# Probability and its Applications

*A Series of the Applied Probability Trust*

# Probability and its Applications

*Anderson:* Continuous-Time Markov Chains.
*Azencott/Dacunha-Castelle:* Series of Irregular Observations.
*Bass:* Probabilistic Techniques in Analysis.
*Choi:* ARMA Model Identification.
*Gani (Editor):* The Craft of Probabilistic Modelling.
*Grandell:* Aspects of Risk Theory.
*Gut:* Stopped Random Walks.
*Guyon:* Random Fields on a Network.
*Last/Brandt:* Marked Point Processes on the Real Line.
*Leadbetter/Lindgren/Rootzén:* Extremes and Related Properties of Random Sequences and Processes.
*Nualart:* The Malliavin Calculus and Related Topics.
*Resnick:* Extreme Values, Regular Variation and Point Processes.
*Shedler:* Regeneration and Networks of Queues.
*Todorovic:* An Introduction to Stochastic Processes and Their Applications.

Xavier Guyon

# Random Fields on a Network

## Modeling, Statistics, and Applications

Translated by Carenne Ludeña

With 28 Illustrations

Springer-Verlag
New York Berlin Heidelberg London Paris
Tokyo Hong Kong Barcelona Budapest

Xavier Guyon
Statistique Appliquée et
Modélisation Stochastique (SAMOS)
Université Paris 1
75634 Paris Cedex 13
*and*
Université Paris-Sud
91405 Orsay Cedex
France

*Series Editors*

J. Gani
Stochastic Analysis Group, CMA
Australian National University
Canberra ACT 0200
Australia

C.C. Heyde
Stochastic Analysis Group, CMA
Australian National University
Canberra ACT 0200
Australia

T.E. Kurtz
Department of Mathematics
University of Wisconsin
Madison, WI 53706
USA

Mathematics Subject Classification (1991): 60G60

Library of Congress Cataloging-in-Publication Data
Guyon, Xavier.
Random fields on a network: modeling, statistics, and applications/Xavier Guyon.
p. cm. – (Probability and its applications)
Includes bibliographical references and index.
ISBN 0-387-94428-1
1. Random fields. I. Title. II. Series: Springer series in statistics. Probability and its applications.
QA274.45.G89 1995
003'.76–dc20 94-41485

Printed on acid-free paper.



Production managed by Francine McNeill; manufacturing supervised by Jacqui Ashri.
Photocomposed copy prepared from the author's TeX files using Springer-Verlag's "svplain.-sty" macro.
Printed and bound by Edwards Brothers, Inc., Ann Arbor, MI.
Printed in the United States of America.

9 8 7 6 5 4 3 2 1

ISBN 0-387-94428-1 Springer-Verlag New York Berlin Heidelberg

# Preface

The theory of spatial models over lattices, or random fields, has developed significantly over the last few years. There has been a growing demand from applied research areas like spatial modeling and image analysis. There has also been a growing interaction with well-established areas like statistical mechanics and spectral analysis. This conjunction has led to an autonomous and original development of the theory of stochastic fields and of its applications in statistics and computation. This book is intended as a guide. Our aim is to accompany the reader (student, engineer, researcher) in the discovery of these developments and their applications.

This book is largely inspired by courses given at the University of Orsay and at the University of Paris I-Paris VII, and can be considered as the basis for a graduate-level course. With this idea in mind, we have included exercises and bibliographical comments at the end of each chapter. Required previous knowledge is relatively limited: spectral theory for those interested in second-order processes, basic results from probability and statistics, and rudimentary knowledge of finite Markov chains. Compared to the usual theory of processes, this field is relatively new.

Spatial modeling is essentially different from time modeling because it is not causal. That is, the chronology of time does not exist. Even over $\mathbb{Z}$, it is not reasonable to model the production of a corn plant based only on that of the plant on its left. But structural differences between temporal and spatial models are more than causality and noncausality: They involve the role played by boundary conditions both in statistical mechanics and statistics, and the absence of factorization techniques for several complex variable polynomials. These differences led to the definition of new and specific ideas, models, results, techniques, and algorithms.

Experimental design for agricultural research was undoubtedly the first area to consider the spatial dimension of a problem. However, after that, many other fields, such as geography, cartography, economics, epidemiology, biology, ecology and meteorology, have called upon spatial modeling, static or dynamic, defined over regular or irregular lattices. The mathematical development of this area and of its realm of applicability is due in great part to Whittle, Besag, Ripley and Künsch, among others.

After the Geman brothers' fundamental article, Bayesian methodology gave a strong impulse to pattern recognition and image analysis: object reconstruction, detection of movement, and segmentation for cartography or for data reduction. Markovian techniques, with a coherent mathematical setup, can be readily applied and are usually algorithmically efficient. They are an improvement over traditional ad hoc methods. This algorithmic applicability illustrated by the Gibbs sampler has given new strength to the study of conditional graphic models, for a long time not considered because of their analytical and numerical complexity, and to Bayesian statistics.

The first two chapters are devoted to the study of two general families of models. In each case, we discuss numerous examples. The first chapter deals with second-order models, and the second with Gibbs models and Markovian fields. They can be read independently. For notation and basic results concerning statistical mechanics, we have followed Georgii.

Chapter 3 gives limit theorems for fields as well as a general presentation of the asymptotics for minimum contrast statistics in the nonergodic setup. This nonergodic setup, required in order to deal with Gibbs specifications in the presence of a possible phase transition, can be applied in many other modeling problems.

Chapters 4 and 5, the natural extensions of Chapter 3, deal, respectively, with statistics for second-order processes and statistics for Markovian fields. Again, they can be read independently. The last chapter, based on Chapter 2, is devoted to stochastic algorithms for simulation, optimization, and estimation. It also contains a detailed exposition of the principal Bayesian techniques for image analysis and includes examples for each of the given methods.

Taking advantage of the translation of the French version, I added some new results on stochastic algorithms, as well as some new exercises.

This book was possible because of the contributions of all referenced works. Particularly, I would like to thank Julian Besag, who introduced me to this field and who taught me to appreciate the "Anglo-Saxon" way with statistics: motivated, simple, and efficient. Thanks also to Hans Ruedi Künsch, to whom is owed much of the mutual development of statistical mechanics and spatial statistics, for his very precious collaboration. This book would not exist but for the human and scientific qualities of all in the Statistics group in Orsay. In particular, I thank the Image group lead by Robert Azencott; without possibly mentioning everybody, thanks to B. Chalmond, C. Graffigne, L. Younes, J.-M. Dinten, J.-F. Yao, and C. Hardouin, whose work, discussions, and experience made possible many of the ideas presented here. Thanks also to P. Doukhan and F. Comets for their helpful remarks and their careful reading of the manuscript.

Finally, thanks to Monique Le Bronnec for her careful typing of the French version, and to Carenne Ludeña for the translation and English adaptation. Both jobs were always done with smiles and much competence. Any remaining error is mine!

Paris, France

Xavier Guyon

# Contents

CHAPTER 1

# Second-Order Stationary Models on $\mathbb{Z}^d$

Let $X = \{X_t, t \in \mathbb{Z}^d\}$ be a second-order stationary and centered process with complex values

$$(1.1) \qquad \begin{cases} X_t \in L^2(\Omega, \mathcal{A}, P), & X_t(\omega) \in \mathbb{C}; \\ E(X_t) = 0, & t \in \mathbb{Z}^d; \\ R(t-s) = E(X_t \overline{X}_s), & s,\ t \in \mathbb{Z}^d. \end{cases}$$

To the Abelian group $G = (\mathbb{Z}^d, +)$ we associate $T^d$, $T = [0, 2\pi[$ the group of continuous characters on $G$; to $\lambda = (\lambda_1, \lambda_2, \ldots, \lambda_d)$ in $T^d$ we associate the group homomorphism $\lambda : G \longrightarrow (\mathbb{C}, X)$

$$k \longmapsto \lambda(k) = \exp i < \lambda, k >, \quad \text{where } < \lambda, k >= \sum_1^d \lambda_i k_i = \lambda.k.$$

To $X$ we associate the Hilbert subspace $H_X$, the closure of $\{X_t, t \in \mathbb{Z}^d\}$ under the scalar covariance product.

## 1.1. Spectral Representations

There are no fundamental differences between the case $d = 1$ and $d \geq 2$ (for $d = 1$, cf. Azencott and Dacunha–Castelle [5] ; for $d \geq 1$, cf. Rosenblatt [139]).

*Spectral Measure.* Assume $R = (R(n), n \in \mathbb{Z}^d)$ is a nonnegative definite function since for all finite subset $\Lambda$,

$$E\Big|\sum_{i\in\Lambda} c_i X_i\Big|^2 = \sum_{i,j\in\Lambda} c_i\bar{c}_j \; R(i-j) \geq 0.$$

Bochner's theorem (Rudin [143]) gives the following representation of the covariance:

$$R(n) = \int_{T^d} e^{i<n,\lambda>} F(d\lambda), \quad n \in \mathbb{Z}^d, \tag{1.2}$$

where $F$ is a nonnegative, regular, and bounded measure over $T^d$.

$F$ is the *spectral measure* of the process $X$. If $F$ is absolutely continuous with respect to Lebesgue measure on $T^d$ (with total mass $(2\pi)^d$), the density $f$ of $F$ is called *the spectral density* of the process. If $X$ is real, $f$ is even.

The density is in $L^2(d\lambda)$ if and only if $R = (R(n), n \in \mathbb{Z}^d)$ is in $\ell^2(\mathbb{Z}^d)$. Then $f$ allows the Fourier representation:

$$f(\lambda) = \frac{1}{(2\pi)^d} \sum_{\mathbb{Z}^d} R(n) e^{-i<\lambda,n>}. \tag{1.3}$$

If $R \in \ell^1$, the convergence in the expansion is uniform and $f$ is continuous. If $k \in \mathbb{N}^d$ and if $f$ is of class $\mathcal{C}^k$, then the covariance satisfies

$$\limsup_{n\to\infty} \; n^k |R(n)| < \infty,$$

where $n^k = n_1^{k_1} \times \cdots \times n_d^{k_d}$; $n \to \infty$ means that at least one coordinate tends to infinity.

*The Spectral Random Field.* The concepts of uncorrelated random fields on $T^d$ and of stochastic integrals with respect to this field are analogous to those introduced for $d = 1$. For $X$, we have the representation

$$X_n = \int_{T^d} e^{i<n,\lambda>} \phi(d\lambda) \; ,$$

where $\phi : \mathcal{B}(T^d) \times \Omega \longrightarrow \mathbb{C}$ is an uncorrelated random field with basis $F$; that is to say:

- $A \longrightarrow \phi_A \in L^2(\Omega, \mathcal{A}, P)$ and is centered.
- If $A \cap B = \phi$, $\quad E(\phi_A \cdot \overline{\phi_B}) = 0$.
- If $A_n \searrow \phi$, $\quad \phi_{A_n} \xrightarrow{L^2} 0$.
- $E|\phi_A|^2 = F(A)$.

$f \longrightarrow \int_{T^d} f d\phi$ is an isometry of $L^2(F(d\lambda))$ on $H_X$.

REMARKS: The results are still true for a process $X$ defined on a locally compact Abelian group $G$ (here, $(\mathbb{Z}^d, +) = G$). $T^d$ is identified with the group $\Gamma$ of the continuous characters on $G$, equipped with its Haar measure

$\lambda$. If $G$ is discrete, $\Gamma$ is compact; if $G$ is compact, $\Gamma$ is discrete; and if $G = G_1 \times G_2$, $\Gamma = \Gamma_1 \times \Gamma_2$. In particular, we have the matching $G \longrightarrow \Gamma : \mathbb{Z}$ and $T, T$ and $\mathbb{Z}, \mathbb{R}$ and $\mathbb{R}$, $Z/pZ \longrightarrow Z/pZ$; this last situation refers to that of the finite torus with $p$ points. Let us notice that the triangular lattice of $(\mathbb{R}^2, +)$

$$G_T = \{n + mj,\ (n,m) \in \mathbb{Z}^2\},\ j = \exp \tfrac{1}{3} i\pi\},$$

is isomorphous to $\mathbb{Z}^2$ and that the hexagonal regular lattice $G_H$ used in mathematical morphology is not a group:

$$G_H = \{n + mj,\ (n,m) \in \mathbb{Z}^2,\ j = \exp \tfrac{1}{3} i\pi,\ n - m \not\equiv 0(3)\}.$$

## 1.2. Various Regularity Concepts

**(1.2.1) Definition.** *Let $\mathcal{I}$ be a translation invariant and nonempty family of $\mathbb{Z}^d$ subsets. We will say that $X$ is $\mathcal{I}$-regular if:*

$$\bigcap_{T \in \mathcal{I}} H(X, T^c) = \{0\},$$

*where $H(X,T^c)$ is the Hilbert space generated by the $X_t$, $t \notin T$.*

Among the four definitions of regularity as defined below, the first two don't require any order on $\mathbb{Z}^d$, whereas the last two are related to some order. On $\mathbb{Z}$, only $R3 \equiv R4$ is usually studied.

R1: *Noncausal regularity.* $\mathcal{I} = \{T \subseteq \mathbb{Z}^d, T \text{ finite}\}$. Theorem 1.2.1 (Rozanov [141]) characterizes such fields.

R2: *Basic process.* $\mathcal{I} = \{\{t\},\ t \in \mathbb{Z}^d\}$.

R3 : *Lexicographic regularity.* Let $\leq$ be a total order relation on $\mathbb{Z}^d$ compatible with respect to the addition and let $\mathcal{I}$ be the associated family, $\mathcal{I} = \{\{s : s \leq t\},\ t \in \mathbb{Z}^d\}$. On $\mathbb{Z}$, this is the common definition of regularity. Theorem 1.2.3 (cf. Helson and Lowdenslager [91]) characterizes such regularity for a process on $\mathbb{Z}^2$.

A total order relation on $\mathbb{Z}^2$ is characterized by a subset $S \subseteq \mathbb{Z}^2$ which is a *half space*, that is:

$S$ stable for the addition, $S \cup \{-S\} = \mathbb{Z}^2$, $S \cap \{-S\} = \{0\}$.

The lexicographic order corresponds to

$$S_{lex} = \{(m,n) \in \mathbb{Z}^2,\ m > 0 \quad \text{or} \quad m = 0,\ n > 0\}.$$

If $\alpha$ is irrational, the subsets $S_\alpha = \{(m,n) \in \mathbb{Z}^2, \quad m\alpha + n \geq 0\}$ also define a total order (cf. [101], [115], [143]).

R4: *Quarter plane regularity.* If $<<$ is the strong partial order, $t = (t_1, \ldots, t_d) << s = (s_1, \ldots, s_d)$ if $t_j \leq s_j$, $j = 1, d$, it is the regularity notion associated to:

$$\mathcal{I} = \{\{s, t << s\},\ t \in \mathbb{Z}^d\} \text{ ([143], theorem 1.2.4, [101] and [162]).}$$

To each regularity corresponds a linear representation of $X$ which is the basis of second-order modeling of spatial phenomena: $R1$ for the noncausal Markovian representation, and $R3$ and $R4$ for the causal representations.

We will say that the process is $\mathcal{I}$*–singular* if and only if

$$H(X) = \bigcap_{\mathcal{I}} H(X, T^c) = H(X, T^c), \quad \text{for every } T \in \mathcal{I}.$$

Following standard Hilbertian arguments, we have the following decomposition into regular and singular parts:

DECOMPOSITION THEOREM. — *Let $X$ be a second-order stationary process (1.1), and let $\mathcal{I}$ be a translation invariant family of subset of $G$. Then there exist two orthogonal second-order stationary processes, one $Y$, $\mathcal{I}$–regular, the other $Z$, $\mathcal{I}$–singular, such that $X_t = Y_t + Z_t, t \in \mathbb{Z}^d$. Such a decomposition is unique.*

## *1.2.1. Noncausal Regularity*

**(1.2.1) Theorem.** *$X$ is noncausal regular ($R1$) if and only if its spectral measure $F$ is absolutely continuous and if its spectral density verifies for a nonzero trigonometric polynomial $\delta_0(\lambda) = \sum_T c_t \exp i < \lambda, t >$, $T$ finite*

$$\int_{T^d} |\delta_0(\lambda)|^2 f^{-1}(\lambda) d\lambda < \infty\ .$$

The result is a consequence of the following Lemma, describing, for finite $T$, the space $\Delta_T$, orthogonal complement of $H(X, T^c)$ in $H_X$.

**(1.2.1) Lemma.** *$\Delta_T$ is isometric to the Hilbert space of trigonometric polynomials $\delta(\lambda) = \sum_T c_t \exp i < \lambda, t >$ which verify almost everywhere (a.e.):*

$$\delta(\lambda) F(d\lambda) = \delta(\lambda) f(\lambda) d\lambda, \tag{1}$$

$$\int |\delta(\lambda)|^2 f^{-1} d\lambda < \infty, \tag{2}$$

*for the scalar product* $< \delta_1,\delta_2 > \; = \int \delta_1(\lambda)\overline{\delta_2(\lambda)} f^{-1}(\lambda)d\lambda$ *where* $f$ *is the density of the absolutely continuous part of* $F$.

*Proof of the Lemma:* $H_X$ is isometric to $L^2(F)$; let $\varphi \in L^2(F)$ be associated to $h \in \Delta_T$; if $t \notin T$, we have

$$0 = < h, X_t > = \int \exp(-i < \lambda, t >)\varphi(\lambda)F(d\lambda). \tag{1.4}$$

As $T$ is finite, there exists a cone $C$ of angular opening strictly larger than a half space such that (s.t.) for $t \in C$ :

$$\int \exp(-i < \lambda, t >) \; G(d\lambda) = 0 \quad \text{where} \quad G(d\lambda) = \varphi(\lambda)F(d\lambda).$$

We can then apply the Riesz–Bochner Theorem ([91], [143]), which says that if a measure on $T^d$ has its Fourier coefficients equal to zero on a cone of opening larger than the half plane, this measure is absolutely continuous with respect to (w.r.t.) Lebesgue measure,

$$\delta(\lambda) = \frac{\varphi(\lambda)F(d\lambda)}{d\lambda} = \varphi(\lambda)f(\lambda). \tag{1.5}$$

According to (1.4), $\delta$ is a trigonometric polynomial on $T^d$ and for $h_1$ and $h_2$ of $\Delta_T$, we associate $\delta_1$ and $\delta_2$ s.t., from (1.5):

$$\int h_1(\lambda)\bar{h}_2(\lambda)F(d\lambda) = \int \delta_1(\lambda)\delta_2(\lambda)f^{-1}(\lambda)d\lambda.$$

□

*Proof of the Theorem:* The regularity condition is equivalent to

$$\lim_{T \to \mathbb{Z}^d} \Delta_T = H_X; \tag{1.6}$$

that is, every element $h \in H_X$ can be arbitrarily approximated by $\delta \in \Delta_T$, $T$ finite.

*Necessity:* Since $H_X \neq \{0\}$, there exists a finite $T$ and $\delta_0 \in \Delta_T$, $\delta_0 \neq 0$ s.t. $\int |\delta_0|^2 f^{-1}(\lambda)d\lambda < \infty$ because of the Lemma. Let $\psi$ be an a.e. $\lambda$ nonzero variable. Then for $\varphi = \frac{\delta}{f}$, $\delta$ is a trigonometric polynomial:

$$\int \varphi\overline{\psi}F(d\lambda) = \int \delta(\lambda)\overline{\psi}(\lambda)d\lambda = 0.$$

And now it follows from (1.6) that $F$ is absolutely continuous.

*Sufficiency:* Let the finite measure $\mu(d\lambda)$ be defined by $\mu(d\lambda) = \frac{|\delta_0|^2}{f}d\lambda$, $\varphi \in L^2(F)$ and $\psi = \frac{\varphi \cdot f}{\delta_0}$. Since, by hypothesis, $\psi \in L^2(\mu)$ and

$$0 = \inf_{c,T} \int \Big|\frac{\varphi \cdot f}{\delta_0} - \sum_T c_t e^{i<\lambda,t>}\Big|^2 \frac{|\delta_0|^2}{f} d\lambda = \inf_{c,T} \int \Big|\varphi - \frac{\delta_0}{f}\sum_T c_t e^{i<\lambda,t>}\Big|^2 f(\lambda)d\lambda$$

it follows that $X$ is $(R1)$ regular. □

### *1.2.2. Basic Processes: Noncausal Linear Representations*

We will say that a process $X$ is basic if and only if it has noncausal regularity and if $X_t \notin H(X, \{t\}^c)$.

**(1.2.2) Theorem.** (on $\mathbb{Z}$, cf. [140] Theorem 11.1) *$X$ is basic if and only if $F$ is absolutely continuous and if $f^{-1}$ is integrable.*

*Proof:* It follows from Lemma 1.2.1 and from the equivalence

$$X_0 \notin H(X, \{0\}^c) \Longleftrightarrow \Delta_{\{0\}} \neq \{0\} \Longleftrightarrow \int f^{-1} d\lambda < \infty .$$

□

*Linear Noncausal Decomposition of a Basic Process.* Let $X$ be a basic process. To the decomposition $H_X = H(X_s,\ s \neq t) \oplus \Delta_{\{t\}}$, we can associate the Conditional Auto Regression (CAR) infinite linear representation [135] given by

$$X_t = \sum_{s \neq 0} c_s X_{t-s} + e_t, \quad c \in \ell^2(\mathbb{Z}^d), \quad e_t \perp X_s \quad \text{if } s \neq t . \tag{1.7}$$

Here, $e = \{e_t, t \in \mathbb{Z}^d\}$ is the *conditional errors* process. In the Gaussian setting, $\sum_{s \neq 0} c_s X_{t-s}$ is the conditional expectation of $X_t$ given $X_s$, $s \neq t$. Since $e_0$ is orthogonal to every $X_u$, $u \neq 0$, this translates into:

$$\int e^{-i<\lambda,u>} [1 - \sum_{s \neq 0} c_s e^{-i<\lambda,s>}] f(\lambda) d\lambda = 0, \quad u \neq 0.$$

Here, $F(d\lambda) = f(\lambda)d\lambda$ is the spectral measure of $X$. From Plancherel's Theorem, one has

$$f(\lambda) = \frac{\sigma_e^2}{1 - \sum_{s \neq 0} c_s \exp i < \lambda, s >}, \quad c_s = c_{-s}, \text{ and } \sigma_e^2 = Var(e_t). \tag{1.8}$$

So that because of (1.7), $e$ is stationary and regular, and its spectral density is:

$$f_e(\lambda) = \sigma_e^2 (1 - \sum_{s \neq 0} c_s \exp i < \lambda, s >).$$

In other words, the conditional errors are a *colored noise* with covariance:

$$cov(e_t, e_{t+s}) = \begin{cases} \sigma_e^2 & \text{if } s = 0. \\ -\sigma_e^2 c_s & \text{if } s \neq 0. \end{cases} \tag{1.9}$$

### *1.2.3. Regularity and Causal Representations*

We shall consider $\mathbb{Z}^2$ with the total order relationship associated to the half plane:

$$(i, j) \leq (k, \ell) \iff (k - i, \ell - j) \in S \quad \text{(cf. definition R3)}.$$

We have the following results [91] :

THEOREM (SZEGÖ). — *Let $S$ be a half plane of $\mathbb{Z}^2$. Let $F$ be a finite positive measure over $T^2$. Let*

$$F(d\lambda) = f(\lambda)d\lambda + \mu_s(d\lambda)$$

*be its decomposition into regular and singular components. Then if $\widehat{X}_0$ is the projection of $X_0$ onto $H_0^- = H(X_s,\ s < 0,\ s \neq 0)$, one has*

$$\|X_0 - \widehat{X}_0\|^2 = \exp\Big\{\frac{1}{4\pi^2}\int_{T^2} \log f(\lambda)d\lambda\Big\}. \tag{1.10}$$

FACTORIZATION THEOREM. — *Let $f \geq 0$ be a spectral density over $T^2$. Consider a half plane $S$. Then $f$ can be written as*

$$f(\lambda) = \Big|b_0 + \sum_S b_s e^{i<\lambda,s>}\Big|^2, \quad b_0 \neq 0, \quad (b_s) \in \ell^2 \tag{1.11}$$

*if and only if*

$$\int \log f(\lambda)d\lambda > -\infty. \tag{1.12}$$

Both theorems are nontrivial generalizations of analogous one-dimensional results.

*Causal Regularity Criteria.* Consider the order relationship associated to $S$ and the innovation process $\varepsilon_s = B_{-s}\varepsilon_0$, $\varepsilon_0 = X_0 - \hat{X}_0$. Here, $B_s$ is the unitary operator defined over $H_X$ such that $B_s X_t = X_{t-s},\ \ t \in \mathbb{Z}^2$. If $X$ has causal regularity, then $H_X = H_\varepsilon$ and for all $t$, $H(\varepsilon_s,\ s \leq t) = H(X_s,\ s \leq t)$. In other words one has the following infinite Moving Average $(MA)$ representation:

$$X_t = \sum_{s \geq 0} b_s \varepsilon_{t-s}. \tag{1.13}$$

**(1.2.3) Theorem.** *X is S–regular if and only if*

(*a*) *F is absolutely continuous, with density f such that*

(*b*) $\int \log f(\lambda) d\lambda > -\infty$.

*Proof:* If $X$ is $S$–regular, then (1.13) is true and therefore $F$ is absolutely continuous. Because of Szegö's Theorem, $\log \sigma_\varepsilon^2 = \frac{1}{4\pi^2} \log f(\lambda) > -\infty$. Conversely, because of the Factorization Theorem, we can define a white noise $\eta = \{\eta_s\}$ :

$$\eta_s = \int \exp(i\lambda \cdot s) \overline{\varphi}^{-1}(\lambda) \Phi(d\lambda)$$

such that

$$H^-_{\eta,t} \subseteq H^-_{X,t}, \quad \text{and therefore} \quad X_t = \sum_{s \geq 0} b_s \eta_{t-s}.$$

□

OBSERVATIONS:

(1) The characterization given by (*a*) and (*b*) is intrinsic and does not depend on the order defining half-plane $S$.

(2) When dealing with lexicographical order, it is possible to give a simple Hilbertian characterization of regularity. We define

- $H_s = \text{esp}\{X_t,\ t \leq s\}$, and $H_{-\infty,\ell} = \bigcap_{k \in \mathbb{Z}} H_{k,\ell}$;
- $H^1_\ell = \text{esp}\{X_{i,j},\ j \leq \ell\}$, $I^1_\ell = H_{-\infty,\ell} \ominus H^1_{\ell-1}$, the line jump innovation, where $\ominus$ stands for orthogonal difference;
- $I_{k,\ell} = H_{k,\ell} \ominus H_{k-1,\ell}$, $H_{-\infty,-\infty} = \bigcap_{s \in \mathbb{Z}^2} H_s$.

And now one has $H_{k,\ell} = H_{-\infty,-\infty} \oplus \left(\sum_{j \leq \ell} I^1_j\right) \oplus \left(\sum_{(i,j) \leq (k,\ell)} I_{i,j}\right)$.

One has lexicographical regularity if $H_{-\infty,-\infty} = I^1_0 = \{0\}$, in other words, if there is no memory in $(-\infty, -\infty)$ or line jump innovation. It is not enough that $H_{-\infty,-\infty} = \{0\}$. Indeed, consider $X_{s,t} = \eta_t$, where $\eta$ is a one-dimensional white noise ($I_0 = \text{esp}\{\eta_0\}$; then $4\pi^2 F(d\lambda, d\mu) = \sigma^2_\eta \delta_0(d\lambda) d\mu$).

(3) If $X$ is basic, it has lexicographical regularity but it also has noncausal regularity: $X$ admits the following three representations:

| Representation | of $X$ | of $f$ |
|---|---|---|
| $MA(\infty)$ causal (1.13) | $X_t = \sum_{s\geq 0} b_s \varepsilon_{t-s}$ | $f(\lambda) = \sigma_\varepsilon^2 \left\| \sum_{s\geq 0} b_s e^{i\lambda\cdot s}\right\|^2$ |
| $AR(\infty)$ causal | $X_t = \sum_{s>0} a_s X_{t-s} + \varepsilon_t$ | $f(\lambda) = \sigma_\varepsilon^2 \left\| 1 - \sum_{s>0} a_s e^{i\lambda\cdot s}\right\|^{-2}$ |
| Markovian or $CAR$ (1.7) | $X_t = \sum_{s\neq 0} c_s X_{t-s} + e_t$ | $f(\lambda) = \sigma_e^2 (1 - 2\sum_{s>0} c_s \cos \lambda \cdot s)^{-1}$ |

The coefficients $(a_s, s > 0)$ of the $AR$ causal representation are defined by

$$\widehat{X_0} = \sum_{s>0} a_s X_{-s}.$$

## *1.2.4. Quarter Plane Regularity ([101], [143], [162])*

Let $s = (s_1, s_2)$, and let $t = (t_1, t_2)$. We shall call $s \ll t$ the strong order relationship defined by $s_1 \leq s_2$ and $t_1 \leq t_2$. We now define

$$H_1(t) = \text{esp}\{X_s,\ s_1 \leq t_1\},\ H_2(t) = \text{esp}\{X_s,\ s_2 \leq t_2\},$$

$$H_1((s,t]) = H_1(t) \ominus H_1(s), \quad \text{if } s < t,$$

$$H_2((s,t]) = H_2(t) \ominus H_2(s), \quad \text{if } s < t.$$

The quarter plane innovation space is defined by

$$H(t) = H_1((i-1,i]) \cap H_2((j-1,j]), \quad t = (i,j).$$

If $s \neq t$, $H(s) \perp H(t)$. If $X$ is regular in the strong order

(1.14) $$H_X = \oplus_t H(t).$$

Here, (1.14) is equivalent to $(R4)$ regularity. If $(R4)$, $H_k(t) = \bigoplus_{s_k \leq t_k} H(s)$, $k = 1, 2$. In particular, the orthogonal projection operators $E_t^1$ and $E_t^2$ over $H_1(t)$ and $H_2(t)$ commute. This allows us to define the white noise $\nu$,

$$\nu_{i,j} = (1 - E_{i-1}^1)(1 - E_{j-1}^2) X_{ij},$$

for which one has the quarter plane decomposition

(1.15) $$X_{i,j} = \sum_{k\leq i, \ell \leq j} d_{k,\ell} \nu_{i-k,j-\ell}.$$

A sufficient condition to assure quarter plane regularity (1.14) is that $F$ be absolutely continuous with bounded, strictly positive and lower semicontinuous (lsc) density $f$ [162].

## 1.3. ARMA Models and Markovian Models

In this paragraph, $\leq$ stands for lexicographical order over $\mathbb{Z}^2$:

$$(i,j) \leq (k,\ell) \iff (j < \ell) \text{ or } (j = \ell,\ i \leq k)\ .$$

Let $\varepsilon = \{\varepsilon_t, t \in \mathbb{Z}^2\}$ be a white noise on $L^2(\Omega, \mathcal{A}, P)$.

### *1.3.1. Linear Processes*

Let $c = \{c_s, s \in \mathbb{Z}^2\}$ be a sequence of $\ell^2(\mathbb{Z}^2)$. To $\varepsilon$ and $c$ we associate the linear process:

$$X_t = \sum_{s \in \mathbb{Z}^2} c_s \varepsilon_{t-s}$$

with spectral density:

$$f(\lambda) = \frac{\sigma^2}{4\pi^2} \left| \sum_{\mathbb{Z}^2} c_s e^{i\lambda \cdot s} \right|^2 .$$

**(1.3.1) Example.** Generalized ARMA. Consider $F(z_1, z_2)$ and $G(z_1, z_2)$ two holomorphic functions on a neighborhood of $T^2$ such that $F$ is not zero over $T^2$. Then $\frac{G}{F}$ has a Laurent series in a neighborhood of $T^2$, with exponentially decreasing coefficients $c = (c_s, s \in \mathbb{Z}^2)$. We define

$$B_1 X_{k,\ell} = X_{k-1,\ell}, \quad B_2 X_{k,\ell} = X_{k,\ell-1},$$

the backshift operators in each direction. We will say that $X$ is a $(F, G, \varepsilon)$–*Generalized ARMA* if

$$F(B_1, B_2) X_{k,\ell} = G(B_1, B_2) \varepsilon_{k,\ell}, \quad (k,\ell) \in \mathbb{Z}^2. \tag{1.16}$$

The formal definition (1.16) is the $L^2$ development:

$$\begin{cases} X_{k,\ell} = \sum_{u,v} c_{u,v} \varepsilon_{k-u,\ell-v} \\ f(\lambda_1, \lambda_2) = \frac{\sigma^2}{4\pi^2} \left| \frac{G}{F} \right|^2 (e^{i\lambda_1}, e^{i\lambda_2}). \end{cases} \tag{1.17}$$

Since there exist $a$ and $b$ s.t. $0 < a \leq f \leq b < \infty$, $X$ is basic, has lexicographical regularity, and $H_X = H_\varepsilon$.

**(1.3.2) Example.** ARMA Fields. If $F = P$, $G = Q$ are two polynomials:

$$P(z_1, z_2) = \sum_R a_{k,\ell} z_1^k z_2^\ell, \quad a_{00} = 1, \ R \subseteq \mathbb{Z}^2 \text{ finite;}$$

$$Q(z_1, z_2) = \sum_S b_{k,\ell} z_1^k z_2^\ell, \quad b_{00} = 1, \ S \subseteq \mathbb{Z}^2 \text{ finite;}$$

we say we have an ARMA $(P, Q, \varepsilon)$ and write

$$P(B)X = Q(B)\varepsilon. \tag{1.18}$$

If $Q \equiv 1$, we have an *AR* field. If $P \equiv 1$, we have an *MA* field.

As with time series, ARMA models are interesting because any stationary process can be approximated by an ARMA of relatively small parametric dimension.

As with time series, in order to be identifiable, an ARMA must be well specified (e.g., indicating that $\varepsilon$ is the innovation process for $X$). Canonical ARMA representation over $\mathbb{Z}$ ([5]) is assured if $P$ is nonzero over $|z| \leq 1$. In the spatial context we refer to *causal representation.*

### *1.3.2. Causal ARMA Representation*

We will say that a polynomial is unilateral (or analytical) if its support belongs to $(\mathbb{Z}^2)^+$ (the zero or positive half plane). Define $H_t(Y) = H(Y_s, s \leq t)$.

**(1.3.1) Definition.** *The* ARMA $(P,Q,\varepsilon)$ *is causal if*

$$H_t(\varepsilon) = H_t(X), \quad \textit{and then} \quad \varepsilon_t \perp X_s \quad \textit{for all } s < t. \tag{1.19}$$

If the representation is causal, $\varepsilon$ is the innovation process of $X$. It is easy to check that the process $MA(Q, \varepsilon)$ is causal if and only if $Q$ is unilateral. For an *AR* process, we have the following characterization:

**(1.3.1) Theorem.** $(P,\varepsilon)$ *is an AR causal representation if one of the following conditions is true:*

(*a*) $\int \log |P| d\lambda = 0$.

(*b*) (*b*1): $P$ *is unilateral.*

(*b*2): $P(z_1,0) \neq 0$ *if* $|z_1| \leq 1$ *and* $P(\exp i\lambda_1, z_2) \neq 0$ *if* $|z_2| \leq 1$, *for all* $\lambda_1 \in \mathbb{R}$.

*Proof:*

(*a*) That (*b*) is equivalent to (*a*) results from the fact that the spectral density can be written as $(4\pi)^{-2}\sigma_\varepsilon^2 |P(\lambda)|^{-2}$ and from Szegö's Theorem.

(*b*) Suppose $P$ is not unilateral. Then $\varepsilon_0$ can be decomposed as

$$\varepsilon_0 = V^- + V^+, \quad V^- = \sum_{s\in R, s\geq 0} a_u X_{-u}, \quad V^+ = \sum_{s\in R, s<0} a_u X_{-u} \neq 0,$$

and thus $\varepsilon_0 \notin H_0(X)$. However, then $\varepsilon$ cannot be the innovation process of $X$.

Consider the algebra of analytic functions over $D^2 = \{z \in \mathbb{C}, |z| \leq 1\}^2$,

$$\mathcal{A} = \{f : D^2 \longrightarrow \mathbb{C},\ f(z_1, z_2) = \sum_{(k,\ell)\geq(0,0)} a_{k\ell} z_1^k z_2^\ell\}.$$

A general result for uniform algebras ([61]) states that $f \in \mathcal{A}$ is invertible over $\mathcal{A}$ if and only if $f$ isn't zero over $\mathcal{A}$'s spectrum (the set of continuous homomorphisms of $\mathcal{A}$ onto $\mathbb{C}$). Here the spectrum identifies with $D^2$ and therefore $P$ is not zero over $\Delta$:

$$\Delta = \{(z_1, 0),\ |z_1| \leq 1\} \cup \{(e^{i\lambda}, z_2),\ \lambda \in T,\ |z_2| \leq 1\}\ .$$

Conversely, let $P$ be analytical and not zero over $\Delta$. Since the map $z_2 \longrightarrow P(e^{i\lambda}, z_2)$ is not zero over $D$, there exists a holomorphic representation of its logarithm in a neighborhood of $D$: $\operatorname{Log} P(e^{i\lambda}, z) = \operatorname{Log}|P(e^{i\lambda}, z)| + i\ arg(P(e^{i\lambda}, z))$.

Integrating along $\mathcal{C}$ defined by:

$$\mathcal{C} = \left\{ \begin{array}{lll} |z| = 1, \text{ positive sense}, & & z = [\varepsilon, 1] \\ |z| = \varepsilon, \text{ negative sense}, & & \text{in both senses} \end{array} \right\}$$

and observing that $\mathcal{C}$ belongs to an open set where $\operatorname{Log} P(e^{i\lambda}, \cdot)$ is holomorphic, Cauchy's Theorem yields for all $\varepsilon \in ]0, 1]$

$$\int_0^{2\pi} \operatorname{Log}|P(e^{i\lambda}, e^{i\mu})| d\mu = \int_0^{2\pi} \operatorname{Log}|P(e^{i\lambda}, \varepsilon e^{i\mu})| d\mu.$$

So that because of the Lebesgue Theorem,

$$\int_0^{2\pi} \operatorname{Log}|P(e^{i\lambda}, e^{i\mu})| d\mu = 2\pi\ \operatorname{Log}|P(e^{i\lambda}, 0)|.$$

Since $P(z_1, 0)$ is not zero over $D$, a similar argument gives

$$\int_0^{2\pi} \operatorname{Log}|P(e^{i\lambda}, 0)| d\lambda = 0 \text{ and so, } \operatorname{Log}\sigma_\varepsilon^2 = \frac{1}{4\pi^2}\int_{T^2} \operatorname{Log} f(\lambda, \mu) d\lambda d\mu,$$

so that $\varepsilon$ is the innovation process of $X$. □

OBSERVATIONS:

(1) A consequence of characterization (*b*) is that a sufficient condition for causality is

$$\sum_{u \in R, u \neq 0} |a_u| < a_0 = 1.$$

This condition is not necessary, however (cf. 4.3.1, Observation 3 of Corollary 4.3, and Exercise 1).

(2) Condition (*a*) is intrinsic and does not depend on the half plane $S$ chosen to define the order relationship. The result is then true of every ARMA relative to any order relationship.

(3) Numerically (*b*) is not easy to check whereas (*a*) is, approximating the integral by a Riemman sum.

We have the following Corollary:

**(1.3.1) Corollary.** *$(P,Q,\varepsilon)$ is a causal* ARMA *representation if $P$ and $Q$ are unilateral and each one satisfies either condition (a) or (b) of Theorem* 1.3.1.

*Proof:* Let $Y_t = Q(B)\varepsilon_t$, $H_t(Y) \subseteq H_t(\varepsilon)$, and $Q$ satisfy either $(a)$ or $(b)$; $H_t(\varepsilon) \subseteq H_t(Y)$ because $Q^{-1}$ is analytical. In much the same way for $X$, $P(B)X_t = Y_t$. So one has $H_t(X) = H_t(Y)$. □

### *1.3.3. Some Differences Between $\mathbb{Z}$ and $\mathbb{Z}^d$, $d \geq 2$*

*AR Models and Causal Representation.* Over $\mathbb{Z}$ it is known that any $AR$ model, $P(B)X_t = \varepsilon_t$, $\partial^o P = m$, admits $2^m$ equivalent $AR$ representations, two of which are causal: one for the natural order and the other for the inverted time order. Indeed, let $z_1$ ($|z_1| \neq 1$) be a root of $P$. Define $Q = T_{z_1} \cdot P$ with

$$T_{z_1}(z) = \frac{z_1(\overline{z} - \overline{z}_1^{-1})}{(z - z_1)}$$

so that $T_{z_1}$ has modulus 1 over the torus. Then $X$ is an $AR(Q, \eta)$ with $\eta = T_{z_1}(B)\varepsilon$ a white noise. In particular, if every zero of $P$ has modulus less than 1, we can transform the $AR$ representation into one where all roots have modulus greater than 1. In other words, into a canonical representation (minimum phase representation). Over $\mathbb{Z}^d$, $d \geq 2$, this transformation is no longer possible.

**(1.3.2) Theorem.** *Contrary to what happens in $\mathbb{Z}$, an $AR$ process over $\mathbb{Z}^d$, $d \geq 2$ does not, in general, admit an equivalent finite causal $AR$ representation.*

*Proof:* Consider the four nearest neighbor isotropic $AR$ model over $\mathbb{Z}^2$:

$$X_s = \alpha \sum_{t:|t-s|=1} X_t + \varepsilon_s, \quad 0 < |\alpha| < \frac{1}{4},$$

$$\sigma_\varepsilon^2 f^{-1}(\lambda, \mu) = [1 - 2\alpha(\cos\lambda + \cos\mu)]^2 = Q(e^{i\lambda}, e^{i\mu}).$$

Assume $Q = |P|^2$, where $P$ is unilateral and finite. A close examination of its support yields $P(z_1, z_2) = a + bz_1 + cz_1^2 + dz_2 + ez_2^2 + fz_1z_2 + gz_1^{-1}z_2$, $c$ and $d \neq 0$. Identifying $|P|^2$ with $Q$ imposes $cd = 0$. Thus, $d$ (or $c$) $= 0$, which is a contradiction. □

**(1.3.3) Example.** $(1+\beta^2)X_{s,t} = \beta(X_{s+1,t} + X_{s-1,t} + X_{s,t-1}) + \varepsilon_{st}$, $|\beta| < 1$, admits the causal infinite representation ([13])

$$X_{s,t} = 2\beta X_{s-1,t} + \beta^2 X_{s-2,t} - \beta X_{s-1,t-1} + \beta(1-\beta^2)\sum_{j=0}^{\infty} \beta^s X_{s+j,t-1} + \varepsilon_{s,t}.$$

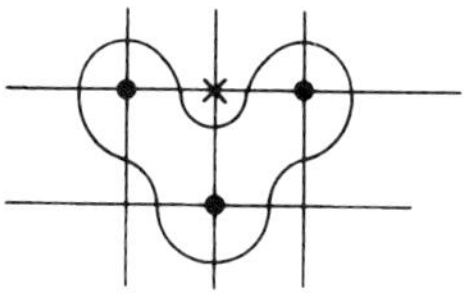
*Noncausal AR*

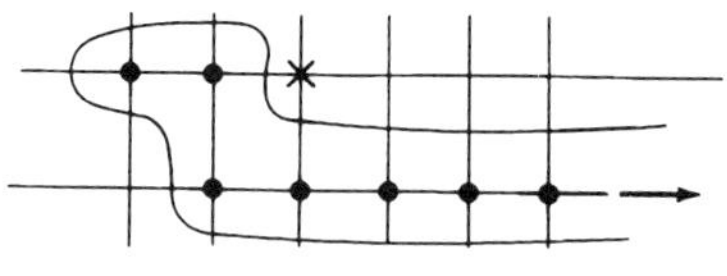
*Equivalent infinite causal AR*

*MA Models and Processes with Bounded Support Covariance.* Every MA process has a bounded support covariance function. On $\mathbb{Z}$, the converse is true because of the following theorem:

THEOREM (FEJER). — *Consider $Q$ a trigonometric polynomial over $T$, with real coefficients such that $Q(e^{i\lambda}) \geq 0$. Then there exists a polynomial $P$ with real coefficients such that $Q = |P|^2$.*

As the proof of Theorem 1.3.2 shows, this property is no longer true over $\mathbb{Z}^2$ : the $\mathbb{Z}^2$ process with correlation $\rho$, $0 < \rho < 1/4$ with four nearest neighbors and zero elsewhere, does not admit a finite MA representation.

However, if we assume that the covariance function can be factorized as $R(s,t) = R_1(s)R_2(t)$, the converse is true. Such processes are called *separable processes* by Martin [118']. Likewise, a factorizing ARMA $f(\lambda, \mu) = f_1(\lambda)f_2(\mu)$ admits a finite causal ARMA representation. Because of this, factorizing models are often used for spatial modelling. The simplest of these AR models is

$$X_{st} = \alpha X_{s-1,t} + \beta X_{s,t-1} - \alpha\beta X_{s-1,t-1} + \varepsilon_{s,t}, |\alpha|, \ |\beta| < 1$$

with covariance:

$$R(s,t) = \sigma_\varepsilon^2 (1-\alpha^2)^{-1}(1-\beta^2)^{-1}\alpha^{|s|} \cdot \beta^{|t|} \ .$$

If additionally sampling is done over a rectangular grid $[1,n] \times [1,m]$, the covariance matrix factorizes as a Kronecker product. This simplifies the numerical problems associated with the statistics of these models.

*Extension of Positive Type Functions.* Let $S$ be a finite subset of $\mathbb{Z}^d$, $\Delta = S - S = \{s-t, s,t \in S\}$ and let $r$ be a positive type function over $\Delta$. That is, $r(s) = \overline{r}(-s)$, and for every sequence $c_t$, $t \in S$, $\sum_{s,t} c_s \overline{c}_t r(s-t) \geq 0$. Is there a positive type function $R$ (a covariance) over all of $\mathbb{Z}^d$ such that $r = R$ over $\Delta$?

If $d = 1$, the answer is yes (Krein [103]). If, however, $d \geq 2$, the problem doesn't always have a solution (Rudin [145]) . In some sense, $r$ can be a covariance over a finite subset $\Delta$ of $\mathbb{Z}^2$ without there being a stationary process $X$ over $\mathbb{Z}^2$ having $r$ as its covariance function over $\Delta$ (cf. Seghier [146] for the characterization of functions $r$ that admit an extension).

## *1.3.4. Second-Order Markovian Modeling*

Consider $X$ a centered process defined over $\mathbb{Z}^d$ with spectral density $f$. Assume $f^{-1}$ is integrable. Then $X$ admits the infinite noncausal representation (1.7).

**(1.3.2) Definition.** *Second-Order L-Markovian Models. Let $L$ be a symmetric finite subset of $\mathbb{Z}^d$, $0 \notin L$. We will say $X$ is second-order L-Markovian if $X$ is such that*

$$\text{(1.20)} \qquad \left\{ \begin{array}{ll} X_t = \sum_{s \in L} c_s \ X_{t-s} + e_t, & c_s = c_{-s} \\ \\ e_t \perp X_s \quad if \ s \neq t, & \sigma_e^2 = var(e_t). \end{array} \right.$$

In this case, $X$'s spectral density is (cf. (1.8)):

$$\text{(1.21).} \qquad f(\lambda) = \sigma_e^2 (1 - \sum_L c_s \cos \lambda \cdot s)^{-1}$$

$\{e_t\}$, the *conditional error,* is a *colored noise* with spectral density

$$\text{(1.22)} \qquad f_e(\lambda) = \sigma_e^2 (1 - \sum_L c_s \cos \lambda \cdot s).$$

This model can also be considered as a $CAR$ model. Over $\mathbb{Z}$ and $\mathbb{Z}^2$, $f \in L^1$ is equivalent to $f^{-1}$ not equal to zero, and thus a sufficient condition is

$\sum_L |c_s| < 1$. Over $\mathbb{Z}^d$, $d \geq 3$, $f^{-1}$ can be zero and yet $f$ be integrable (cf. Exercise 1).

We have the following link between $AR$ models and the Markovian CAR representations:

**(1.3.3) Theorem.** *AR Representation and Markovian CAR Representation.*

*(a) Every AR model is a linear markovian model: if*

$$X_t + \sum_R a_s X_{t+s} = \varepsilon_t, \quad a_0 = 1$$

*then $X$ is L-Markovian such that if $\widetilde{R} = R \cup \{0\}$:*

$$\begin{cases} X_t + \sum_L c_s X_{t+s} = e_t, \ L = (\widetilde{R} - \widetilde{R}) \backslash \{0\}, \\ c_0 = 1, \quad \sigma_e^2 \sum a_u a_{u+s} = \sigma_\varepsilon^2 c_s, \ s \in L \cup \{0\} \ . \end{cases}$$

*(b) Contrary to what happens in $\mathbb{Z}$, the converse is not true over $\mathbb{Z}^d$, $d \geq 2$. In this case, the class of Markovian CAR processes is strictly greater than the class of autoregresive processes.*

*Proof:* $(a)$ is a consequence of the expansion of $|P|^2$ and of the following identity:

$$f(\lambda) = \frac{\sigma_\varepsilon^2}{|P(e^{i\lambda})|^2} = \frac{\sigma_e^2}{Q(e^{i\lambda})}, \quad a_0 = c_0 = 1 \ .$$

$(b)$ The proof was already discussed in Theorem 1.3.2. □

**(1.3.4) Example.** Over $\mathbb{Z}$, the following two representations are equivalent:

$$\begin{cases} AR(1): & X_t = \rho X_{t-1} + \varepsilon_t, & \varepsilon \text{ white noise } |\rho| < 1 \\ CAR(1): & X_t = \alpha(X_{t-1} + X_{t+1}) + e_t, & X_t \bot e_s \text{ if } t \neq s \end{cases}$$

$$\alpha(1 + \rho^2) = \rho, \quad \sigma_\varepsilon^2 = \sigma_e^2(1 + \rho^2) \ .$$

The following three representations are equivalent:

$$\begin{cases} \text{Bilateral } AR: X_t = \alpha(X_{t-1} + X_{t+1}) + \varepsilon_t, \varepsilon \ b.b., \ 0 < \alpha < \frac{1}{2} \\ \text{Canonical } AR: X_t = a_1 X_{t-1} + a_2 X_{t-2} + \eta_t, \ \eta \ b.b. \\ CAR: X_t = c_1(X_{t-1} + X_{t+1}) + c_2(X_{t-2} + X_{t+2}) + e_t \\ \qquad\qquad \text{with } e_t \bot X_s \text{ if } s \neq t \end{cases}$$

$$a_1 = 4\alpha(1+\sqrt{1-4\alpha^2})^{-1}, \quad a_2 = (\sqrt{1-4\alpha^2}-1)(\sqrt{1-4\alpha^2}+1)$$

$$\sigma_\eta^2 \cdot (1+\sqrt{1-4\alpha^2}) = 2\sigma_\varepsilon^2 \quad \text{and} \quad \eta_t = -\frac{B(z_1 B - 1)}{\alpha z_2 (B - z_1)}\varepsilon_t$$

if $|z_1| < 1$, $|z_2| > 1$ are the roots of $\alpha z^2 - z + \alpha = 0$,

$$c_1 = \frac{2\alpha}{1+\alpha^2}, \quad c_2 = -\frac{\alpha^2}{1+\alpha^2}, \quad \sigma_e^2(1+2\alpha^2) = \sigma_\varepsilon^2.$$

**(1.3.5) Example.** Over $\mathbb{Z}^2$:

(1) There is an equivalence between $AR$ and $CAR$ representations

$$\begin{cases} AR: & X_{s,t} = \alpha X_{s-1,t} + \beta X_{s,t-1} + \varepsilon_{s,t}, \ \varepsilon \ b.b., \ |\alpha| + |\beta| < 1 \\ CAR: & X_{s,t} = \frac{\alpha}{1+\alpha^2+\beta^2}(X_{s-1,t} + X_{s+1,t}) \\ & \quad + \frac{\beta}{1+\alpha^2+\beta^2}(X_{s,t-1} + X_{s,t+1}) \\ & \quad - \frac{\alpha\beta}{1+\alpha^2+\beta^2}(X_{s-1,t+1} + X_{s+1,t-1}) + e_{s,t} \\ & \quad e_{s,t} \perp X_{u,v} \text{ if } (s,t) \neq (u,v) \\ & \text{and} \quad \sigma_e^2(1+\alpha^2+\beta^2) = \sigma_\varepsilon^2 \end{cases}$$

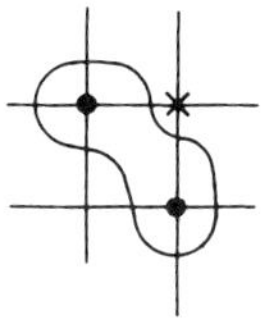

*AR support* : $R$

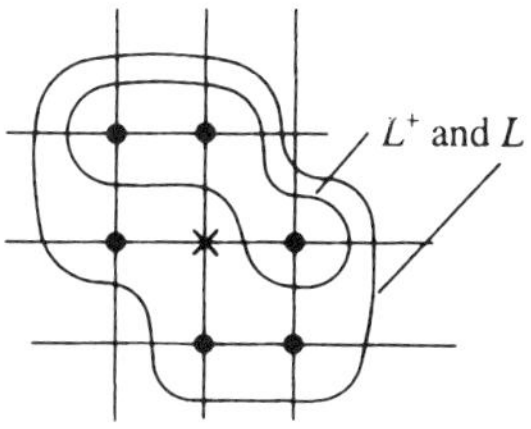


*CAR Support* : $L^+$ *and* $L$

(2) Bilateral $AR$ model and associated $CAR$ models:

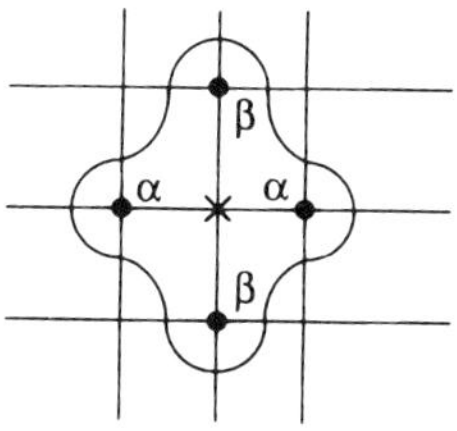


*AR Support* : $R$

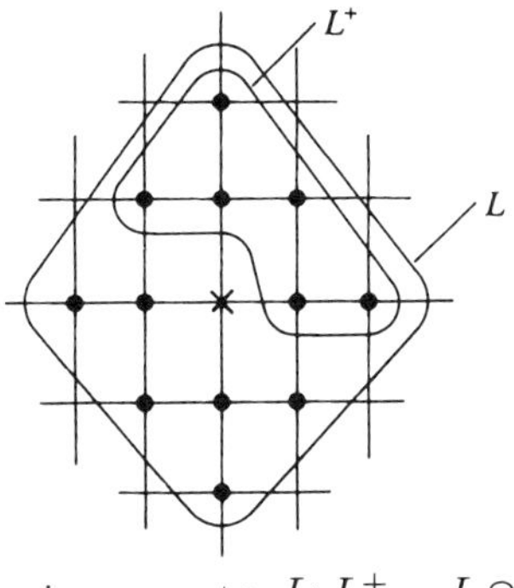


*Markovian support* : $L; L^+ = L \cap (\mathbb{Z}^2)^+$

$$c_{1,0} = 2\alpha A,\ c_{0,1} = 2\beta A,\ c_{2,0} = 2\alpha^2 A,\ c_{0,2} = 2\beta^2 A$$

$$c_{1,1} = c_{-1,1} = -2\alpha\beta A,\ \sigma_e^2 = \sigma_\varepsilon^2 A,\ A = (1 + 2\alpha^2 + 2\beta^2)^{-1}.$$

For the $AR$ representation, symmetric coefficients are not required.

# 1.4. Identification of Processes with Rational Spectra

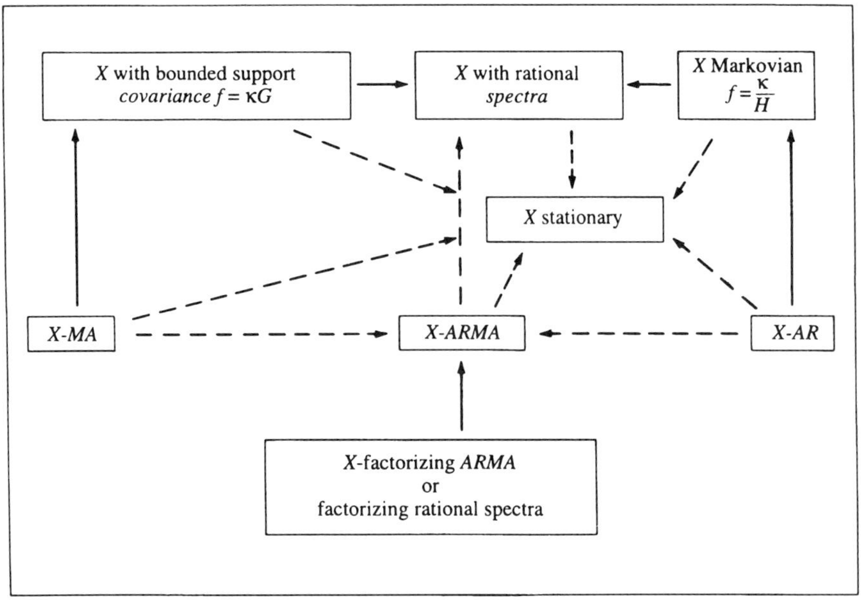


⟶ *for inclusion (the converse is never true over* $\mathbb{Z}^2$*);*

−− → *for "dense subset of".*

*Inclusion of principal rational spectra models over* $\mathbb{Z}^2$.

Consider $X$ a real stationary process over $\mathbb{Z}^d$ with rational spectral density
(1.23)

$$\begin{cases} f = \kappa \frac{G}{H} \\ G(\lambda) = \sum_M d_s \exp i < \lambda, s > \geq 0, \qquad M \text{ finite},\ d_s = d_{-s},\ d_0 = 1, \\ H(\lambda) = \sum_L c_s \exp i < \lambda, s > \geq 0, \qquad L \text{ finite},\ c_s = c_{-s},\ c_0 = 1. \end{cases}$$

We asume that $G$ and $H$ don't have any common factors. If $d = 1$, $X$ is an ARMA (with a canonical representation). However, if $d \geq 2$, this spectra family is strictly greater than that of ARMA models (cf. § 1.3.3).

### *1.4.1. Identification Properties*

*Identification of Rational Spectra.*

**(1.4.1) Theorem.** *Let $X$ have spectral density* (1.23) *with covariance $R$.*

*(a) Identification of $M$ (support of $G$) and of $H$: The following equations are true and allow the identification of $M$ and $H$*

$$\sum_{s \in L} c_s R(t-s) = 0 \quad \text{if} \quad t \notin M.$$

*(b) Identification of $G$: $Y = HX$ has spectral density $\kappa HG$ which allows the (linear) identification of $\kappa$ and $G$.*

*Proof:* Because of the isometry between $H_X$ and $L^2(f)$, one has:

$$Cov(X_u, H(B)X_0) = \int \exp -i < \lambda, u > G(\lambda) d\lambda = 0 \quad \text{if } u \notin M.$$

Conversely, suppose there exists a trigonometric polynomial $H$ whose support is in $L$ s. t. $cov(X_u, H(B)X_0) = 0$ if $u \notin M$. Then

$$\int \exp -i < \lambda, u > H(\lambda) f(\lambda) d\lambda = 0, \quad u \notin M$$

and thus, $H(\lambda)f(\lambda) = G(\lambda)$, $G$'s support in $M$. This gives $(a)$; $(b)$ follows immediately. □

*Identification of a Markovian L-field* [63]. The covariance satisfies

$$\sum_{L} c_s R(t+s) = \delta_{s,t}\sigma_e^2 \quad \text{if} \quad t \in L \cup \{0\}.$$

Now consider $R^L$, the covariances conditional to $\{X_s, s \in L\}$. Then $L$ is characterized by the following

**(1.4.2) Theorem.** *$X$ is a Markovian L-field if and only if*

$$R^L(X_0, X_t) = 0 \quad if \quad t \notin L \cup \{0\} \ .$$

*Proof:* If $X$ is an $L$-field, then $R^L(X_0, X_t) = R^L(e_0, X_t) = 0$ if $t \notin L \cup \{0\}$. Conversely, consider the following linear decomposition of $X_0$ over $H(X_t,\ t \neq 0)$,

$$X_0 = X_0(L) + X_0(\overline{L}) + e_0, \quad e_0 \perp X_t \quad \text{if } t \neq 0.$$

Here, $X_0(S)$ stands for the component over $H(S)$. Then, if $t \notin L \cup \{0\}$,

$$0 = R^L(X_0, X_t) = r(X_0(\overline{L}), X_t), \quad \text{i.e.,} \quad X_0(\overline{L}) = 0.$$

□

OBSERVATION: Theorem 1.4.2 is analogous to the classical result for the identification of the order $p$ of an $AR$ on $\mathbb{Z}$: if $k > p$, the covariance of $X_0$ and $X_k$, conditional to $X_1, \ldots, X_{k-1}$, is zero.

*Identification of Causal* ARMA *Representations.*
Consider the Causal ARMA $X$:

$$\begin{cases} P(B)X_t = Q(B)\varepsilon_t, & \varepsilon \text{ is the lexicographical innovation of } X; \\ P(z) = \sum_R a_s z^s, & R \subseteq (\mathbb{Z}^d)^+, \ R \text{ finite}, \ a_0 = 1; \\ Q(z) = \sum_S b_s z^s, & S \subseteq (\mathbb{Z}^d)^+, \ S \text{ finite}, \ b_0 = 1 \end{cases}$$

Then, as over $\mathbb{Z}$, one has the following identification property:

**(1.4.3) Theorem.** *Let* $s_0 = \text{Max}\{s, s \in S, b_s \neq 0\}$.

(a) *If* $t > s_0$, $\sum_R a_s r(t-s) = 0$. *We thus obtain the linear identification of* $P$ *and its order* $s_0$.

(b) $Y = P(B)X = Q(B)\varepsilon$, *which permits the quadratic identification of* $Q$.

We remark that identification of an ARMA can be done in two steps: in the first place, identification of the rational spectrum (Theorem 1.4.1), and in the second, that of the realization, unique up to a global sign, of $G = |Q^2|$ (resp., $H = |P|^2$) for the $MA$ part (resp., $AR$).

### *1.4.2. Covariance Calculus and Resolution of the Yule–Walker Equations*

If $f$ is known, its covariances can be calculated by numerical integration. If $f$ is rational, then its covariances satisfy linear recurrences after a certain order.

Over $\mathbb{Z}(d = 1)$, these recurrences can be solved analytically: the $R(k)$ can be expressed in terms of a finite number of initial covariances and of basic functions $k^n \lambda^k$, $|\lambda| < 1$ ([5], [23]).

Over $\mathbb{Z}^d$, $d \geq 2$, the situation is not the same. We do not know how to solve analytically a multi-indexed recurrence. On the other hand, we would have to know the covariance over an initial infinite subset in order to extend it to all of $\mathbb{Z}^d$ ([13], [17], [81], [120], cf. Exercise 8).

**(1.4.1) Example.** The simplest $AR$ model is

$$X_{st} = \alpha X_{s-1,t} + \beta X_{s,t-1} + \varepsilon_{s,t}, \ |\alpha| + |\beta| < 1, \text{ with covariance}$$

$$R(k,\ell) = \lambda^{-k}\mu^{-\ell} + \beta^{\ell}\sum_{i=0}^{k}\binom{-k}{i}(-\alpha)^i(\lambda^{k-i} - \lambda^{i-k}), \ k \geq 0, \ \ell > 0$$

$$\text{and} \quad \alpha\lambda + \beta\mu = 1 = \alpha\lambda^{-1} + \beta\mu^{-1} .$$

**(1.4.2) Example.** The Nearest-Neighbor Markovian Model:

$$X_{st} = \alpha(X_{s-1,t} + X_{s+1,t} + X_{s,t-1} + X_{s,t+1}) + e_{s,t}, \quad |\alpha| < 1/4, \quad \kappa = Var\ e.$$

Recurrences over $R$ are given by

$$R(k,\ell) = \delta_{k,\ell}\ \kappa + \alpha(R(k-1,\ell) + R(k+1,\ell) + R(k,\ell-1) + R(k,\ell+1)).$$

So we would need to know $R(k,0)$, $k \geq 0$ in order to extend the model to all of $\mathbb{Z}^2$ ($R(k,0)$ and $R(0,\ell)$, $k \geq 0$, $\ell \geq 0$ if the model is not isotropic). $R(0,0)$ can be expressed by the elliptic integral $R(0,0) = \frac{2\kappa}{\pi}K(16\alpha^2)$, $R(1,0) = \frac{R(0,0)-\kappa}{4\alpha}$, $K(x) = \int_0^{\pi/2}(1 - x\sin^2\theta)^{-1/2}$, $0 \leq x < 1$.

**(1.4.3) Example.** $f(\lambda,\mu) = \kappa(1 + 2\gamma\cos\mu)(1 - 2\alpha\cos\lambda - 2\beta\cos\mu)^{-1}$.

We would need to know $R(k,0)$, $R(0,\ell)$, $k,\ell \geq 0$, and $R(s,1)$, $s \in \mathbb{Z}$ in order to extend $R$ by recurrence over all of $\mathbb{Z}^2$.

If $R$ is the covariance associated to a rational spectrum, then the linear recurrences can be identified, theoretically, over $\mathbb{Z}^d$, $d \geq 2$, as with $\mathbb{Z}$. Based on the corner method ([5], [10]), we shall define two enumerations: one, $\{s(i),\ i \geq 0\}$, beginning with the $n_0$ points of $L$; the other, $\{t(j), j \geq 0\}$, beginning with the $m_0$ points of $M$. We now apply the corner method to the $n \times n$ matrix:

$$\mathcal{R}^{n,m} = (R(t(m) + s(i), s(j)), 1 \leq i,j \leq n) .$$

Because the method is not numerically stable, it is advised to consider statistical methods based on penalized quasi-likelihood methods (cf. 3.4.3).

We remark that the covariances associated to a rational spectrum model have exponential decay: if $H$ is not zero over a neighborhood of

$\{\delta \leq |z| \leq \delta^{-1}\}^d$ of $T^d$, $0 < \delta < 1$, and if $f$ is bounded by $M$ over $T^d$, then the Laurent development of $f(z)$ yields:

$$|R(k)| \leq M\ \delta^{|k|}, \quad \text{where } |k| = \sum_1^d |k_\ell|.$$

# 1.5. Nonstationary Representations

## 1.5.1. Finite Collection of Sites: $S = \{1, 2, \ldots, n\}$

Let $X = \{X_1, \ldots, X_n\}$ centered with covariance $\Sigma$.

*AR Representations:* For a given enumeration of $S$, there exists a unique causal representation in terms of $\varepsilon_p$, with $\varepsilon$ a white noise,

$$\sum_{j=1}^{p} a_{pj} X_j = \varepsilon_p,\ p = 1,\ n,\ \sigma_\varepsilon^2 = 1, \quad \text{i.e., } AX = \varepsilon. \tag{1.24}$$

Here, the lower triangular matrix $A$ is defined by the factorization $\Sigma^{-1} = {}^tA \cdot A$. Both $\Sigma$ and $A$ depend on $\frac{1}{2}n(n+1)$ parameters. This assures the identification of the causal $AR$ model.

All the equivalent $AR$ representations (noncausal and nonidentifiable in general) are obtained as

$$BX = \eta, \quad cov\ \eta = I_n, \quad B = PA, \quad \eta = P\varepsilon,$$

$P$ orthogonal. More specifically, in order for the model to be well specified, $\frac{1}{2}n(n-1)$ constraints have to be fixed: the causal choice corresponds to $B_{ij} = 0$, $1 \leq i < j \leq n$.

*MA Representations:* A causal $MA$ representation is $X = A^{-1}\varepsilon$. Other equivalent representations are given by $X = B^{-1}\eta$.

*Markovian Representation:*

$$X_i = \sum_{j:j\neq i} c_{ij} X_j + e_i, \quad Var\ e_i = \sigma_i^2, \quad e_i \perp X_j \quad \text{si } i \neq j. \tag{1.25}$$

If we assume $c_{ii} = 0$, $C = (c_{ij})$, and $D = (\delta_{ij}\sigma_i^2)$, the Yule–Walker equations yield $\Sigma = (I - C)^{-1}D$. In other words, (1.25) shall be well specified if $D^{-1}(I - C)$ is symmetric and positive definite. In particular, the constraints $c_{ij}\sigma_j^2 = c_{ji}\sigma_i^2$ must be satisfied.

### *1.5.2. Infinite Collection of Sites*

Let $X = \{X_i,\ i \in S\}$, a centered process with $\sup_i (Var(X_i)) < \infty$, such that for all $i$, $X_i \notin H(X_j,\ j \neq i)$. For any given enumeration $S = \{1, 2, \dots, n, \dots\}$ of $S$, we define intrinsically a white noise $\varepsilon = \{\varepsilon_i, i \in S\}$, a causal $AR$ representation, and a causal $MA$ representation by

$$\sum_{j=1,n} a_{n,j} X_j = \varepsilon_n, \quad n \geq 1, \quad X_n = \sum_{j=1,n} b_{n,j} \varepsilon_j, \quad n \geq 1.$$

If $\Sigma_n = Cov(X_1, X_2, \dots, X_n)$, $A_n$ is the lower triangular matrix such that ${}^tA_n \cdot A_n = \Sigma_n$ and $B_n = A_n^{-1}$.

*Second-Order Linear Processes* (cf. 1.6.2). Let $\varepsilon = (\varepsilon_j)$ be a centered non-correlated process. Define $\sigma_j^2 = Var\ \varepsilon_j$. Then the $MA$ representation

$$X_i = \sum_{j \in S} b_{i,j} \varepsilon_j\ , \quad b_{ii} = 1,\ i \in S, \tag{1.26}$$

is well defined if

$$M = \sup_i \Big( \sum_{j \in S} b_{i,j}^2 \sigma_j^2 \Big) < \infty. \tag{1.27}$$

Assume that additionally

$$\alpha = \sup_i \Big( \sum_{j : j \neq i} |b_{i,j}| \Big) < 1 \tag{1.28}$$

and that $S$ is metric, and $|i - j|$ is the distance from $i$ to $j$. The operator $B = (b_{ij})$ over $\ell_\infty(S)$ is invertible and bicontinuous, and $\|B\| = \alpha$. Also, if $\Delta = I - B$, $A = B^{-1} = \sum_{n \geq 0} \Delta^n$, $\|A\| \leq (1 - \alpha)^{-1}$, since one has, by recurrence, that

$$\sum_j |\Delta^n{}_{ij}| \leq \sum_{j,k} |\Delta^{n-1}{}_{ik}|\ |\Delta_{kj}| \leq \alpha^n.$$

This gives sense to the $AR$ representation:

$$\sum_j a_{i,j} X_j = \varepsilon_i, \quad i \in S. \tag{1.29}$$

Indeed, defining $\varepsilon_{i,N} = \sum_{j:|j-i| \leq N} a_{ij} X_j$, since $A \cdot B = I$, one has $\varepsilon_{i,N} = \varepsilon_i - \sum_{k \in S} \varepsilon_k \Big( \sum_{j:|j-i|>N} a_{ij} b_{jk} \Big)$. Thus, for every $i$:

$$\|\varepsilon_{i,N} - \varepsilon_i\|^2 \leq \sum_{\substack{j:|j-i|>N \\ j':|j'-i|>N}} |a_{ij}|\ |a_{ij'}| \sum_k |b_{jk}|\ |b_{j'k}| \sigma_k^2$$

$$\leq M\Big( \sum_{j:|j-i|>N} |a_{ij}|^2 \Big) \underset{N \to \infty}{\longrightarrow} 0.$$

**(1.5.1) Theorem.** *Under condition* (1.27), *the* $MA$ *representation* (1.26) *is well defined. If additionally condition* (1.28) *is true,* $X$ *admits the* $AR$ *representation* (1.29).

Conversely, consider $\varepsilon = \{\varepsilon_j\}$ a noncorrelated noise, $\sigma_j^2 = Var\ \varepsilon_j$, and

(1.30) $$\sup_i \sigma_j^2 < \infty.$$

Let $A = (a_{ij})$, with ones in its diagonal, satisfy

(1.31) $$\beta = \sup_i \sum_{j:j \neq i} |a_{ij}| < 1.$$

Then $B = A^{-1}$ exists, with norm $\sup_i \sum_j |b_{ij}| \leq (1-\beta)^{-1} < \infty$.

**(1.5.2) Theorem.** *Under* (1.30) *and* (1.31), *the* $AR$ *representation*

(1.32) $$\sum_j a_{ij} X_j = \varepsilon_i, \quad i \in S,$$

*is well defined.* $X$ *is the* $MA$ *process:* $X_i = \sum_j b_{ij}\varepsilon_j$ *with* $B = A^{-1}$.

*Conditional Autoregressions.* If the distribution of $\varepsilon$ and the $MA$ (1.26) or $AR$ (1.32) representations are valid, $X$ has a well-defined distribution. If, on the other hand, $X$ is specified by its conditional distributions and $S$ is not finite, this in general is not enough to determine $X$'s distribution. Conditional distributions are generally not enough to determine the joint distribution. These issues shall be considered in Chapter 2 and § 2.3 will deal with Markovian Gaussian specifications, that is, with conditional Gaussian autoregressions.

## 1.6. Multivariate Markovian Gaussian Fields

$X = \{X_i, i \in S\}$, $X_i \in \mathbb{R}^p$ is a multidimensional, centered Gaussian field.

### 1.6.1. Case $S = \{1, 2, \ldots, n\}$: Contextual Discriminant Analysis

Assume $X = \{X_1, \ldots, X_n\}$ has the following conditional distributions:
(1.33)

$$E(X_i \mid X_j, j \neq i) = \sum_{j:j \neq i} C_{ij} X_j, \ Var(X_i \mid X_j, j \neq i) = V_i \quad i = 1, n.$$

Under which conditions does (1.33) specify a global distribution, and what is this global distribution? Define $X_c = {}^t({}^tX_1, \ldots, {}^tX_n)$, the "concateneted" form of $X$ with covariance $\Sigma$. We shall write

$$\Sigma^{-1} = (\Sigma^{ij}), \tag{1.34}$$

the $p \times p$ block matrices associated to $\Sigma^{-1}$. Define $C_{ii} = -I_p$.

**(1.6.1) Theorem.**

(*a*) (1.33) *specifies a centered Gaussian distribution X under the following conditions*:

(*i*) $C_{ij} V_j = V_i \, {}^tC_{ji}$, $i,j = 1,n$.

(*ii*) $( - V_i^{-1}, i,j = 1,n)$ *or* $( - C_{ij}, \, i,j = 1,n)$ *are* $np \times np$ *positive definite matrices. Then if* $\Sigma = ( - V_i^{-1} C_{ij})$, *X is* $\mathcal{N}(0,\Sigma)$.

(*b*) *Conversely, if X is* $\mathcal{N}(0, \Sigma)$, *X satisfies the conditional Gaussian model* (1.33) *with* $C_{ij} = -(\Sigma^{ii})^{-1} \Sigma^{ij}$, $V_i = (\Sigma^{ii})^{-1}$, $i, j = 1, n$.

*Proof:* Joint densities are related to conditional ones by

$$\frac{f(x)}{f(y)} = \prod_{i=1}^{n} \frac{f(x_i | x_1, \ldots, x_{i-1}, y_{i+1}, \ldots, y_n)}{f(y_i | x_1, \ldots, x_{i-1}, y_{i+1}, \ldots, y_n)},$$

$$\frac{f(x)}{f(y)} = \prod_{i=1}^{n} \frac{f(x_i | y_1, \ldots, y_{i-1}, x_{i+1}, \ldots, x_n)}{f(y_i | y_1, \ldots, y_{i-1}, x_{i+1}, \ldots, x_n)}$$

(cf. [25]) so that because of (1.33),

$$-2 \ \log\Big(\frac{f(x)}{f(0)}\Big) = \sum_{i=1}^{n} {}^t x_i V_i^{-1} x_i - 2 \sum_{i=2}^{n} \sum_{j=1}^{i-1} {}^t x_i V_i^{-1} C_{ij} x_j$$

$$= \sum_{i=1}^{n} {}^t x_i V_i^{-1} x_i - 2 \sum_{i=1}^{n-1} \sum_{j=i}^{n} {}^t x_i V_i^{-1} C_{ij} x_j.$$

Since both expressions are identical, one has

$$-2 \ \log f(x) = \text{Const} + \sum_{i=1}^{n} {}^t x_i V_i^{-1} x_i - \sum_{i=1}^{n} {}^t x_i V_i^{-1} C_{ij} x_j.$$

This gives $(a)$; $(b)$ is standard. □

**(1.6.1) Corollary.** *Assume $i \neq j$. Then $X_i$ and $X_j$ are conditionally independent with respect to $\{X_k, k \neq i, k \neq j\}$ if and only if $C_{ij} = 0$. $\partial i = \{j \mid j \neq i$ and $C_{ij} \neq 0\}$ is the Markov neighborhood of $i$.*

• *Spatial $\times$ Multispectral Covariance Factorization* ([116], [158], [171]). Assume that for all $i, j$, $C_{i,j} = c_{ij} I_p$ with $c_{i,j} \in \mathbb{R}$; one has $p$ real Markovian marginal models, identical up to the residual variance:

$$X_i(k) = \sum_{j \neq i} c_{ij} X_j(k) + e_i(k) \quad i = 1, n, \ k = 1, p \ .$$

Dependence among them is given by matrices $V_i = cov(e_i)$, $i = 1, n$.

Assume also that $V_i = V$, $i = 1, p$. Then if $\otimes$ stands for the Kronecker product and defining $c_{ii} = -1$, $C = (c_{ij})$, one has $\Sigma = (-C) \otimes V$; that is, $\Sigma_{ij} = -c_{ij} V$ is the $(i, j)$–th $p \times p$ block of $\Sigma$.

Conditions $(i)$ and $(ii)$ of Theorem 1.6.1 are assured if $c_{ij} = c_{ji}$ and $-C$ is positive definite. We shall see why this factorization leads to interesting analytical simplifications for the problem of classification using contextual discriminant analysis.

• *Contextual Discriminant Analysis Under Covariance Factorization* (cf. also 6.8.3). One wishes to classify a site $s$ (we will also say a *pixel*: picture element) among $K$ different Gaussian populations with different means.

*Context Blind Classification* $(BC)$. We only consider $X_s \in \mathbb{R}^p$, the observation at $s$, and not what happens around $s$. If $s$ is in class $k$, $X_s \sim \mathcal{N}_p(\mu_k, V)$ and one chooses, given $x_s$,

$$(BC) \qquad \widehat{k} = \underset{k}{\text{Arg Max}}\, \ell_k, \quad \ell_k = {}^t\mu_k V^{-1}(x_s - \tfrac{1}{2}\mu_k).$$

*Simple Contextual Classification* $(SCC)$. One has observations over a neighbourhood $S$ of $s$, and we shall assume that if $s$ belongs to class $k$, then all points in $S$ also belong. Consider $S$ an $n$-point set (traditionally $n = 5$ or 9). Define $X_S = \{X_t, t \in S\}$; then, if $s$ belongs to class $k$,

$$X_S \sim \mathcal{N}_{np}(M_k, \Sigma), \quad M_k = 1\!\!1_n \otimes \mu_k$$

and classification is done according to:

$$(SCC) \qquad \widehat{k} = \underset{k}{\text{Arg Max}}\, L_k, \quad L_k = {}^tM_k \Sigma^{-1}(x_S - \tfrac{1}{2}M_k).$$

If one has two populations $(K = 2)$, classification errors are, respectively, given by:

$$e_{BC} = \phi\Big(-\tfrac{1}{2}d\Big), \quad e_{SCC} = \phi\Big(-\tfrac{1}{2}D\Big).$$

Here, $\phi$ stands for the distribution function of the reduced normal, and $d$, and $D$ are the Mahalanobis' distances between two populations under $BC$ or $SCC$. It is intuitive that $D \geq d$ and thus that contextual clasification is to be preferred. This gain, however, increments the amount of numerical calculation needed. If the covariance factorizes, this problem disappears and the $SCC$ method can be explicitly calculated.

• *SCC in a Factorizing Covariance Setting.* Define $C$ as the partial correlation matrix and $V$ as the multispectral variance matrix. Then $\Sigma = C \otimes V$, $M_k = 1_n \otimes \mu_k$, and one has

$$\begin{cases} {}^tM_k(C\otimes V)^{-1} = ({}^t1_nC^{-1})\otimes({}^t\mu_kV^{-1}), \\ {}^tM_k(C\otimes V)^{-1}x_S = {}^t\mu_kV^{-1}z_s \quad \text{with} \\ z_s = {}^t1_nC^{-1}x_S \sim \mathcal{N}_p(\nu^2\mu_k, \nu^2V) \quad \text{under } k, \\ {}^tM_k(C\otimes V)^{-1}M_k = \nu^2\ {}^t\mu_k\mu_k \quad \text{with } \nu^2 = {}^t1_nC^{-1}1_n. \end{cases}$$

The discriminating function thus becomes $L_k = {}^t\mu_kV^{-1}(z_s - \nu^2\mu_k)$; we point out:

(*i*) the discriminant vector $z_s$ has the same dimension as $x_s$: incorporating context doesn't make the algorithm much more complicated.

(*ii*) $\nu^2$ appears as a gain factor of the $SCC$ method over the $BC$. For example, if $K = 2$, one has $D = \nu d$ with

$$\nu^2 = {}^t1_nC^{-1}1_n \geq n/\lambda \geq 1.$$

Here, $\lambda$, the greatest eigenvalue (g.e.v.) of $C$, is less than $n$; $\nu$ varies from 1 to $\sqrt{n}$ and then from $\sqrt{n}$ to infinity when the spatial correlation varies from 1 to 0 and then from 0 to $-1$, i.e., is attractive, independent or repulsive. If we consider a window that for each $s$ picks out $s$ and its four nearest neighbors ($n = 5$), and if $\rho(u,v) = \rho^{|u|+|v|}$, then

$$\nu^2 = \frac{5-3\rho}{\rho+1} = \begin{cases} 1 & \text{if } \rho \longrightarrow 1_-, \\ 5 & \text{if } \rho = 0, \\ \infty & \text{if } \rho \longrightarrow (-1)_+, \end{cases}$$

and the discriminant vector is

$$z_s = \frac{1-3\rho}{1+\rho}x_s + \frac{4}{1+\rho}\overline{x}_S \quad \text{with} \quad \overline{x}_S = \frac{1}{4}\sum_{t\in S:t\neq s} x_t.$$

### *1.6.2. Stationary Markovian Gaussian Fields over $S = \mathbb{Z}^d$*

Consider $L$ a symmetric neighborhood of the origin, $0 \notin L$. The stationary form of (1.33),

$$E(X_i|\cdot) = \sum_L C_\ell X_{i+\ell}, \quad Var(X_i|\cdot) = V, \tag{1.35}$$

must satisfy

$(i)^*$ $C_{-s}V = V\ {}^tC_s\ V^{-1}$, $s \in L$, and

$(ii)^*$ $\{(-C_{i-j}),\ i,j \in \Lambda\}$ is positive definite (p.d.)

for every $\Lambda$ finite subset of $\mathbb{Z}^d$. Here, $C_0 = -I$.

*Yule–Walker Equations:* If $X$ is a real, stationary Markovian field with covariance $R_h = E(X_t\ {}^tX_{t+h})$, one has

$$\begin{cases} R_0 = V + \sum_L C_\ell R_\ell = V + \sum_L C_\ell\ {}^tR_{-\ell}, \\ R_h = \sum_L C_s R_{h-s} \quad h \neq 0. \end{cases} \tag{1.36}$$

The first equation is a consequence of (1.35) and the second is true because of the following variance decomposition:

$Var\,(X_s) = E_{X_{\bar{s}}}(Var(X_s|X_{\bar{s}})) + Var_{X_{\bar{s}}}(E(X_s|X_{\bar{s}}))$ where $X_{\bar{s}} = \{X_t, t \neq s\}$

Condition $(ii)^*$ is not practical because it must be verified for every $\Lambda \in \mathcal{S}$. Define $\|A\| = (g.e.v.({}^t\overline{A}A))^{1/2}$, $\overline{A}$ the conjugate of $A$, as the spectral norm of a square matrix.

**(1.6.2) Theorem.** [*117*] *Assume that $(C_\ell, \ell \in L)$ and $V$ satisfy $(i)^*$, that $V$ is positive definite, and that*

$$\sum_L \lambda_\ell < 1, \tag{1.37}$$

*where $\lambda_\ell = \|V^{-1/2}C_\ell V^{1/2}\|$, and $V^{1/2}$ is the symmetric square root of $V$. Then there exists a stationary Gaussian Markovian field over $\mathbb{Z}^d$ that satisfies* (1.35) *with spectral density matrix:*

$$F_X(\lambda) = (I - \sum_{\ell \in L} C_s \exp(i < \lambda, \ell >))^{-1} V.$$

*Proof:* Assume that the local conditions (1.35) are true. We will show that the moment generating function satisfies

$$G_X(z) = (I - \sum_L C_\ell z^\ell)^{-1} V \tag{1.38}$$

if $z \in T^d$, $\ell \in \mathbb{Z}^d$, and $z^\ell = z_1^{\ell_1} \times \cdots \times z_d^{\ell_d}$. Indeed, $G_X(z) = \sum_{\mathbb{Z}^d} R_h z^h$. Replacing $R_0$ and $R_h$, $h \neq 0$, by (1.36), one has

$$G_X(z) = V + \sum_{\ell \in L} \sum_{h \in \mathbb{Z}^d} C_\ell R_h z^{\ell+h} = V + \Big(\sum_{\ell \in L} C_\ell z^\ell\Big) G_X(z),$$

which yields (1.38). If it exists, the spectral density matrix is given by

$$F_X(\lambda) = G_X(e^{-i\lambda_1}, \ldots, e^{-i\lambda_d}) = (I - \sum_{\ell \in L} C_\ell \, e^{-i<\lambda,\ell>})^{-1} V.$$

Under $(i)^*$, $F_X$ is Hermitian and defines a Gaussian process over $\mathbb{Z}^d$ if the following two conditions are also satisfied [77]:

$$(a) \;\; F_X(\lambda) \text{ is } d.p.; \quad (b) \;\; \int_{T^d} \varphi(\lambda) d\lambda < \infty.$$

Here, $\varphi(\lambda)$ stands for the g.e.v. of $F_X(\lambda)$. However, $F_X = V^{1/2} A^{-1} V^{1/2}$, where

$$A = I - B, \quad B = \sum_{\ell \in L} C_\ell^* e^{i<\lambda,\ell>}, \quad C_\ell^* = V^{-1/2} C_\ell V^{1/2}.$$

If $\alpha_1 \leq \alpha_2 \leq \cdots \leq \alpha_p$ are the eigenvalues of $A$, and $\delta_1 \leq \delta_2 \leq \cdots \leq \delta_p$ are the absolute values of the eigenvalues of $B$, then $\alpha_1 \geq 1 - \delta_p > 0$ since $\delta_p = \|B\| \leq \sum_L \|C_\ell^*\| = \sum_L \lambda_\ell < 1$. So that $\alpha_1 > 0$, $A$ and thus $F_X$ is d.p. and

$$\varphi(\lambda) = \|F_X(\lambda)\| < \|V\| \; \|A^{-1}\| = \|V\|/\alpha_1 < \|V\| \Big(1 - \sum_L \lambda_\ell\Big)^{-1}$$

is bounded. □

REMARKS:

(1) If the process factorizes, (1.37) becomes $\sum_L |c_\ell| < 1$.

(2) The field is *reversible* if for all $n$ and $\{t_1, \ldots, t_n\}$, $\{X_{t_1}, \ldots, X_{t_n}\}$ and $\{X_{s-t_1}, \ldots, X_{s-t_n}\}$ have the same law . Now this means $F_X(\lambda)$ is real and either $F_X(\lambda)$ $(\lambda \in T^d)$ or $R_h$ $(h \in \mathbb{Z}^d)$ is symmetric. In the case of a Markovian field, this is equivalent to

$$C_s = C_{-s} \quad \text{and} \quad C_s = V \, {}^tC_s \, V^{-1}, \; s \in L.$$

So that the spectral density can be written as

$$F_X(\lambda) = (I - \sum_L C_s \cos < \lambda, s >)^{-1} V$$

and $\lambda_s = \{g.e.v.(C_s {}^tC_s)\}^{1/2}$ does not depend any longer on $V$. If, in addition, $C_s = {}^tC_s$, then $\lambda_s = g.e.v.(C_s)$.

(3) Let $X$ be an $AR$ process: $X_t = \sum_R A_s X_{t+s} + \mathcal{E}_t$, $\mathcal{E}_t$ centered, independent and identically distributed (i.i.d.), $Cov\ \mathcal{E}_t = \Gamma$. Under stationary conditions, the spectral density matrix of $X$ is $[I - \sum_R A_s e^{i<\lambda,s>}]^{-1}\Gamma[I - \sum_R A'_s e^{-i<\lambda,s>}]^{-1}$. If $d = 1$, there is a biunivocal correspondence between $AR$ models and Markovian models.

## 1.7. Mixing Properties for Linear Spatial Processes

Let $X$ be a random field over $S$, with $S$ a given metric, and $A$ and $B$ two subsets of $S$. We define the *strong mixing coefficient* of $X$ over $A$, $B$ by

(1.39)

$$\alpha^X(A,B) = \sup\{|P(U \cap V) - P(U)P(V)|,\ U \in \mathcal{F}(X(A)),\ V \in \mathcal{F}(X(B))\}$$

where $\mathcal{F}(Z(E))$ is the $\sigma$-algebra generated by $Z$ over $E$. We shall also define the following coefficients, which depend on $a$, $b \in \mathbb{N} \cup \{\infty\}$ and $k \in \mathbb{R}$:

$$\alpha^X_{a,b}(k) = \sup\{\alpha^X(A,B),\ |A| \le a,\ |B| \le b,\ d(A,B) \ge k\}. \tag{1.40}$$

We shall examine mixing properties first for stationary Gaussian fields and then, more generally, for linear spatial processes. For references, see [52], [53], [73] and [169].

### 1.7.1. Regularity and Mixing for Stationary Gaussian Processes

Assume $X = \{X_t, t \in \mathbb{Z}^d\}$, $X_t \in \mathbb{R}$ is Gaussian and stationary with spectral density $f$. Let

$$\rho_X(A,B) = \sup\{Corr(\varphi,\psi), \varphi \in L^2(X(A)), \psi \in L^2(X(B))\} \tag{1.41}$$

stand for the maximum correlation between $L^2(X(A))$ and $L^2(X(B))$.

**(1.7.1) Theorem.** (Kolmogorov–Rozanov).

$$\rho_X(A,B) = \sup\{Corr(\varphi,\psi), \varphi \in H_X(A),\ \psi \in H_X(B)\}.$$

*Proof:* We can reduce the problem to $A$ and $B$ finite and define a linear transformation which leaves invariant the spaces generated by $X_A$ and $X_B$ and such that the covariance matrix of the new variables has the following reduced form: if $|A| = n \le |B| = m$, then

$$Cov\ X_A = I_n,\ Cov X_B = I_m, Cov(X_A, X_B) = \begin{pmatrix} \rho_1 & & 0 & \mathbf{0} \\ & \ddots & & \vdots \\ 0 & & \rho_n & \mathbf{0} \end{pmatrix}.$$

Here, $\rho_1 \geq \rho_2 \geq \cdots \geq \rho_n \geq 0$. Spaces $H_X(A)$ and $L^2(X(A))$ are also invariant under this transformation.

Let $H_\ell(x)$ be the Hermite polynomial of degree $\ell$. If $i = (i_1, \ldots, i_n)$ is a $n$-multindex of $\mathbb{N}^n$, $x_A = (x_1, \ldots, x_n)$, we write $H_i(x_A) = \prod_{\ell=1,n} H_{i_\ell}(x_\ell)$. A reduced element of $g_A \in L^2(X(A))$ can be written as

$$g_A(x_A) = \sum_{i \in \mathbb{N}^A} a_i H_i(x_A), \text{ where}$$

$$a_0 = Eg_A(X_A) = 0,\ E(g_A(X_A))^2 = 1 = \sum_{\mathbb{N}^A} \frac{a_i^2}{i!},\ i! = \prod_{\ell=1}^{n} (i_\ell!).$$

With analogous normalization conditions for the $b_j$, one has

$$g_B(x_B) = \sum_{j \in \mathbb{N}^B} b_j H_j(x_B) = g_{B,1}(x_B) + g_{B,2}(x_B)$$

$$\text{where} \qquad g_{B,1}(x_B) = \sum_{j=(j_1,\ldots,j_n;0,\ldots,0) \in \mathbb{N}^A \times 0} b_j H_j(x_B)$$

depends only on $x_{B,1} = (x_1, x_2, \ldots, x_n; 0, \ldots, 0)$. Finally,

$$E[g_A(X_A) g_B(X_B)] = E[g_A(X_A) g_{B,1}(X_{B,1})]$$

$$= \sum_{i \in \mathbb{N}^A} a_i b_i \frac{\rho_1^{i_1} \times \rho_n^{i_n}}{i_1! \cdots i_n!} \leq \rho_1 \sum_i \frac{|a_i b_i|}{i!} \leq \rho_1 = \rho(X_1, Y_1)\ .$$

□

Since $\mathbb{1}_U \in L^2(X(A))$ when $U \in \mathcal{F}(X(A))$, one has $\alpha^X(A,B) \leq \rho_X(A,B)$. The following result, which was communicated to us by I. Ibragimov, relates the *regularity coefficient* $\rho_X$ to the spectral density; known results over $\mathbb{Z}$ [94] are thus generalized to $\mathbb{Z}^d$.

**(1.7.2) Theorem.** *Assume* $f(\lambda) \geq a > 0$ *over* $T^d$. *Then*

$$\rho_X(A,B) \leq a^{-1} \Delta_K(f),$$

*where* $K = \text{dist}(A{,}B)$ *and*

$$\Delta_K(f) = \inf\{\|f - P\|_\infty, P(\lambda) = \sum_{|t|<K} c_t e^{i<\lambda,t>}\}$$

*stands for the best trigonometric approximation of* $f$ *by a polynomial of degree* $K-1$. *In particular,*

$$\alpha^X_{\infty,\infty}(k) \leq a^{-1} \Delta_k(f).$$

*Proof:* Consider the following two inner products over $L^2(X)$ defined by the norms

$$\|\varphi\|_f^2 = \int |\varphi|^2 f d\lambda, \quad \|\varphi\|_2^2 = \int |\varphi|^2 d\lambda.$$

Because of $f$'s lower bound, $\|\varphi\|_2 \leq a^{-1/2}\|\varphi\|_f$. Applying Theorem 1.7.1, it is enough to study the correlation between $\varphi \in H_X(A)$ and $\psi \in H_X(B)$, such that $\|\varphi\|_f = \|\psi\|_f = 1$.

$$\rho(\varphi,\psi) = \int \varphi\overline{\psi} f d\lambda = \int \Big[\sum_{s\in A-B} \gamma_s e^{i<\lambda,s>}\Big]\Big[P_{K-1} + (f - P_{K-1})\Big] d\lambda$$

$$= \int \Big[\sum_{s\in A-B} \gamma_s e^{i<\lambda,s>}\Big](f - P_{K-1}) d\lambda$$

$$(\text{if } s \in A - B,\ |s| \geq K \text{ and } \partial^0 P_{K-1} < K)$$

$$= \int \varphi \cdot \overline{\psi}(f - P_{K-1}) d\lambda \leq \Delta_K(f)\|\varphi\|_2\|\psi\|_2 \leq a^{-1}\Delta_\kappa(f).$$

□

As over $\mathbb{Z}$, we can control $\rho$ and $\alpha$ due to the properties of polynomial approximations of $f$:

- • If $f$ is continuous, $\Delta_K(f) \longrightarrow 0$ if $K \longrightarrow \infty$ (Stone–Weirstrass' Theorem).
- •• If $f$ is analytical (e.g. rational), $\Delta_K(f) \longrightarrow 0$ exponentially in $K$.
- ••• Let's examine an upper bound for $\Delta_K(f)$ considering the linear representation of $X$,

(1.42)
$$X_t = \sum_{s\in\mathbb{Z}^d} b_{t-s}\varepsilon_s, \quad \varepsilon \text{ is a white noise.}$$

Since $f(\lambda) = \sum_{s\in\mathbb{Z}^d} c_s e^{i<\lambda,s>}$, $c_s = \sum_j b_j b_{j+s}$, we get:

$$\Delta_k(f) \leq \sum_{|s|\geq k} |c_s| \leq 2\|b\|_\infty \sum_{|s|\geq \frac{k}{2}} |s|\ |b_s|\ .$$

**(1.7.2) Corollary.** *If $X$ is the linear Gaussian process* (1.42), *and* $f(\lambda) \geq a > 0$, *then*

$$\alpha^X_{\infty,\infty}(k) \leq \frac{2}{a}\|b\|_\infty\Big(\sum_{|s|\geq\frac{k}{2}} |s|\ |b_s|\Big),\ |s| = \sum_1^d |s_i|\ .$$

REMARKS:

(1) If $X$ has a spectral density bounded from below by $a > 0$ and if $Y_t = P(B)X_t$, $P$ is a polynomial (which eventually has zeros over $T^d$), and if $r$ is greater than both the degree and the number of nonzero coefficients of $P$, then

$$\alpha^Y_{a,b}(k) \le \alpha^X_{a+r,b+r}(k-r), \quad k \ge r .$$

(2) Let $R$ be $X$'s covariance. If

$$\begin{aligned} c(A,f) &= \inf\{\|\varphi\|_2,\ \varphi \in H_X(A),\ \|\varphi\|_f = 1\} \\ &= \inf\Big\{\textstyle\sum_A c_t^2 : \sum_{s,t\in A} c_s c_t R(s-t) = 1\Big\} < \infty. \end{aligned}$$

The Theorem's proof shows that if $\lambda\{f > 0\} > 0$

$$\rho_X(A,B) \le c(A,f)c(B,f)\Delta_K(f).$$

## *1.7.2. Mixing Properties for Linear Fields*

Let $Z$ be a real field over $S = \mathbb{Z}^d$, and let $X$ be the following linear field over $S$

$$X_i = \sum_j b_{i,j} Z_j, \quad b_{i,i} = 1,\ i \in S \quad (X = BZ). \tag{1.43}$$

For $\delta > 0$, that $X$ belongs to $L^\delta$ is guaranteed if

$$(L1)\quad \begin{cases} \sup_i E|Z_i|^\delta < \infty, & (1.44) \\ \gamma = \sup_i \Big(\sum_j |b_{i,j}|^\rho\Big) < \infty, \quad \rho = \inf\{1,\delta\}. & (1.45) \end{cases}$$

Indeed, the sequence $X_i^N = \sum_{j:|j-i|\le N} b_{ij} Z_j$ is Cauchy in $L^\delta$. This space is complete if $\delta < 1$ under the metric $d(X,Y) = E|X-Y|$ and is Banach if $\delta \ge 1$. Notice that (1.45) is more restrictive than the second order condition

$$\sup_i \sum_j b_{i,j}^2 < \infty, \tag{1.46}$$

which permits us to define $X$, for $\delta = 2$, if $Z$ is a noncorrelated process that satisfies (1.44). Under (1.45) it is not necessary to ask any condition other than (1.44) in order to define $X$.

*Weak Dependence and Regularity of Z.*

$(M1)$ For $b < \infty$, $\alpha^Z_{a,b}(k) \longrightarrow 0$ exponentially when $k \to \infty$.

$(M2)$ If $V \subset S$ is finite, $Z_V = \{Z_j, j \in V\}$ has a conditional density $p_V(z|y)$ which satisfies uniformly in $V$,

$$\int_{\mathbb{R}^V} |p_V(z+x|y) - p_V(z|y)|dz \leq C \sum_{i \in V} |x_i|$$

for a certain constant $C$. The exponential condition in (M1) can be weakened (cf. [52]).

**(1.7.1) Example.** Independent $Z$ Processes. If $Z_i$ with density $p_i$ satisfies

$$\int_{\mathbb{R}} |p_i(z+x) - p_i(z)|dz \leq C|x|,$$

then defining $V = \{1, 2, \ldots, n\}$, $(M2)$ is a consequence of

$$|p_V(z+x) - p_V(z)| \leq \sum_{k=1}^{n-1} \Big(\prod_{i=1}^{k} p_i(z_i+x_i)\Big)\Big(\prod_{k+1}^{n} p_i(z_i)\Big)|p_k(z_k+x_k) - p_k(x_k)|.$$

**(1.7.2) Example.** $Z$ is a Markovian Field (cf. Chapter 2). $Z$ is specified by its conditional distributions $\pi_V$, and $V$ is a finite subset of $S$. We shall assume

$$\int_{\mathbb{R}^V} \sup_{y \in \mathbb{R}^{\partial V}} \Big|\pi_V(z+x|y) - \pi_V(z|y)\Big| dz \leq C \sum_{i \in V} |x_i|$$

where $\partial V$ stands for the Markov boundary of $V$.

*Inversibility Hypothesis for Linear Representations.*

$(L2)$ There exists $A = (a_{i,j}, i, j \in S)$ such that $\sup_i \sum_j |a_{i,j}| < \infty$ and $B \cdot A = I$ in the sense that for all $i, j \in S$, $\sum_k b_{i,k} a_{k,j} = \delta_{i,j}$. Consider the following examples:

(1) *Nonstationary Representation.*

If $\alpha = \sup_i \sum_{j:j \neq i} |b_{i,j}| < 1$, $(L2)$ is guaranteed. Writing $B = I - \Delta$, $A = \sum_{n \geq 0} \Delta^n$ exists and satisfies the assumptions. Indeed, $|\Delta^n_{i,j}| \leq \alpha^n$, which gives the first part of $(L2)$. Clearly $BA = I$.

(2) *Stationary Representation* $(b_{i,j} = b_{i-j})$.

Consider $b(z) = \sum_i b_i z^i$, $z \in T^d$, $b(z) \neq 0$. Assume that for a certain $k > \frac{d}{2}$, one has $\sum_i |i|^k\ |b_i| < \infty$. Then $a(z) = b^{-1}(z) = \sum_i a_i z^i$ has an absolutely convergent Fourier Series and $(L2)$ is true for the associated matrices $A$ and $B$. We can control the mixing coefficient of $X$ in terms of the two following bounds for the tails of the linear representations:

$$\begin{cases} a_\tau(m) = & \sup_i \Big( \sum_{j:|j-i|>m} |b_{i,j}|^\tau \Big), \\ \\ \ell_\tau(m) = & \max\Big\{ a_\tau(m)^{\frac{1}{1+\tau}},\ a_2(m)^{\frac{\tau}{2(1+\tau)}} \Big\}. \end{cases}$$

Indeed, we have the following Theorem:

**(1.7.3) Theorem.** *Let $X$ be the linear process* (1.43) *under conditions* $(L1-L2)$ *and* $(M1-M2)$*. Then, for all $k = 2m+p$ and for $\tau < \delta$ if $\delta > 1$, $\tau = \delta$ if not, one has*

$$\alpha^X_{a,b}(k) \leq C\ b\ \ell_\tau(m) + \alpha^Z_{u,v}(p). \tag{1.47}$$

*Here, $u = a(2m+1)^d$ and $v = b(2m+1)^d$. Constant $C$ depends only on the mixing coefficient of $Z$, $\delta$, $\tau$ and $\gamma$.*

**(1.7.3) Corollary.** *Let $X$ be a linear process of $L^2$ defined by* (1.43)*. Assume $b = (b_{i,j})$ satisfies* (1.46)*, and that $Z$ is centered, independent noise which satisfies:*

(*a*) *there exists $\delta > 2$ such that $\sup_i E|Z_i|^\delta < \infty$;*

(*b*) *for each $i$, $Z_i$ has density $p_i$ such that, uniformly in $i$,*

$$\int_{\mathbb{R}} |p_i(z+x) - p_i(z)|dz \leq C|x|.$$

*Then, under the inversibility condition* $(L2)$,

$$\alpha^X_{a,b}(2m) \leq C'b \cdot \Big[ \sup_i \Big( \sum_{j:|j-i|>m} b^2_{i,j} \Big) \Big]^{1/3}.$$

If $|b_{i,j}| < C\rho^{|i-j|}$, $|\rho| < 1$, the brackets behave like $m^d \rho^{2m}$ and $\alpha^X_{a,b}(m)$ decays exponentially.

OBSERVATIONS:

(1) In Theorem 1.7.3, $u$ is an upper bound for the cardinal of the $m$-neighborhood of $A$ (of cardinal $a$). If $S \neq \mathbb{Z}^d$, the proposition is still true with an adequate modification of $u, v$.

(2) If $S = \mathbb{Z}$, dependence on $b = |B|$ disappears. If $B = [1,n]^d \subset \mathbb{Z}^d$, if $|b_{i,j}| \leq C \prod_{\ell=1}^{d} g(|i_\ell - j_\ell|)$ with $g(x) = x^{-k}$, for big enough $k$, dependence on $b$ is like $b^{\frac{d-1}{d}}$.

(3) Condition $(L2)$ can be relaxed if $X = BZ'$ and $Z'_i = P_i Z$ where $P_i$ is a family of polynomials with bounded degree applied to the backward

shift operator. In this case, the mixing properties of $Z'$ are those of $Z$ and under conditions $(L, M)$ for $B$ and $Z$, the above results are still true.

*Proof:* Consider $C \subset S$ and $Z_C$ the marginal distribution of $Z$ over $C$, and $\mathcal{Z}_C$ the $\sigma$-algebra defined by $Z_C$. Fix $m \in \mathbb{N}$ and consider $C^m$ the following $m$-neighborhood of $C$:

$$C^m = \{i \in S : \exists j \in C \ s.t. \ |i - j| \leq m\}.$$

Let $W_i^C$, $R_i^C$ be defined by $W_i^C = \sum_{j \in C^m} b_{i,j} Z_j$, $X_t = W_t^C + R_t^C$. $W_i^C$ is $\mathcal{Z}_{C^m}$-measurable. We will write $R_C = R_C^C$. Let $E \subset \mathbb{R}^A$, and $F \subset \mathbb{R}^B$ be two Borel sets. Define

$$\mu = \mathbb{P}(X_A \in E, X_B \in F) - \mathbb{P}(X_A \in E)\mathbb{P}(X_B \in F).$$

We can write $\mu = \mu_1 + \mu_2 + \mu_3 + \mu_4 + \mu_5$, where

$$\begin{aligned}
\mu_1 &= \mathbb{P}(X_A \in E, X_B \in F) - \mathbb{P}(W_A \in E, X_B \in F), \\
\mu_2 &= \mathbb{P}(W_A \in E, X_B \in F) - \mathbb{P}(W_A \in E, W_B \in F), \\
\mu_3 &= \mathbb{P}(W_A \in E, W_B \in F) - \mathbb{P}(W_A \in E)(W_B \in F), \\
\mu_4 &= [\mathbb{P}(W_A \in E) - \mathbb{P}(X_A \in E)]\mathbb{P}(W_B \in F), \\
\mu_5 &= \mathbb{P}(X_A \in E)[\mathbb{P}(W_B \in F) - \mathbb{P}(X_B \in F)].
\end{aligned}$$

Except for $\mu_3$, which may be bounded by $\alpha^Z(A^m, B^m)$, all other terms can be written, for some event $U$, as

$$\mathbb{P}((X_C \in S) \cap U) - \mathbb{P}((W_C \in S) \cap U).$$

For a famlily of positive real numbers $(\eta_t)_{t \in C}$, we define the set

$$H_C = \{x \in \mathcal{R}^C : \ |x_t| \leq \eta_t, \ t \in C\}.$$

Consider $\xi = \mathbb{P}(W_B^B \in F - r_B | y) - \mathbb{P}(W_B^B \in F | y)$ for some $r_B \in H_B$. Then $(L2)$ implies that there exist $B^m = (b_{i,j}, i \in B^m, j \in B)$, an operator of $\mathbb{R}^{B^m} \longrightarrow \mathbb{R}^B$, and $A^m = (a_{i,j}, i \in B, j \in B^m)$, an operator of $\mathbb{R}^B \longrightarrow \mathbb{R}^{B^m}$, that satisfy

$$B^m \cdot A^m = I_B + U^m, \quad U^m : \mathbb{R}^B \longrightarrow \mathbb{R}^B,$$

where, if we write $\|C\| = \sup_i \sum_j |c_{i,j}|$, then

$$\|U^m\| \leq \|A\| \sup_i \sum_{j \notin B^m} |b_{i,j}| \leq \|A\| \sup_i \sum_{j : |j-i| > m} |b_{i,j}|.$$

If $m$ is big enough, this expression is bounded by $\frac{1}{2}$. Thus, $I_B + U^m$ is invertible with

$$B^m \widehat{A}^m = I_B, \quad \widehat{A}^m = A^m(I_B + U^m)^{-1}, \quad \|\widehat{A}^m\| \leq 2\|A\|.$$

This means that $\widehat{r} = \widehat{A}^m r_B$ satisfies $B^m \widehat{r} = r_B$, and one has, using (M2),

$$\begin{aligned} &|\mathbb{P}(W_B^B \in F - r_B|y) - \mathbb{P}(W_B^B \in F|y)| \\ &\qquad = |\mathbb{P}(B^m(Z_{B^m} + \widehat{r}) \in F|y) - \mathbb{P}(B^m Z_{B^m} \in F|y)| \\ &\qquad \leq \int_{\mathbb{R}^{B^m}} |p_{B^m}(z + \widehat{r}|y) - p_B(z|y)| dz \leq C \cdot \|\widehat{A}^m\| \sum_{i \in B} |r_i| \end{aligned}$$

so that

$$\xi \leq 2 \cdot C \|A\| \sum_B |r_i|. \tag{1.48}$$

*Estimation of* $\xi = P((x_c \in S) \cap U) - P((W_C \in S) \cap U)$: Recalling the above definition of $H_C$, we have

$$\begin{aligned} |\xi| &\leq P(R_c \notin H_C) \\ &+ \int_{H_C} |P(W_C \in S - r|R_C = r) - P(W_C \in S|R_C = r)| P_{R_C}(dr) \\ &\leq P(R_C \notin H_C) + \kappa \sum_C |\eta_t|, \text{ from (1.5.1).} \end{aligned}$$

*An upper bound for* $\mathbb{P}(R_B \notin H_B)$: $H_B = \{x \in \mathbb{R}^B : |x_t| \leq \eta_t, t \in B\}$

$$\mathbb{P}(R_B \notin H_B) \leq \sum_{i \in B} \mathbb{P}(|R_i^B| > \eta_i) \leq \sum_{i \in B} E(|R_i^B|^\tau) \eta_i^{-\tau}$$

for all $\tau \leq \delta$. Under the exponential mixing condition for $Z$ and because of the Nagaev–Fuk inequalities which bound the moments of a sum (cf. Doukhan [52]) one has

$$E|R_i^B|^\tau \leq C \; S_i(\tau, B^m), \quad C < \infty,$$

$$S_i(\tau, D) = \max\Big\{\sum_{j \notin D} |b_{i,j}|^\tau, \; \Big(\sum_{j \notin D} b_{i,j}^2\Big)^{\tau/2}\Big\}, \; \tau = \delta \text{ if } \delta \leq 1, \; \tau < \delta \text{ if not.}$$

Combining the above results and optimizing for each $t \in B$, the contribution of $(CS_i(\tau, B^m)\eta_i^{-\tau} + \eta_i)$ in the upper bound for $|\mu|$, we get

$$|\mu| \leq C' \sum_{i \in B} (S_i(\tau, B^m))^{\frac{1}{1+\tau}} + \alpha^Z(A^m, B^m) \;, \; C' < \infty,$$

which completes the proof. □

For the Corollary, since the $Z_i$ are noncorrelated, condition (1.46) enables us to define $X$. Now it is enough to choose $\tau = 2$ in the above proof.

## 1.8. Bibliographical Comments

Spectral representations can be achieved over $\mathbb{Z}^d$ as over $\mathbb{Z}$: Rudin's book (1962) gives a general version of Bochner's Theorem and Azencott and Dacunha–Castelle's presentation can be directly generalized to $\mathbb{Z}^d$ (cf. also Rosenblatt, 1985). Characterization of the different kinds of spatial regularities is quite another problem for higher dimensions: noncausal regularity is considered in Rozanov (1967); Helson and Lowdenslager (1958) generalized Szegö's Theorem, the Factorization Theorem, as well as Riesz's result (which assures absolute continuity of a measure on the torus) to dimension two. The quarter plane representations are studied in Rudin (1969); Tjøstheim considers certain ARMA representations; a detailed study on the factorization of ARMA processes was done by Korezlioglu and Loubaton (1986) (cf. also Loubaton). Besag (1981) gives a complete account of the recurrence relationships satisfied by the covariance of the nearest neighbor Markovian models, and of the differences between $\mathbb{Z}$ and $\mathbb{Z}^2$ which had already appeared in Whittle (1954) and Besag (1972). The only result we know on the identification of rational spectra is due to Garber (1981). This work generalizes to $L$-fields the identification criteria of an $AR$ process over $\mathbb{Z}$ in terms of the partial correlations.

The results on multidimensional fields are attributed to Mardia (1988). Contextual classification by discriminant analysis and its particularly simple form when the covariance can be factorized is due to Switzer (1980) (cf. also Mardia, 1984). The most complete review on this topic can be found in Yao (1990).

The linear process' mixing results (cf. Doukhan and Guyon 1991) generalize those of Gorodetskii (1977): for fields, for bilateral representations, and in the case of a not-necessarily independent generator process. For the mixing properties and their applications, the reader can check the very complete review of Doukhan (1994). The generalization to $\mathbb{Z}^d$ of the regularity coefficient of a stationary Gaussian process was communicated personally to us by Ibragimov.

## 1.9. Exercises

(1) (a) Check that if $a > 0$, $P(\lambda) = 1 - \cos\lambda + a\cos 2\lambda > 0$. Deduce the existence of the following Markovian model over $\mathbb{Z}$:

$$X_t = \frac{1}{2}(X_{t-1} + X_{t+1}) + \frac{a}{2}(X_{t-2} + X_{t+2}) + e_t.$$

(b) Check that $f(\lambda) = \left(1 - \frac{1}{d}\sum_{j=1,d} \cos\lambda_j\right)^{-1}$ is integrable for $d \geq 3$ but not for $d \leq 2$.

(c) Show that there exists $\gamma > 0$ s.t. $(1-\frac{1}{2}(\cos\lambda+\cos\mu)+\gamma\cos(\lambda+\mu))^{-1}$ is integrable over $[0, 2\pi[^2$.

(2) Find the equivalent $CAR$ representations of the following $AR$:

(a) $X_{st} = \alpha\ X_{s-1,t} + \beta\ X_{s,t-1} + \gamma\ X_{s+1,t-1} + \varepsilon_{s,t}$.

(b) $X_{st} = \alpha(X_{s-1,t} + X_{s+1,t}) + \beta(X_{s,t-1} + X_{s,t+1}) + \gamma(X_{s-1,t-1} + X_{s+1,t+1}) + \varepsilon_{st}$.

(3) Among the following $AR$, find those that are causal:

(a) $X_{st} = \frac{1}{4}(X_{s-1,t} + X_{s,t-1} + X_{s+1,t-1}) + \varepsilon_{s,t}$.

(b) $X_{st} = \frac{1}{4}(X_{s-1,t} + X_{s+1,t} + X_{s,t-1}) + \varepsilon_{s,t}$.

(c) $X_{st} = \frac{1}{2}(X_{s-1,t} + X_{s,t-1}) - \frac{1}{4}X_{s-1,t-1} + \varepsilon_{s,t}$.

(4) Write the general form of a factorizing $AR$ with autoregression domain $R = \{0,1,2\}^2 \backslash \{(0,0)\}$. Describe its spectral density and its covariance.

(5) Under what conditions does one get $H = |P|^2$, where $P$ is a trigonometric polynomial, $H = 1 + 2(a\ \cos\lambda + b\ \cos\mu + c\ \cos(\lambda+\mu))$?

(6) Over $\mathbb{Z}$, find the correspondences between the following three equivalent representations:

($AR$): $X_t = \alpha(X_{t-1} + X_{t+1}) + \eta_t$, $\eta$ a white noise $(0, \sigma_\eta^2)$.

($AR$-causal): $X_t = a_1 X_{t-1} + a_2 X_{t-2} + \varepsilon_t$, $\varepsilon$ a white noise innovation $(0, \sigma_\varepsilon^2)$.

($CAR$): $X_t = c_1(X_{t-1} + X_{t+1}) + c_2(X_{t-2} + X_{t+2}) + e_t$, $Var(e_t) = \sigma_e^2$.

(7) $f = G \cdot H^{-1}$ is the rational spectral density of a field over $\mathbb{Z}^2$, and $H$ is never zero over $\{\delta < |z| < \delta^{-1}\}^2$, $0 < \delta < 1$. If $f$ is bounded by $M$, show that the field's covariance satisfies $|R(k_1, k_2)| \leq M\delta^{k_1+k_2}$ (use the Laurent development of $f$ in a neighborhood of the torus).

(8) Write the Yule–Walker equations for the following models:

(a) $f(\lambda,\mu) = \sigma_e^2(1 - 2\alpha\ \cos\lambda - 2\beta\ \cos\mu)^{-1}$.

(b) $g(\lambda,\mu) = (1 + 2\gamma\cos\lambda)f(\lambda,\mu)$; $|\alpha| + |\beta| < \frac{1}{2}$.

Determine the set of initial conditions under which the recurrent equations can be used to calculate the covariance everywhere.

(9) Identify, as a series, the variance of the isotropic nearest neighbor Markovian model over $\mathbb{Z}^2$. Give the covariance at distance 1.

(10) Consider the following $CAR$ models over $\mathbb{Z}$ and over $\mathbb{Z}^2$, with correlation $\rho(\cdot)$:

(1) $X_t = a(X_{t-1} + X_{t+1}) + e_t$, $\varepsilon = \left(\frac{1}{2} - a\right) \longrightarrow 0_+$.

(2) $X_{st} = \alpha(X_{s-1,t} + X_{s+1,t} + X_{s,t-1} + X_{s,t+1}) + e_{st}$, $\varepsilon = \left(\frac{1}{4} - \alpha\right) \longrightarrow 0_+$

(a) Show that for (1), $1 - \rho(1) \sim \sqrt{2\varepsilon}$.

(b) With the results of § 1.4.2 and the equivalence $K(x) \sim \frac{1}{2}\log\left(\frac{1-\sqrt{x}}{8}\right)$ if $x \to 1_-$, show that for (2), $1 - \rho(1,0) \sim -\frac{\pi}{2}(\log\varepsilon)^{-1}$.

(c) For which values of $\varepsilon$ does $\rho(1) = \rho(1,0) = 0.9$ ; 0.99 ; 0.999?

(11) *Skeleton of an* $\mathbb{R}^d$ *process over* $\mathbb{Z}^d$ *and vice versa.*

(a) Let $X$ be a stationary, second-order process over $\mathbb{R}^d$, with spectral density $f(\lambda)$, $\lambda \in \mathbb{R}^d$. For $n \in \mathbb{N}^*$, consider $X^{(n)} = \left\{X_{\frac{i}{n}}, i \in \mathbb{Z}^d\right\}$ the skeleton of $X$ over $(\mathbb{Z}/n)^d$, $f_n(\mu)$ the spectral density of $X^{(n)}$, $\mu \in T^d$. Show that $\lim_{n\to\infty} \frac{1}{n^d} f_n(\frac{\lambda}{n}) \to f(\lambda)$, $\lambda \in \mathbb{R}^d$.

(b) Over $\mathbb{Z}$, consider the $AR(1)$: $X_i^{(n)} = \rho_n X_{i-1}^{(n)} + \varepsilon_i^{(n)}$. Give an equivalent $CAR$ representation. Show that if $n(1-\rho_n) = \alpha > 0$, $n\sigma^2_{\varepsilon^{(n)}} = 2\alpha\sigma^2$, $(X^{(n)}) \to X$, the Ornstein–Uhlenbeck diffusion process which satisfies $dX_t = -\alpha X_t dt + \sigma\sqrt{2\alpha} dW_t$.

(c) Over $\mathbb{Z}^2$, we consider the isotropic four nearest neighbor Markovian model: $X_{ij}^{(n)} = a_n(X_{i-1,j}^{(n)} + X_{i+1,j}^{(n)} + X_{i,j-1}^{(n)} + X_{i,j+1}^{(n)}) + e_{ij}^{(n)}$. Show that, in order for $\frac{1}{n^2} f_n(\frac{\lambda}{n})$ to have a limit, it is necessary that $1 - 4a_n \sim \frac{a^2}{n^2}$ and $\sigma^2_{e^{(n)}} \sim K$. Is the limit integrable? Deduce that $(X^{(n)})$ can't converge toward any field in $\mathbb{R}^2$.

(12) Let $X$ be the isotropic four nearest neighbor Markovian model over $\mathbb{Z}^2$. What is the spectral density of the process over $\mathbb{Z}$, $Y = \{X_{s,0}, s \in \mathbb{Z}\}$? Is the trace of a Markovian process over $\mathbb{Z}^2$ also Markovian over $\mathbb{Z}$? Does the trace of a process with rational spectrum over $\mathbb{Z}^2$ also have a rational spectrum?

(13) Consider the following two independent $CAR$ models over $\mathbb{Z}$:

$$X_t = \alpha(X_{t-1} + X_{t+1}) + e_t, \qquad Y_t = \beta(Y_{t-1} + Y_{t+1}) + \widetilde{e}_t, \quad t \in \mathbb{Z}.$$

Find the ARMA representation of $Z = X + Y$ (find the spectral density of $Z$). Under which condition is $Z$ Markovian?

(14) If we order $S = \{1,2,3,4\}$, show that $X \in \mathbb{R}^4$ with covariance $\Sigma$ given below admits the $MA$ representation

$$\Sigma = \begin{pmatrix} 1 & 1 & 1 & 1 \\ 1 & 2 & 2 & 2 \\ 1 & 2 & 3 & 3 \\ 1 & 2 & 3 & 4 \end{pmatrix}, \qquad (MA): \begin{cases} X_1 = \varepsilon_1,\ X_2 = \varepsilon_1 + \varepsilon_2, \\ X_3 = \varepsilon_1 + \varepsilon_2 + \varepsilon_3, \\ X_4 = \varepsilon_1 + \varepsilon_2 + \varepsilon_3 + \varepsilon_4, \\ Var(\varepsilon_i) = 1. \end{cases}$$

Deduce the $CAR$ and causal $AR$ representations.

(15) Let $Z$, $Z_t = {}^t(X_t, Y_t) \in \mathbb{R}^2$ be a Gaussian process over $\mathbb{Z}$.

(a) Establish the relationship between the two representations:
*Causal AR(1)*: $Z_t = AZ_{t-1} + \mathcal{E}_t$, $\mathcal{E}_t$ i.i.d. $\mathcal{N}_2(0, \Gamma)$.
*CAR*: $Z_t = C_{-1}Z_{t-1} + C_1 Z_{t+1} + E_t$, *cov* $E_t = V$, $E(Z_s E_t) = 0$ if $s \neq t$.

(b) Identify the reversible submodels and those with factorizing covariance.

(c) If $C_1 = C_{-1} = \begin{pmatrix} a & b \\ b & a \end{pmatrix}$, $V = \sigma_e^2 \begin{pmatrix} 1 & -\alpha \\ -\alpha & 1 \end{pmatrix}$, what is the stationary condition? Find the distribution of $\mathcal{L}(X_t \mid X_s, s \neq t, Y_\ell, \ell \in \mathbb{Z})$.

(16) *Exchangeable Gaussian Model over* $S = \{1, 2, \dots, n\}$.
Let the $X \in \mathbb{R}^n$ be the $AR$ given by $X = \alpha JX + \varepsilon$ with $\varepsilon \sim \mathcal{N}_n(0, \sigma^2 I)$, $J_{ij} = 1$ if $i \neq j$, 0 if not. Recall that the set of matrices of the form $aI + bJ$, $a, b \in \mathbb{R}$ are invariant under multiplication and inversion.

(a) Under what conditions is the model well defined? Show that *Cov* $X = r_0 I + r_1 J$. Find $r_0$ and $r_1$.

(b) Show that the $CAR$ representation can be written as $X = \beta JX + E$ and find $\beta$ and $cov(E)$.

CHAPTER 2

# Gibbs and Markovian Fields

## 2.1. Gibbs Fields

Let $S$ be a numerable set of *sites* (generally $S = \mathbb{Z}^d$), let $E$ be a metric, complete and separable space (polish space) of states, and let $\mathcal{E}$ be its $\sigma$–algebra (generally $E$ is finite, compact, or $\mathbb{R}^k$). A *configuration* over $S$ is an element $\omega$ of the measurable space of configurations $(\Omega, \mathcal{F}) = (E, \mathcal{E})^S$. A *random field* over $S$ is a probability measure $\mu$ over $(\Omega, \mathcal{F})$.

Let $\mathcal{S}$ be the collection of nonempty finite subsets of $S$. One way to define a field $\mu$ is to give a *coherent family of marginal distributions* $\{P_V, V \in \mathcal{S}\}$ which are then coupled via the Kolmogorov Theorem: for a given *coherent marginal specification*, there *exists a unique* field $\mu$ such that its marginal distributions are $\mu_V = P_V$, $V \in \mathcal{S}$.

Another way of defining a random field model is to consider a *conditional specification*, that is, a coherent family of conditional probability kernels indexed by $\mathcal{S}$ and to ask the following question: Does there exist a field $\mu$ such that its conditional distributions are the ones considered? (problem proposed by D.L.R.: Dobrushin–Landford–Ruelle). If yes, is the field unique (absence or presence of phase transition), and what are its characteristics (is it mixing, does it depend continuously on the specification, large deviation properties for translation independent specifications). In this chapter we shall consider the definitions and most important results for Gibbs fields as well as multiple examples, which shall be used later on for the statistical analysis of fields or applications to pattern analysis.

We will write $\lambda$ for a positive reference measure over the state space $(E, \mathcal{E})$. If $V \in \mathcal{S}$, $\lambda_V$ is the product measure over $(E, \mathcal{E})^{\otimes V}$.

## *2.1.1. Conditional Specifications and Gibbs Measures*

*Conditional Kernel.* If $\mu$ is a random field, for each $V \in \mathcal{S}$, we can define the *transition kernel*

$$\mu_V(\cdot \mid \cdot) : \mathcal{F}(V) \times \Omega(S \backslash V) \longrightarrow [0,1]; (B \times y) \longmapsto \mu_V(B|y) = E_\mu(1_B|y).$$

Here, $\mathcal{F}(V)$ stands for the $\sigma$-algebra of configurations over $V$, and $\Omega(S\backslash V)$ is the space of $V$-external configurations, $\mu_V(\cdot|y)$ is a probability over $\mathcal{F}(V)$, conditional to $y$; $\mu_V(B|\cdot)$ is a random variable (r.v.) $\mathcal{F}(S\backslash V)$-measurable. If, in general, this kernel is not well defined, it will be in the Gibbs case.

*Coherence of Kernels.* Let $V \subseteq V'$ be two elements of $\mathcal{S}$, and let $\pi_V$ and $\pi_{V'}$ be two conditional kernels, relative to $V$ and $V'$, $B \in \mathcal{F}(V)$, $B' \in \mathcal{F}(V'\backslash V)$ and $z \in \Omega(S\backslash V')$. Composition of both kernels is another kernel over $V'$ defined by

$$(2.1) \qquad (\pi_{V'}\pi_V)(BB'|z) = \int_{B'} \pi_V(B|yz)\pi_{V'}(E^V, dy|z).$$

For the family $\{\mu_V\}$ given by the field $\mu$, one has

$$(2.2) \qquad \mu_{V'}\mu_V = \mu_{V'}.$$

We will say that the family is *coherent.*

**(2.1.1) Definition.** *Conditional Specifications. We shall call any family of kernels $\pi = \{\pi_V, V \in \mathcal{S}\}$ that satisfies condition* (2.2) *a conditional specification.*

*Gibbs Measure Associated to a Specification $\pi$.* We will call $\mathcal{G}(\pi)$ the set of fields $\mu$ that admit $\pi$ as their conditional specification: $\mu \in \mathcal{G}(\pi) \iff \mu_V \equiv \pi_V \ \mu - a.e.$ for all $V \in \mathcal{S}$. $\mathcal{G}(\pi)$ can be empty (cf. Prum [133]), with only one element (absence of *phase transition*) or have more than one element (the latter is a phase transition setting). If $\mathcal{G}(\pi) \neq \emptyset$, an element of $\mathcal{G}(\pi)$ is called a Gibbs measure, or a Gibbs state associated to $\pi$. We shall be interested in $\mathcal{G}(\pi)$ when $\pi$ is defined by a Gibbs potential.

We observe that even though a specification $\pi$ can be reconstructed by considering *its* conditional distributions at each site $i \in S$, $\pi_{\{i\}}$, a family of conditional distributions $\{\nu_i, i \in S\}$ does not allow in general the construction of a specification $\nu$ (cf. Appendix 1 and Exercise 3). The problem is thus to construct coherent specifications: the "natural" answer is to perturb i.i.d. fields by local interactions of 2 bodies, of 3 bodies, ..., which leads to a Gibbs specification.

### *2.1.2. Gibbs Specifications*

*Interaction Potential* is defined by a family $\phi = (\phi_A, A \in \mathcal{S})$ of applications

$$\phi_A : \Omega(A) \longmapsto \mathbb{R} \quad s.t.$$

(*i*) For every $A$, $\phi_A$ is $\mathcal{F}(A)$-measurable.

(*ii*) If $\Lambda \in \mathcal{S}$ and $\omega \in \Omega$, then the sum

$$U_\Lambda^\phi(\omega) = \sum_{A \in \mathcal{S}: A \cap \Lambda \neq \emptyset} \phi_A(\omega) \quad \text{exists.} \tag{2.3}$$

In Statistical Mechanics, $-U_\Lambda^\phi(\omega)$ is called the *energy* of $\omega$ in $\Lambda$, and $-U_\Lambda^\phi$ is the Hamiltonian on $\Lambda$. We will say that $\phi$ is $\lambda$-*admissible* if for all $\Lambda \in \mathcal{S}$, $\omega \in \Omega$,

$$Z_\Lambda^\phi(\omega) = \int_{\Omega(\Lambda)} \exp U_\Lambda^\phi(\omega_\Lambda, {}_\Lambda\omega) \lambda^\Lambda(d\omega_\Lambda) < \infty.$$

Here, $\omega_\Lambda$ and ${}_\Lambda\omega$ are the configurations over $\Lambda$ and $S \backslash \Lambda$.

*Gibbs Specification Associated to a Potential* $\phi$. If $\phi$ is admissible, the family

$$\pi_\Lambda^\phi(\omega_\Lambda | {}_\Lambda\omega) = Z_\Lambda^\phi(\omega) \exp U_\Lambda^\phi(\omega_\Lambda, {}_\Lambda\omega), \Lambda \in \mathcal{S} \tag{2.4}$$

is *coherent.* We give the proof for finite $E$. For $x \in E^V$, $y \in E^{V' \backslash V}$, $z \in E^{S \backslash V'}$, we have the following:

$$\begin{aligned}
&\frac{\pi_{V'}(x, y|z)}{\pi_{V'}(E^V, y|z)} \\
&= \frac{\exp\Big( \sum_{A \cap V' \neq \phi} \phi_A(x, y, z) \Big)}{\sum_{u \in E^V} \exp\Big( \sum_{A \cap V' \neq \phi} \phi_A(u, y, z) \Big)} \\
&= \frac{\exp\Big( \sum_{A \cap V \neq \phi} \phi_A(x, y, z) \Big) \exp\Big( \sum_{A \cap V = \phi, A \cap V' \neq \phi} \phi_A(y, z) \Big)}{\Big( \sum_{u \in E^V} \exp\Big( \sum_{A \cap V \neq \phi} \phi_A(u, x, y) \Big) \Big) \exp\Big( \sum_{A \cap V = \phi, A \cap V' \neq \phi} \phi_A(y, z) \Big)} \\
&= \pi_V(x \mid y, z).
\end{aligned}$$

$\{\pi_V^\phi, V \in \mathcal{S}\}$ is called the *Gibbs specification associated to potential* $\phi$. The normalization constant $Z_\Lambda^\phi(\cdot)$ is called the *partition function* on $\Lambda$ for potential $\phi$. Finally, the set of Gibbs measures associated to $\pi^\phi$ is called $\mathcal{G}(\phi)$.

*Potentials with Bounded Support, Mean Energy by Site.* The space of *summable* potentials is given by

$$\mathcal{B} = \Big\{\phi : |||\phi|||_i = \sum_{A\in\mathcal{S}:A\ni i} \|\phi_A\|_\infty < \infty, \text{ for all } i \in S\Big\}.$$

A summable potential is admissible if and only if $\lambda(E) < \infty$: indeed, the condition is sufficient since $\|U_\Lambda^\phi(\omega)\|_\infty \leq \sum_{i\in\Lambda} |||\phi|||_i$ and it is necessary because $\lambda(E)\exp(-|||\phi|||_i) \leq \int \exp U_{\{i\}}^\phi(\omega)\lambda(d\omega_i) < \infty$. In what follows, we shall consider that $\lambda$ *is finite positive measure over* $E$. Consider the following two fundamental examples of summable potentials:

(1) Potentials with *bounded support*: For each $i \in S$, there exists $\Delta_i \in \mathcal{S}$ s.t. $\phi_A \equiv 0$ except if $A \subseteq \Delta_i$ for a given $i$.

(2) Potentials *with range* $R$: When $S$ is a metric space, $\phi_\Delta \equiv 0$ except if the diameter of $A \leq R$.

*Potentials Invariant Under Translations over* $S = \mathbb{Z}^d$. These are potentials that satisfy for each $j \in S$,

$$\phi_{A+j}(\omega) = \phi_A(\tau_j\omega), \quad \text{with} \quad (\tau_j(\omega))_i = \omega_{i-j},\ i \in S.$$

We will call $\mathcal{B}_s \subseteq \mathcal{B}$ the subset of summable, invariant under translation potentials: $\|\phi\| = |||\phi|||_0$ is a norm over $\mathcal{B}_s$ and $\mathcal{B}_s^b$, and the subset of bounded support potentials is dense in $\mathcal{B}_s$.

*Mean Energy by Site.* This concept shall be required for the study of maximum likelihood estimation and for the assertion of the variational principle.

The *Mean Energy* at site $i$ is given by

(2.5)
$$\overline{\phi}_i = \sum_{A\ni i} |A|^{-1}\phi_A.$$

If the potential is invariant under translations, one has $\overline{\phi}_i(\omega) = \overline{\phi}_0(\tau_i\omega)$ and

(2.6)
$$\begin{cases} U_\Lambda^\phi(\omega) &= \displaystyle\sum_{i\in\Lambda}\overline{\phi}_i(\omega) + \varepsilon_\Lambda^\phi(\omega), \\ \varepsilon_\Lambda^\phi(\omega) &= \displaystyle\sum_{A:A\cap\Lambda\neq\phi \text{ et } A\not\subseteq S}\Big(1 - \frac{|A\cap\Lambda|}{|A|}\Big)\phi_A(\omega). \end{cases}$$

REMARK: If $(\Lambda_n)$ is an increasing sequence of domains, and if $\phi$ is an invariant under translation, summable potential supported over $R$, then the residual term $\varepsilon_{\Lambda_n}^\phi(\omega)$ is $O(|\partial_R\Lambda_n|)$ in the supremum norm (here, $\partial_R\Lambda_n$ is the $R$-neighborhood of $\Lambda_n$):

(2.7)
$$\|\varepsilon_\Lambda^\phi\| \leq |\partial_R\Lambda|\cdot|||\phi|||_0.$$

## *2.1.3. Existence and Uniqueness of the Gibbs Measure and Mixing Properties*

**(2.1.1) Theorem.** Existence of a Gibbs Measure ([50], [71]). *Assume $E$ is a polish space where the reference measure $\lambda$ is positive and finite and the potential $\phi$ is summable; then $\mathcal{G}(\phi)$ is not empty and compact.*

The topology on $\mathcal{P}(\Omega, \mathcal{F})$, probability space over $\Omega$, is that given by local convergence (cf. [71], Chapter 4). The proposition is true if $\phi$ is of bounded support as each $\phi_A$ is bounded: this is the case of Markovian potentials. The case $E$ compact (or finite) is a classical situation for the application of this result. The argument used for proving the Theorem is that the sequence $(\pi^{\phi}_{\Lambda_n}(\cdot|\omega), n \in \mathbb{N})$, $\Lambda_n \nearrow S$ is relatively compact for all $\omega \in \Omega$. Thus, there exists a converging subsequence in $\mathcal{P}(\Omega, \mathcal{F})$.

**2.1.3.1. Unicity: Dobrushin–Simon's Conditions.**

A function $f : \Omega \longrightarrow \mathbb{R}$ is *local* (or Markovian) if there exists finite $A$ *s.t.* $f(\omega) = f(\omega_A)$. A function $f$ is *quasi-local* (or quasi-Markovian) if there exist local functions $(f_n,\ n \geq 0)$ *tq* $\|f - f_n\|_\infty \longrightarrow 0$. If $E$ is a separable, metric space, every uniformly continuous function $f$ is quasi-local. If $E$ is finite, $f$ is quasi-local if it is continuous.

*Dobrushin's Influence Measure* (cf. [50], [107]). Let $a$ and $b$ be two sites of $S$, $a \neq b$. Define for a specification $\pi$

$$\gamma_{a,b}(\pi) = \sup \tfrac{1}{2}\|\pi_b(\cdot|\omega) - \pi_b(\cdot|\omega')\|. \tag{2.8}$$

Here, $\|\mu\| = \sup\{|\mu(f)|,\ \|f\|_\infty = 1\}$ is the total variation norm of $\mu$ and the sup is taken over all configurations $\omega$, $\omega'$ identical except at site $a$. If $a = b$, we define $\gamma_{a,a}(\pi) = 0$. $\gamma_{a,b}$ is a measure of the influence of site $a$ over the conditional distribution $\pi_b(\cdot|\cdot)$ in $b$. We will write $\gamma_{a,b}(\phi)$ for $\gamma_{a,b}(\pi^\phi)$.

*Dobrsuhin's Condition* $(D)$. We will say that a Gibbs potential satisfies the Dobrushin condition if each potential is quasi-local and if

$$(D) \quad \alpha(\phi) = \sup_{a \in S} \sum_{b \in S} \gamma_{a,b}(\phi) < 1. \tag{2.9}$$

**(2.1.2) Theorem.** Unicity of the Gibbs State. *Under* $(D)$, $|\mathcal{G}(\phi)| \leq 1$. *Thus, under the existence conditions of Theorem* 2.1, *we have existence and uniqueness of Gibbs measure.*

In this case we say that there isn't a phase transition. If $E$ is compact, we have existence and uniqueness if $\phi$ is bounded and of bounded support and if we have condition $(D)$. As we shall see later, condition $(D)$ allows us to describe the mixing properties of the unique measure $\mu$ of $\mathcal{G}(\phi)$. However, as the next example shows, $(D)$ is a sufficient but not a necessary condition for uniqueness.

**(2.1) Example.** Ising Model over $\mathbb{Z}^2$. $\pi$ is the specification associated to the interaction potential of four nearest neighbors, with $\beta \geq 0$

$$\phi_{\{s,t\}}(x_s, x_t) = \begin{cases} \beta x_s x_t & \text{if } \|s-t\|_1 = 1, \\ 0 & \text{if not} \end{cases}$$

$\phi_{\{s\}} \equiv 0$ (we will say that there isn't an exterior field) and $\phi_A \equiv 0$ if $|A| > 2$. The state space is $E = \{-1, +1\}$. The conditional distribution at $s = (i,j)$ is $\pi(x|\cdot) = (2ch(\beta v))^{-1} \exp(\beta x v)$, where $v$ stands for the sum of the values at the four nearest neighbors. Define $u = x_{i-1,j} + x_{i+1,j} + x_{i,j-1} \in \{-3, -1, 1, 3\}$; $w = x_{i,j+1}$. Then

$$\pi(-1|u,w) = (1 + \exp(2\beta(u+w)))^{-1}, \ \pi(1|u,w) = 1 - \pi(-1|u,w).$$

Setting $w = -1$ and $w' = 1$, we get,

$$|\pi(-1|u,-1) - \pi(-1|u,1)| = |\pi(1|u,1) - \pi(1|u,-1)| = \frac{sh2\beta}{ch2\beta + ch2\beta u},$$

which attains its upper bound, $\frac{1}{2}th2\beta$, at $u = 1$. Condition $(D)$ is thus

*Dobrushin* : $\beta < \frac{1}{4}\log 2 \simeq 0.275.$

This Gibbs model is *one of the few* examples for which the exact uniqueness condition is known. It was obtained by Onsager [126] following the analycity condition for the pressure:

*Onsager* : $\beta < \beta_c = \frac{1}{2}\log(1+\sqrt{2}) \simeq 0.441.$

Simon's sufficient condition (see later) is

*Simon* : $\beta < \frac{1}{2}\log\frac{5}{3} \simeq 0.255.$

The Dobrushin condition is not always easy to obtain. A stricter condition, based directly on the potentials, was given by Simon [149]. It is obtained as a consequence of the following property:

**(2.1.1) Lemma.** *Let $\lambda$ be finite over $(E,\mathcal{E})$, let $h$ and $g$ be two real, bounded functions over $E$, and let $\mu_h$ and $\mu_g$ be the associated measures*

$$d\mu_f = \left(\int e^f d\lambda\right)^{-1} e^f \cdot d\lambda.$$

*Then, for the total variation norm, $\|\mu_h - \mu_g\| \leq \frac{1}{2}\delta(h-g) \leq \|h-g\|_\infty$, where $\delta(u) = \sup_x u(x) - \inf_x u(x)$ stands for the oscillation of $u$.*

*Proof:* Let $\nu_\theta = \mu_{\theta h+(1-\theta)g}$, $q = h - g$, $f \in \mathcal{C}(\Omega)$ such that $\|f\|_\infty = 1$. It can be shown directly that $\frac{d}{d\theta}\nu_\theta(f) = \nu_\theta(fq) - \nu_\theta(f)\nu_\theta(q)$. Therefore, since $\int_0^1 \frac{d}{d\theta}\nu_\theta(f)d\theta = \nu_1(f) - \nu_0(f) = \mu_h(f) - \mu_g(f)$, we have the following series of inequalities

$$\begin{aligned}
|\mu_h(f) - \mu_g(f)| &= \left|\int_0^1 \frac{d}{d\theta}\nu_\theta(f)d\theta\right| = \left|\int_0^1 \nu_\theta[f(q - \nu_\theta(q))]d\theta\right| \\
&\leq \int_0^1 |\nu_\theta[f(q - \nu_\theta(q))]|d\theta \leq \int_0^1 \nu_\theta(|q - \nu_\theta(q)|)d\theta \\
&\leq \int_0^1 (\nu_\theta([q - \nu_\theta(q)]^2))^{\frac{1}{2}}d\theta \leq \int_0^1 (\nu_\theta([q - m]^2))^{\frac{1}{2}}d\theta
\end{aligned}$$

for all real $m$. Choose $m = \frac{1}{2}\left\{\sup_{x\in E} q(x) + \inf_{x\in E} q(x)\right\}$. Then $\|q - m\|_\infty \leq \frac{1}{2}\delta(q)$ and $|\mu_h(f) - \mu_g(f)| \leq \|q - m\|_\infty \leq \frac{1}{2}\delta(q) \leq \|q\|_\infty$. □

Examples: *Dobrushin's uniqueness condition provided by Simon's conditions.* Examples (2.1.1) to (2.1.3) correspond to weak interactions. Examples (2.1.4) and (2.1.5) show that (2.9) cannot be reduced to a weak dependence situation. We present these examples following [71]. A sufficient condition is established for (2.1.1).

**(2.1.1) Example.** A Bound for Potential Oscillation. $\lambda$ is a finite measure over $(E,\mathcal{E})$, and $\phi$ is a continuous $\lambda$-admissible potential such that:

$$\sup_{a\in S} \sum_{A\ni a} (|A| - 1)\delta(\phi_A) < 2. \tag{2.10}$$

Then the uniqueness condition (2.9) is satisfied.

*Proof:* Consider $a \neq b$ two sites of $S$, $\xi$ and $\eta$ two configurations, equal everywhere except eventually in $b$; and $h(x)$ and $g(x)$ the following two bounded, continuous functions over $E$:

$$h(x) = U^\phi_{\{a\}}(x, \xi_{S\setminus\{a\}}), \quad g(x) = U^\phi_{\{a\}}(x, \eta_{S\setminus\{a\}}).$$

Because of Lemma 2.1 and the definition of $h$ and $g$ and $\gamma_{a,b}(\phi)$, we have

$$\gamma_{a,b}(\phi) \leq \frac{1}{2} \cdot \frac{\delta(q)}{2}, \quad \text{with} \quad q = h - g.$$

However,

$$\delta(q) = \sup_{x,y \in E} \Big( \sum_{A \ni a} \Big| \phi_A(x, \xi_{S \setminus \{a\}}) - \phi_A(x, \eta_{S \setminus \{a\}}) - \phi_A(y, \xi_{S \setminus \{a\}}) + \phi_A(y, \eta_{S \setminus \{a\}}) \Big| \Big) \leq 2 \sum_{A \supseteq \{a,b\}} \delta(\phi_A),$$

$$\alpha(\phi) \leq \sup_a \frac{1}{2} \sum_{b, b \neq a} \Big( \sum_{A \supseteq \{a,b\}} \delta(\phi_A) \Big) \leq \sup_a \frac{1}{2} \sum_{A \ni a} (|A| - 1) \delta(\phi_A).$$

In (2.10), the constant 2 cannot be improved. □

**(2.1.2) Example.** Binary State Space. If $E = \{0, 1\}$, $\lambda$ is the counting measure and

$$\phi_A(\omega) = \begin{cases} K(A) & \text{if } \omega_A = 1, \\ 0 & \text{if not.} \end{cases}$$

condition (2.10) is then

$$\sup_{a \in S} \sum_{A \ni a} (|A| - 1)\ |K(A)| \leq 4. \tag{2.11}$$

**(2.1.3) Example.** Ising Model, $E = \{-1, 1\}$.
If $A \in \mathcal{S}$, $\omega^A = \prod_A \omega_i$; $\phi_A(\omega) = -J(A)\omega^A$; a bound for (2.10) is

$$\sup_{a \in S} \sum_{A \ni a} (|A| - 1)\ th|J(A)| < 1. \tag{2.12}$$

If $S = \mathbb{Z}^d$ and if $\phi$ is a pairwise interaction potential, $\phi_{\{i,j\}}(\omega) = J(i-j)\omega_i\omega_j$ with $J : \mathbb{Z}^d \longrightarrow [0, \infty[$, $J(0) = 0$, the condition becomes

$$\sum_{i \in \mathbb{Z}^d} th(J(i)) < 1. \tag{2.13}$$

**(2.1.4) Example.** Significant Exterior Field. $E = \{-1, 1\}$, $\phi_{\{i\}}(\omega) = -h\omega_i$, $h \in \mathbb{R}$. Then, if

$$e^{|h|} > \sup_{i \in S} \Big( \exp \frac{1}{2} \sum_{\substack{A \ni a \\ |A| > 1}} \delta(\Phi_A) \cdot \sum_{A \ni i} (|A| - 1) \delta(\phi_A) \Big), \tag{2.14}$$

(2.9) is satisfied. Observe that conditions (2.11, 2.12, 2.13) don't consider the exterior field since $|A| - 1 = 0$ if $A = \{i\}$. For the Ising model in the presence of an exterior field $h$, with conditional distribution at a given site

$$\pi(x \mid \cdot) = \frac{\exp x(h+\beta v)}{2ch(h+\beta v)},$$

v is the sum of the four neighboring values, and then condition (2.14) can be writen as $|h| > 4|\beta| + \text{Log}(8|\beta|)$.

**(2.1.5) Example.** Low Temperatures. Consider $\phi$ such that

$$a_\phi = \sup_{i\in S} \sum_{A\ni i} (|A|-1)\delta(\phi_A) < \infty.$$

Then the field associated to $\beta\phi$ satisfies (2.9) if $\beta$ is sufficiently small (cf. (2.10)). The condition is also satisfied if $\beta$ is sufficiently big and if $\phi$ admits a unique maximum in the following sense: there exists an $\omega$ *s.t.*

$$b_\phi = \inf_{i\in S} \inf_{\zeta\in\Omega:\zeta_i\neq\omega_i} (-H^\phi_{\{i\}}(\zeta) + H^\phi_{\{i\}}(\omega_i, \zeta_{S\setminus\{i\}})) > 0.$$

In example (2.1.4), the condition becomes $|h| > \sup_i \frac{1}{2}\sum_{A\ni i,|A|>1} \delta(\phi_A)$. In this case, and if $E$ is finite,

$$\alpha(\beta\phi) \leq \beta a_\phi(|E|-1)^{1/2}\exp(-\beta b_\phi),$$

and thus, $\alpha(\beta\phi) < 1$ if $\beta$ is sufficiently big.

**(2.1.6) Example.** Antiferromagnetic Model. $E = \{1, 2, \ldots, N\}$, $\partial i$ is the set of neighbors of $i$, and there are only pairwise interaction potentials

$$\phi_{\{i,j\}}(\omega) = \begin{cases} -J\delta(\omega_i,\omega_j) & \text{if } i \text{ and } j \text{ are neighbors,} \\ 0 & \text{if not.} \end{cases}$$

Here, $J > 0$ and $\delta$ is the Dirac function. Then

$$\alpha(\phi) \leq \sup_i 2|\partial i|(N - |\partial i|)^{-1} \tag{2.15}$$

so that $\alpha(\phi) < 1$ if for all $i$ the number of states $N$ is greater than $3|\partial i|$ (independently of the value of $J > 0$).

**(2.1.7) Example.** Uniqueness in the One-Dimensional Case, $S = \mathbb{Z}$. We have $\phi$, an admissible potential, $E$, a polish space, and $\lambda$ finite. Then $|\mathcal{G}(\phi)| = 1$ if

$$\sup_{i\in S} \sum_{A:\min A\leq i<\max A} \delta(\phi_A) < \infty. \tag{2.16}$$

This condition is much less restrictive than (2.10). In particular, we have uniqueness if $\phi$ is of bounded support over $R$.

If $\phi$ is invariant under translations (2.16) becomes

$$\sum_{A \ni 0} \frac{\text{diam } A}{|A|} \delta(\phi_A) < \infty. \tag{2.17}$$

For example, consider

$$\phi_A(w) = \begin{cases} |i-j|^{-p} \varphi(\omega_i, \omega_j) & \text{if } A = \{i,j\},\ i \neq j, \\ 0 & \text{if not,} \end{cases}$$

with $\varphi$ symmetric, bounded, and measurable. Then (2.17) is equivalent to $p > 2$, which is also a necessary condition for uniqueness.

### 2.1.3.2. Mixing Properties.

When condition (2.9) is true, we can define $\Gamma^n$, $n \geq 0$, the n-*th* power of $\Gamma = (\gamma_{a,b}(\phi), a, b \in S)$, and $\Gamma^0$ the identity matrix. Since each term $\Gamma^n_{a,b}$ of $\Gamma^n$ is bounded by $\alpha^n$, we can define $\chi = \sum_{n \geq 0} \Gamma^n$.

**(2.1.3) Theorem.** Mixing Under the Uniqueness Condition $(D)$ ([107], [71], [80]). *Assume $\phi$ is a bounded potential with bounded range $\ell$ that satisfies $(D)$, where E is a polish space. Then the (unique) measure $\mu \in \mathcal{G}(\phi)$ satisfies the uniform exponential mixing condition, for $A \in \mathcal{S}$:*

$$\varphi(A, B) \leq C \cdot |A| \alpha^{d(A,B)},$$

*where* $\varphi(A{,}B) = \sup\{|\mu(E \mid F) - \mu(E)|,\ E \in \mathcal{F}_A,\ F \in \mathcal{F}_B,\ \mu(F) > 0\}$.

*Proof:* It follows from the upper bound (Property 2.5, [71])

$$\varphi(A, B) \leq \sum_{b \in A} \Big( \sum_{c \in B, a \notin B} \gamma_{ca} \chi_{ab} \Big).$$

More explicitly, if $d(a,b) > \ell$, $\gamma_{ab} = 0$, and thus $\Gamma^n_{ab} = 0$ if $d(a,b) > n\ell$. Since

$$\chi_{a,b} = \sum_{n > k\ell} (\Gamma^n)_{ab} \quad \text{if } d(a,b) > k\ell,$$

we have

$$\sum_{a:\ d(a,b) > l\ell} \chi_{ab} = \sum_a \sum_{n > k\ell} \Gamma^n_{a,b} \leq \frac{\alpha^{k\ell}}{1-\alpha}.$$

Let $k$ be an integer *s.t.* $d(A,B) \geq k\ell$. In order to estimate $\varphi$, it is enough to consider $a \notin B$ *s.t.* $d(a,B) \geq (k-1)\ell$. Then

$$\varphi(A,B) \leq \alpha |A| \sup_{b \in A} \Big( \sum_{\substack{a \notin B \\ d(a,A) \geq (k-1)\ell}} \chi_{ab} \Big) \leq K |A| \alpha^{d(A,B)}.$$

□

REMARK: Under the same conditions, and if $f$ and $g$ are two quasi-local functions (i.e. continuous functions if $E$ is compact), we can control the $\mu$-covariances by

$$|\mu(fg) - \mu(f)\mu(g)| \leq \frac{1}{4} \sum_{i,j \in S} \delta_i(f)\chi_{ij}(\phi)\delta_j(f)$$

$$\text{with } \delta_i(f) = \sup_{\zeta \equiv \eta \text{ over } S\backslash\{i\}} |f(\zeta) - f(\eta)|.$$

Other (indirect) controls of the covariances can be obtained via the mixing inequalities ([52], [86]).

*Regularity of the Mapping $\phi \longrightarrow \mu_\phi$ in Terms of $\phi$.* This result shall be useful when considering the estimation of a Gibbs field under the uniqueness condition. The result is given in [71] for the case $S = \mathbb{Z}^d$ and $\phi$ invariant under translation. We define $\mathcal{D} = \{\phi : \sum_{A \ni 0} |A| \; \|\phi_A\| < 1\}$. $\mathcal{D}$ is a Banach space. Let $g$ be a quasi-local bounded function, summable in the sense that $\sum_{i \in \mathbb{Z}^d} \delta_i(g) < \infty$. Then $\phi \longrightarrow \mu_\phi(\text{g})$ is of class $\mathcal{C}^1$ over $\mathcal{D}$ under the uniqueness condition. This result can be directly generalized to the non-invariant case if we assume

$$\sup_{i \in S} \Big( \sum_{A \ni i} |A| \; \|\phi_A\| \Big) < 1$$

and if we replace, in the $\psi$-directional derivative form, $\overline{\psi} = \sum_{A \ni 0} |A|^{-1} \psi_A$ by $\overline{\psi}_i = \sum_{A \ni i} |A|^{-1} \psi_A$.

## *2.1.4. Variational Principle*

We shall assume that $S = \mathbb{Z}^d$ and that the potential $\phi \in \mathcal{B}_s$ is invariant under translations. $(\Lambda_n)$ stands for a sequence of cubes of $\mathbb{Z}^d$ whose volume $|\Lambda_n|$ tends to infinity.

*Pressure.* For such a potential and for any sequence $(\omega_n)$ of $\Omega$ (e.g. a sequence of constant configurations), the limit

$$p(\phi) = \lim_{n \to \infty} |\Lambda_n|^{-1} \log Z_{\Lambda_n}(\phi, \omega_n)$$

exists and only depends on $\phi$ (and on $\lambda$) [71]: $p(\phi)$ is the *pressure* associated to potential $\phi$. The analytical behavior of the pressure is the classical indicator of the presence or absence of phase transition. We shall examine more closely the convexity and $\mathcal{C}^2$ regularity properties of $p$ under the uniqueness condition $(D)$.

*Relative and Specific Entropy.* Let $\mu$ and $\nu$ be two finite measures of $\mathcal{M}(\Omega, \mathcal{F})$, $\Lambda \in \mathcal{S}$. Define

$$H_\Lambda(\mu|\nu) = \begin{cases} E_{\mu_\Lambda}\left(\log \dfrac{d\mu_\Lambda}{d\nu_\Lambda}\right) & \text{if } \mu_\Lambda \ll \nu_\Lambda, \\ \infty & \text{if not.} \end{cases}$$

The *relative entropy* $H(\mu|\nu)$ of $\mu$ with respect to $\nu$, two stationary probability measures, is defined by the following property:

$$H(\mu,\nu) = \lim_n H_{\Lambda_n}(\mu|\nu) = \sup_n H_{\Lambda_n}(\mu|\nu).$$

The opposite of the relative entropy is called the Kullback information of $\mu$ relative to $\nu$. The *specific entropy* of $\mu$ is the relative entropy with $\nu = \lambda^S$:

$$h(\mu) = H(\mu|\lambda^S) = \sup_n H_{\Lambda_n}(\mu|\lambda^S).$$

Let $\mu \in \mathcal{P}_s(\Omega)$ be the set of stationary probabilities over $(\Omega, \mathcal{F})$, and let $\nu \in \mathcal{G}_s(\phi)$ be the set of stationary Gibbs measures associated with $\phi$.

**(2.1.4) Theorem.** [71] *Consider $\mu \in \mathcal{P}_s(\Omega)$, $\nu \in \mathcal{G}_s(\phi)$. Then*

(1) $h(\mu|\nu) = h(\mu|\phi) = p(\phi) + E_\mu(\overline{\phi}) + h(\mu) \geq 0$, and

(2) $h(\mu|\phi) = 0 \iff \mu \in \mathcal{G}_s(\phi)$.

As $\inf_{\mu\in\mathcal{P}_s(\Omega)} h(\mu|\phi) = p(\phi) + \inf_{\mu\in\mathcal{P}_s(\Omega)} \{E_\mu(\overline{\phi}) + h(\mu)\}$, (2) transforms into the *variational principle*,

$$-p(\phi) = \inf_{\mu\in\mathcal{P}_s(\Omega)} \{E_\mu(\overline{\phi}) + h(\mu)\},$$

and the infimum is reached for all $\mu \in \mathcal{G}_s(\phi)$ and *only* for these states.

REMARK: This result is interesting only if $\mathcal{G}_s(\phi) \neq \emptyset$. This will be the case if $\mathcal{G}(\phi) \neq \emptyset$ is compact and if $\phi$ is invariant under translations, in particular if $\mathcal{G}(\phi) = \{\mu\}$ with $\mu$ stationary.

In Chapter 5 we shall use the variational principle to establish the consistency of the Maximum Likelihood, based on the following property:

**(2.1.5) Theorem.** *Let $\phi$ and $\psi$ be two translation invariant potentials with bounded support, $\nu \in \mathcal{G}_s(\phi)$, and define $K(\nu,\psi) = -p(\psi) - E_\nu(\overline{\psi})$. Then*

(*a*) $K(\nu,\phi) = h(\nu) \geq K(\nu,\psi)$, and

(*b*) *if $\mathcal{G}_s(\phi) \cap \mathcal{G}_s(\psi) = \emptyset$, the inequality is strict: $K(\nu,\phi) > K(\nu,\psi)$.*

*Proof:* We have $K(\nu,\psi) = \inf_{\mu\in\mathcal{P}_s(\Omega)} \{E_\mu(\overline{\psi}) + h(\mu)\} - E_\nu(\overline{\psi})$. Since the infimum is reached over $\mathcal{G}_s(\psi)$ and since $\nu \notin \mathcal{G}_s(\psi)$, $K(\nu,\psi) < h(\nu)$. But $h(\nu)$ is precisely $K(\nu,\phi)$. □

REMARK: $K$ is the contrast function associated to the log-likelihood calculated for $\psi$ when the true model is $\nu$:

$$|\Lambda_n|^{-1}\log \pi^{\psi}_{\Lambda_n}(\omega) = -|\Lambda_n|^{-1}(\log Z_{\Lambda_n}(\psi,\omega) + \sum_{\Lambda_n} \overline{\psi}(\tau_i\omega) + \varepsilon_{\Lambda_n}(\omega))$$

because of (2.6), and thus under $\nu$ and if $\nu$ is ergodic:

$$|\Lambda_n|^{-1}\log \pi^{\psi}_{\Lambda_n}(\omega) \xrightarrow[\nu-a.s.]{} -p(\psi) - E_\nu(\overline{\psi}) = K(\nu,\psi).$$

**(2.1.6) Theorem.** Convexity and Differentiability of the Pressure.

(1) $\phi \longmapsto p(\phi)$ *is convex over* $\mathcal{B}_s$ *and satisfies*

$$|p(\phi) - p(\psi)| \le \|\phi - \psi\|,$$

*where* $\|\phi\| = \sum_{A\ni 0} \|\phi_A\|$.

(2) *Define* $\widetilde{\mathcal{B}}_s = \{\phi \in \mathcal{B}_s : |||\phi||| = \sum_{A\ni 0} |A|\ \|\phi_A\| < \infty\}$

$$\mathcal{D} = \{\phi \in \widetilde{\mathcal{B}}_s : |||\phi||| < 1 \text{ and } \phi \text{ is quasi-local}\}.$$

*If $E$ is a polish space, and $\mu_\phi$ is the only measure in $\mathcal{G}(\phi) = \mathcal{G}_s(\phi)$, then the pressure is differentiable in all directions $\psi$ of $\widetilde{\mathcal{B}}_s$, and*

$$\frac{\partial}{\partial t} p(\phi + t\psi)\Big|_{s=0} = -\mu_\phi(\overline{\psi})$$

$$\frac{\partial^2}{\partial s\partial t} p(\phi + s\psi + t\widetilde{\psi})\Big|_{s=t=0} = \sum_{i\in\mathbb{Z}^d} [\mu_\phi(\overline{\psi}\cdot\overline{\widetilde{\psi}}\circ\tau_i) - \mu_\phi(\overline{\psi})\mu_\phi(\overline{\widetilde{\psi}})].$$

*Proof of the convexity:* Let $\phi,\psi \in \mathcal{B}_s$, $\Lambda \in \mathcal{S}$ and $\omega \in \Omega$, $0 < s < 1$

$$\begin{aligned}\log Z^{s\phi+(1-s)\psi}_{\Lambda}(\omega) &= \log \lambda_\Lambda((h^\phi_\Lambda)^s (h^\psi_\Lambda)^{1-s}|\omega)\\ &\le s\ \log Z^\phi_\Lambda(\omega) + (1-s)\log Z^\psi_\Lambda(\omega)\end{aligned}$$

with $h^\phi_\Lambda(\omega) = \exp(H^\phi_\Lambda(\omega))$, $H^\phi_\Lambda(\omega) = \sum_{A\in\mathcal{S}:A\cap\Lambda\neq\phi} \phi_A(\omega)$. This gives the convexity. On the other hand,

$$Z^\phi_\Lambda(\omega) = \lambda_\Lambda(h^\psi_\Lambda h^{\phi-\psi}_\Lambda|\omega) \le Z^\psi_\Lambda(\omega)\exp\|H^{\phi-\psi}_\Lambda\|_\infty \le Z^\psi_\Lambda(\omega)\exp(|\Lambda|\ \|\phi-\psi\|).$$

Taking logarithms and dividing by $|\Lambda_n|$, and taking limits and finally interchanging the roles of $\phi$ and $\psi$ we get the announced bound. To check the differentiability of $p$, cf. [71]. □

*Calculating the Derivatives of the Pressure: An Example.* Let $A_\ell$, $\ell = 1, p$, be finite subsets such that 0 belongs to all of them. Define $\phi_{A_\ell}$ a bounded potential supported over $A_\ell$ for each $\ell = 1$, $p$ and $\phi$ the following potential defined by translation:

$$\phi_A(\omega) = \begin{cases} \phi_{A_\ell}(\tau_i \omega) & \text{if for some } i, \ell, \ A = \tau_{-i} A_\ell, \\ 0 & \text{if not.} \end{cases}$$

Consider the exponential family associated to the Hamiltonian

$$H_\Lambda^\phi(\omega) = \sum_{\ell=1}^{p} \theta_\ell \Big( \sum_{i \in S : \tau_{-i} A_\ell \cap \Lambda \neq \phi} \phi_{A_\ell}(\tau_i \omega) \Big), \quad \overline{\phi}(\omega) = \sum_{\ell=1}^{p} \theta_\ell \phi_{A_\ell}(\omega).$$

Once $\phi$ is fixed, $p$ becomes a function of $\theta \in \mathbb{R}^p$. Assume $\theta$ is small enough so that $\mathcal{G}(\theta)$ has a sole element (necessarily stationary and ergodic) $\mu_\theta$, for example,

$$\sum_{\ell=1,p} |\theta_\ell| \, |A_\ell| \, \|\phi_{A_\ell}\| < 1.$$

Then

$$\frac{\partial p}{\partial \theta_\ell} = -E_\theta(\phi_{A_\ell}), \quad \frac{\partial^2 p}{\partial \theta_\ell \partial \theta_k} = \sum_{s \in S} cov_\theta(\phi_{A_\ell}, \ \phi_{A_k + s}).$$

## *2.1.5. Ergodicity and Representation of the Elements of $\mathcal{G}(\phi)$*

Set $S = \mathbb{Z}^d$ and for all $i \in S$, $\tau_i$ is the translation over $\omega$ defined by

$$(\tau_i(\omega))_j = \omega_{j-i}, \quad j \in S.$$

The *invariant's $\sigma$-algebra* is the sub $\sigma$-algebra of $\mathcal{F}$

$$\mathcal{I} = \{A \in \mathcal{F} : \tau_i A = A \quad \text{for all } i \in S\}.$$

The *tail $\sigma$-algebra* is the sub $\sigma$-algebra of $\mathcal{F}$ defined by

$$\mathcal{T} = \bigcap_{\Lambda \in \mathcal{S}} \mathcal{F}(S \backslash \Lambda).$$

A stationary probability measure $\mu \in \mathcal{P}_s(\Omega, \mathcal{F})$ is said to be *ergodic* if $\mu$ is trivial over $\mathcal{I}$. An element $\mu$ of a convex $\mathcal{C}$ of a real vector space is an *extremal point* of $\mathcal{C}$ if $\mu \neq s\nu + (1-s)\nu'$ with $0 < s < 1$, $\nu \neq \nu'$ in $\mathcal{C}$. We define $ex(\mathcal{C})$ as the set of all extreme elements of $\mathcal{C}$. We have the following properties:

(1) $\mu \in ex(\mathcal{P}_s(\Omega)) \Longleftrightarrow \mu$ is trivial over $\mathcal{I} \Longleftrightarrow \mu$ is ergodic.

(2) $\mu \in \mathcal{P}_s(\Omega)$ is characterized (in $\mathcal{P}_s$) by its restriction over $\mathcal{I}$.

(3) If $\mu, \nu \in ex(\mathcal{P}_s(\Omega))$, $\mu \neq \nu$, then they are mutually singular.

(4) For all $\mu \in \mathcal{P}_s(\Omega)$, there exists a linear bijection of $\mathcal{P}_s(\Omega)$ into $\mathcal{P}(ex(\mathcal{P}_s(\Omega)))$, $\mu \longmapsto W_\mu$ *s.t.* $\mu = \int_{ex(\mathcal{P}_s(\Omega))} \nu \, W_\mu(d\nu)$.

Relative to $\mathcal{G}(\pi)$ we have the analogous properties:

(1)' $\mu \in ex(\mathcal{G}(\pi)) \Longleftrightarrow \mu$ is trivial over $\mathcal{T}$.

(2)' $\mu \in \mathcal{G}(\pi)$ is characterized (over $\mathcal{G}(\pi)$) by its restriction over $\mathcal{T}$.

(3)' Two distinct elements of $ex(\mathcal{G}(\pi))$ are mutually singular.

(4)' *Representation of a Gibbs measure in the case of a translation invariant specification:*

(*i*) $ex(\mathcal{G}_s(\pi)) = \mathcal{G}_s(\pi) \cap ex(\mathcal{P}_s(\Omega))$; that is, the extremal elements of $\mathcal{G}_s(\pi)$ coincide with the stationary ergodic Gibbs measures.

(*ii*) Consider $\mu \in \mathcal{G}_s(\pi)$ and $W_\mu$ the measure over $ex(\mathcal{P}_s(\Omega))$ which represents $\mu$ as an element of $\mathcal{P}_s(\Omega)$ (cf. (4)). Then $W_\mu$ is supported over $ex(\mathcal{G}_s(\pi))$:

$$\mu = \int_{ex(\mathcal{G}_s(\pi))} \nu \, W_\mu(d\nu).$$

This representation of a stationary, nonnecessarily ergodic measure will be used in the proof of the consistency of classical estimation methods of $\mu$. On the other hand, it is worth noting that if the specification of $\pi$ is invariant and if $|\mathcal{G}(\pi)| = 1$, then the only element $\mu$ of $\mathcal{G}(\pi)$ is stationary and ergodic.

(5) *Description of* $\mathcal{G}_s(\pi)$:

- • $\mu \in ex(\mathcal{G}_s(\pi)) \Longleftrightarrow \mu$ is ergodic.
- •• If $\nu \in \mathcal{P}_s(\Omega)$ and $\nu \ll \mu \in \mathcal{G}_s(\pi)$, then $\nu \in \mathcal{G}_s(\pi)$.
- ••• $\mathcal{G}_s(\pi)$ is a side of $\mathcal{P}_s(\Omega)$: if $\mu, \nu \in \mathcal{P}_s(\Omega)$ and if for an $s$, $0 < s < 1$, $s\mu + (1-s)\nu \in \mathcal{G}_s(\pi)$, then $\mu$ and $\nu \in \mathcal{G}_s(\pi)$.

### *2.1.6. Large Deviations for a Gibbs Field*

The Large Deviations (LD) inequalities for a Gibbs Field were established by Comets ([34]), Föllmer and Orey ([56]) and Olla ([125]). Following Comets, we shall use them to establish the consistency of the estimation of $\theta$ if the observation follows a distribution $\mu \in \mathcal{G}(\pi_\theta)$, $\pi_\theta$ invariant; although $\mu$ needn't be necessarily stationary (see § 5.2.2). This *LD* method for studying the consistency was extended by Comets and Gidas ([37]) to the case of incomplete observations. In order to do this, they consider $LD$ estimations for incomplete empirical processes and a new variational principle for the conditional pressure.

Here, $S = \mathbb{Z}^d$, $\phi \in \mathcal{B}_s$ and $(\Lambda_n)$ is the sequence of cubes $[-n, n]^d$. The empirical field relative to $\omega \in \Omega$ over $\Lambda_n$ is the random measure

$$R_{n,\omega} = |\Lambda_n|^{-1} \sum_{i \in \Lambda_n} \delta_{\tau_i \omega},$$

which satisfies for every bounded function $f$

$$|\Lambda_n|^{-1} \sum_{i \in \Lambda_n} f(\tau_i \omega) = \int_\Omega f(x) R_{n,\omega}(dx).$$

Consider $\nu \in \mathcal{G}(\phi)$ and $B$ a Borelian of $\mathcal{P}(\Omega, \mathcal{F})$. In the large deviations framework, Int $A$ and $\overline{A}$ stand for the interior and the closure of $A$, and one has

**(2.1.7) Theorem.** Large Deviations for a Gibbs Field.
(2.18)

$$-\inf_{\mu \in \mathrm{Int}(B \cap \mathcal{P}_s(\Omega))} \{p(\phi) + E_\mu(\overline{\phi}) + h(\mu)\} \leq \underline{\lim}_n \, |\Lambda_n|^{-1} \log \nu(R_{n,\omega} \in B)$$

$$\leq \overline{\lim_n} \, |\Lambda_n|^{-1} \log \nu(R_{n,\omega} \in B) \leq - \inf_{\mu \in \overline{B \cap \mathcal{P}_s(\Omega)}} \{p(\phi) + E_\mu(\overline{\phi}) + h(\mu)\}.$$

*Example of a Typical Application.* $\mu \longmapsto h(\mu)$ is lower semicontinuous and $\mu \longmapsto E_\mu(\overline{\phi})$ is continuous. If $E$ is compact, $\mathcal{P}_s(\Omega, \mathcal{F})$ is also and thus the infimun in the upper bound (2.18) is achieved at some $\mu \in \overline{B}$; if $\overline{B}$ has no common element with $\mathcal{G}_s(\phi)$, the variational principle assures that this infimum is $> 0$. That is, one has an exponential upper bound for $\nu(R_{n,\omega} \in B)$. In our example, $\nu \in \mathcal{G}(\phi)$, and $E$ is compact. Let $K : \mathcal{P}_s(\Omega) \longrightarrow \mathbb{R}^m$ be continuous and $\overline{B}_{\varepsilon,\phi}$ the closure of $\mathcal{P}_s(\Omega)$: $\overline{B}_{\varepsilon,\phi} = \{\mu \in \mathcal{P}_s(\Omega) : dist(K(\mu), \ K(\mathcal{G}_s(\phi))) \geq \varepsilon\}$. By construction, $\overline{B}_{\varepsilon,\phi} \cap \mathcal{G}_s(\phi) = \emptyset$ and thus one has for some $C < \infty$ and $\delta > 0$:

$$\nu\{\omega : dist(K(R_{n,\omega}), K(\mathcal{G}_s(\phi))) \geq \varepsilon\} \leq C \exp(-\delta |\Lambda_n|).$$

Remark: If $E$ is not compact but polish, this bound is still true because the sets at level $h$ are compact.

## 2.2. Markovian Fields and Associated Gibbs Potentials

The set of sites $S$, finite or numerable, is equipped with a *graph* structure $G$ symmetrical and reflexive, called the neighborhood graph. We define the *neighborhood* of $i$ ($i$ included) as $V(i) = \{j \in S : j \text{ neighbor of } i\}$, and the *neighborhood boundary* of a finite subset $A \subseteq S$ as:

$$\partial A = \{i \in S \ tq \ i \notin A \text{ et } V(i) \cap A \neq \emptyset\}.$$

If $A = \{i\}$, we will write $\partial i$ for $\partial\{i\}$.

**(2.2.1) Definition.** *Let $\mu$ be a field over $(\Omega,\mathcal{F})$. We will say $\mu$ is G-Markovian if for all finite subsets $\Lambda$ of $S$, all event $A$ of $\mathcal{F}_\Lambda$, and all configuration $\omega$ of $\Omega$:*

$$\mu_\Lambda(A|\omega(S\backslash\Lambda)) = \mu_\Lambda(A|\omega(\partial\Lambda)).$$

If $S$ is metric, and if all neighborhoods $V(i)$ have diameters bounded by $R$, we will say that the Markovian field is of *bounded range* $R$.

### *2.2.1. Representation of a Markovian Field as a Gibbs Field*

The following property is important because it characterizes Markovian fields as Gibbs fields:

*Cliques of a graph $G$:* A nonempty set $C \subseteq S$ is a *clique* if $C$ is a single point, or if all elements of $C$ are pairwise neighbors.

**(2.2.1) Theorem.** Hammersley–Clifford (Besag [14]).

(*a*) *Let $\mu$ be a bounded range Markovian field over $(\Omega,\mathcal{F})$. Let the state space at each site $E_i$ be discrete and assume $\mu$ satisfies the positivity condition: for all finite $V$, $x = x(V) \in \Omega(V)$, $y = y(S\backslash V) \in \Omega(S\backslash V)$, and one has*

$$\mu_V(x \mid y) > 0.$$

*Then we can associate to $\mu$ a bounded support interaction potential $\phi = \{\phi_A, A \in \mathcal{C}\}$ where $\mathcal{C}$ is the family of cliques of the Markovian graph, and $\mu_V = \pi_V^\phi$ where $\pi^\phi$ is the specification associated to $\phi$.*

(*b*) *Conversely, let $\phi = \{\phi_A, A \in \mathcal{C}\}$ be a Gibbs potential. Then the associated specification $\pi^\phi$ is Markovian for the graph structure which has as cliques the elements of $\mathcal{C}$.*

*Proof:*

(*a*) Consider $V \in \mathcal{S}$; $\overline{V} = V \cup \partial V$ is finite. Choose $x = x(V)$ in $\Omega(V)$, $y = y(\partial V)$ in $\Omega(\partial V)$. In each state space $E_i$, let $0_i$ be a reference state $0(V) = (0_i, i \in V)$. Write

$$H(x,y) = \log[\mu(x(V) \mid y(V))/\mu(0(V) \mid y(V))].$$

One has

$$H(0,y) = 0. \tag{2.19}$$

For $A \subseteq \overline{V}$, define the potential

$$\begin{cases} \phi_A(x,y) = \sum_{B \subseteq A} (-1)^{|A \backslash B|} H^B(x,y), \quad \text{with} \\ \\ H^B(z) = H(z(B), 0(\overline{V} \backslash B)). \end{cases} \tag{2.20}$$

$\phi_A \equiv 0$ if $A \cap V = 0$. The Moebius Lemma (cf. Appendix 2) gives

$$H(x,y) = \sum_{A \subset V \cup \partial V, A \cap V \neq \emptyset} \phi_A(x,y).$$

We shall now show that if $A \notin \mathcal{C}$, then $\phi_A \equiv 0$.

Let 1 and 2 be two nonneighboring points of $A$, let $z$ be a configuration s.t. $z_1$ and $z_2 \neq 0$ (if not $\phi_A(z) = 0$), and let $z^i$ stand for the configuration equal to $z$ everywhere except in $i$, where it is 0. Because of (2.20),

(2.21)

$$\phi_A(z) = \sum_{D \subseteq A \backslash \{1,2\}} (-1)^{|A \backslash D|} [(H^D(z) - H^{D \cup \{2\}}(z)) - (H^{D \cup \{1\}}(z) - H^{D \cup \{1,2\}}(z))].$$

But

$$\begin{aligned} H^{D\cup\{1,2\}}(z) - H^{D\cup\{1\}}(z) &= \log \frac{\mu(z_{D\cup\{1,2\}}|0)}{\mu(0_{D\cup\{1,2\}}|0)} - \log \frac{\mu(z_{D\cup\{1\}}|0)}{\mu(0|0)} \\ &= \log \frac{\mu(z_{D\cup\{1,2\}}|0)}{\mu(0_{D\cup\{1,2\}}|0)} - \log \frac{\mu(z^2_{D\cup\{1,2\}}|0)}{\mu(0_{D\cup\{1,2\}}|0)} \\ &= \log \frac{\mu(z_{D\cup\{1,2\}}|0)}{\mu(z^2_{D\cup\{1,2\}}|0)} = \log \frac{\mu(z_2 \mid z_{D\cup\{1\}}, 0)}{\mu(0 \mid z_{D\cup\{1\}}, 0)}, \end{aligned}$$

where this last expression is independent of $z_1$. We thus deduce that in the sum over $D$ that defines $\phi_A$, each expression between braces is 0. Thus, $\phi_A \equiv 0$.

(*b*) If $\pi$ is the specification associated to potential $\phi = \{\phi_A, A \in \mathcal{C}\}$, the relationship

$$\frac{\pi_V(x_V \mid x_{S\backslash V})}{\pi_V(O_V \mid x_{S\backslash V})} = \exp\Big( \sum_{A:A\cap V\neq\phi} (\phi_A(x) - \phi_A(x^V))\Big) \tag{2.22}$$

only depends on the configuration of $x$ over $\overline{V} = \cup_{A\in\mathcal{C}, A\cap V\neq\phi} A$. $\pi$ is thus Markovian for the graph defined by $\mathcal{C}$. □

Part (b) of Theorem (2.2.1) is true any time the potential well defines a probability.

We will say a field is *almost Markovian* if for all finite subsets $V$ of $S$, $y \longmapsto \pi_V(x(V) \mid x(S\backslash V) = y)$ is continuous in $y$. A Markovian field is almost Markovian. There is an analogous representation theorem of any almost Markovian field as a Gibbs field (cf. [133]).

*Energy and Potential Identification.* Assume $S$ is finite. To any probability measure over $\Omega = \prod_S E_i$, we can associate an energy $-H$: $\mu(x) = C\ \exp(H(x))$. $H$, defined up to a constant, can be identified by the restriction $H(0) = 0$, in which case $\mu(x) = \mu(0)\ \exp(H(x))$. If $H$ is defined by a potential $\phi = \{\phi_A, A \subseteq S\}$, we shall have to impose restrictions on $\phi$ in order for the potential to be identifiable. There are several possibilities:

*First possibility*: $H(0) = 0$, the Moebius formula defines $\phi_A$ uniquely if

$$A \subseteq S, \quad \phi_A(x^i) = 0 \quad \text{if } i \in A, \quad \text{where } x_i^i = 0,\ x_j^i = x_j \text{ if not.}$$

*Second possibility*: It is the classical choice of analysis of variance:

$$A \subseteq S, \text{ for all } B \subseteq A: \quad \sum_{x_B} \phi(x_B, x_{A\backslash B}) = 0.$$

Calculating the dimension of a model, directly from the energy function or from the restrictions on the potential is a good way to check model identification.

**(2.2.1) Example.**

$$S = \{1,2,3,4,5\}, \quad E = \{1,2,\ldots,K\},$$

$$\mathcal{C}^* = \{\{1\},\{2,3,4\},\{5\}\}, \quad \mathcal{C} = \{C, C \subset C^* \in \mathcal{C}^*\}.$$

The energy restrictions over every element of the partition yield

$$\dim \phi = 2(K-1) + K^3 - 1 = K^3 + 2K - 3.$$

Restrictions on the Moebius formula yield

$$\dim \phi = 5(K-1) + 3(K-1)^2 + (K-1)^3 = K^3 + 2K - 3.$$

*Calculus of the Conditional Distributions Given the Joint Distribution.* Let $V$ be a finite subset of $S$. Fix the configuration outside of $V$ and define $\mu_V$ a distribution with energy $-H$. If $W \subset V$, in order to determine the conditional distribution $\mu_W(x_W|_W x)$, we shall proceed as follows:

(*i*) Determine potentials $\{\phi_A, A \subseteq V\}$ of $H$, $H(x) = \sum_A \phi_A(x)$.

(*ii*) From (2.3) and (2.4), it follows that

$$\begin{cases} \mu_W(x_W|_W x) = Z_W^{-1}(_W x)\exp\Big\{ \sum\limits_{A: A\cap W \neq \emptyset} \phi_A(x)\Big\}, \quad \text{with} \\ Z_W(_W x) = \sum\limits_{z_W \in \Omega_W} \exp \sum\limits_{A: A\cap W \neq \emptyset} \phi_A(z_{W,W} x). \end{cases}$$

In the particular case of finding the conditional distribution at site $i \in S$, one has, defining $x_{\partial i}$ the configuration of the neighborhood of $i$,

$$\begin{cases} \mu_{\{i\}}(x_i|x_{\partial i}) = Z_i(x_{\partial i})\exp \sum\limits_{A \ni i} \phi_A(x_i, x_{\partial i}), \\ Z_i(x_{\partial i}) = \sum\limits_{z\in\Omega_i} \Big(\exp \sum\limits_{A\ni i} \phi_A(z, x_{\partial i})\Big). \end{cases}$$

If $\Omega_i$ is finite, the normalization constant $Z_i(x_{\partial i})$ can be easily calculated, contrary to the normalization constant of the joint distribution $\mu_V$; it is this that leads toward considering the pseudo conditional likelihood functional (cf. Chapter 5) instead of the usual likelihood.

## *2.2.2. Some Examples of Markovian Fields and Their Associated Potentials*

**(2.2.2) Example.** Six Nearest Neighbor Markovian Fields over a Triangular Lattice.

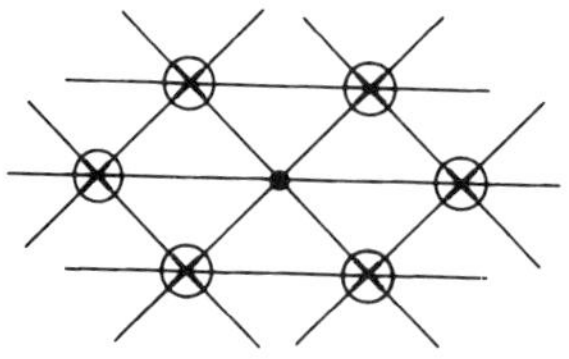

*Triangular lattice and vicinity of six nearest neighbors*

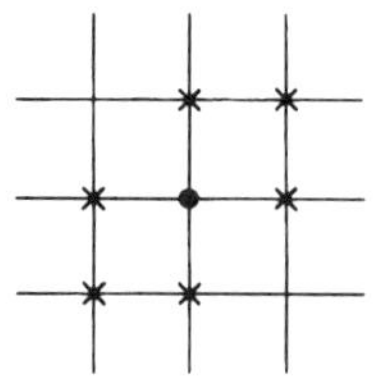

*$\mathbb{Z}^2$ lattice and equivalent neighborhood*

$\mathbb{Z}^2$ with the four nearest neighbors of $(u, v)$ as well as $(u-1, v-1)$ and $(u+1, v+1)$ is isomorphic to the triangular lattice with the neighborhood relationship defined by the six nearest neighbors. The cliques have 1, 2, or 3 points,

$$\mathcal{C}_1 = \{\{i\}, i \in S\}, \quad \mathcal{C}_2 = \{\{i,j\}, \ |i-j| = 1, \ i,j \in S\}$$

$$\mathcal{C}_3 = \{\{i,j,k\}, \ |i-j| = |j-k| = |k-i| = 1\}.$$

If $S$ is finite, we define the field by

$$\mu(x) = Z^{-1} \exp\Big( \sum_{\mathcal{C}_1} \phi_i(x_i) + \sum_{\mathcal{C}_2} \phi_{i,j}(x_i, x_j) + \sum_{\mathcal{C}_3} \phi_{i,j,k}(x_i, x_j, x_k) \Big)$$

with $\phi_A(x^i) = 0$ if $i \in A$. The following table gives the parametric dimension of the model under different hypotheses:

| $E$ with $K$ states | $\mathcal{C}_1$ | $\mathcal{C}_2$ | $\mathcal{C}_3$ |
|---|---|---|---|
| $\mu$ stationary | $K-1$ | $3(K-1)^2$ | $2(K-1)^3$ |
| $\mu$ isotropic | $K-1$ | $\frac{K(K-1)}{2}$ | $\frac{K(K-1)}{4} + \frac{1}{2} \sum_1^{K-1} i^2$ |
| $\mu$ isotropic and binary E | 1 | 1 | 1 |

*Dimension of Markovian models with six nearest neighbors*

In the binary isotropic example, $E = \{0, 1\}$, the distribution is given by

$$\mu(x) = Z^{-1} \exp\Big[ \phi_1 \sum_S x_i + \phi_2 \sum_{\mathcal{C}_2} x_i x_j + \phi_3 \sum_{\mathcal{C}_3} x_i x_j x_k \Big],$$

and the distribution conditional on site $i$ is

$$\mu_i(x_i|\cdot) = \frac{\exp(x_i[\phi_1 + \phi_2 u_i + \phi_3 v_i])}{1 + \exp(\phi_1 + \phi_2 u_i + \phi_3 v_i)}, \ u_i = \sum_{j:|i-j|=1} x_j, \ v_i = \sum_{j,k:\{i,j,k\} \in \mathcal{C}_3} x_j x_k.$$

**(2.2.3) Example.** Nearest Neighbor Ising Model over $\mathbb{Z}^d$.

$E = \{-1, 1\}$. We consider the translation invariant potential:

- $\phi_1(x_i) = h x_i$ ($-h$ is called the exterior field).
- •• If $|i - j| = \sum_{k=1,d} |i_k - j_k| = 1$, $\phi_2(x_i, x_j) = \beta x_i x_j$.

A stationary isotropic $2d$-nearest neighbors Markov field admits as potentials those defined by $\phi_1$ and $\phi_2$. The distributions conditional on $V$, $V \in \mathcal{S}$, and on $i$ are given by

$$\mu_V(x|y) = Z_V^{-1}(y)\exp\Big[h\sum_V x_i + \beta\sum_{\substack{i,j\in V\\|i-j|=1}} x_ix_j + \beta\sum_{\substack{i\in V, j\notin V\\|i-j|=1}} x_iy_j\Big],$$

$$\mu_{\{i\}}(x_i|\cdot) = \frac{\exp x_i(h+\beta v_i)}{2ch(h+\beta v_i)}, \quad v_i = \sum_{j:|j-i|=1} x_j.$$

The pressure was obtained by Onsager [126] for $d = 2$, $h = 0$:

$$p(\beta) = \log 2 + \frac{1}{8\pi^2}\int_{[-\pi,\pi]^2} \log|ch^2 2\beta + sh 2\beta(\cos\lambda + \cos\mu)|d\lambda d\mu$$

Define $\mathcal{G}(h,\beta)$ as the (nonempty) space of Gibbs measures associated to potential $\phi(h,\beta)$. Then:

| | | |
|---|---|---|
| $d = 1$ | any $h$ and $\beta$ | $\lvert\mathcal{G}(h,\beta)\rvert = 1$ |
| $d \geq 2$ | $h \neq 0$ and $\beta \geq 0$ ........ | $\lvert\mathcal{G}(h,\beta)\rvert = 1$ |
| | $h = 0$, $0 < \beta < \beta_c$ .... | $\lvert\mathcal{G}(h,\beta)\rvert = 1$ |
| | $h = 0$, $\beta_c < \beta$ ........ | $\lvert\mathcal{G}(h,\beta)\rvert > 1$ |

*Phase transition or lack thereof: if* $d = 2$, $\beta_c = \frac{1}{2}\log(1+\sqrt{2})$

*Equivalence with Markov Chains.* If $X = (X_t, t \in \mathbb{Z})$ is a homogeneous Markov chain over a finite state space $E$, $X$ is a Markovian field with two nearest neighbors and transition function (cf. 2.2.5):

$$\pi(X_t = y \mid X_{t-1} = x,\ X_{t+1} = z) = \frac{q(x,y)q(y,z)}{q^2(x,z)}$$

if $q(\cdot,\cdot)$ is the transition matrix of $X$. If $E = \{-1, 1\}$ and if we define

$$p = P(X_t = 1 \mid X_{t-1} = 1), \quad q = P(X_t = -1 \mid X_{t-1} = -1),$$

we have the following correspondence between the parameters $(p,q)$ and $(h,\beta)$ of the unilateral and nonunilateral Markov representations:

$$h = \frac{1}{2}\log\frac{p}{q}, \quad \beta = \frac{1}{4}\log\frac{pq}{(1-p)(1-q)},$$

$$p = e^{-h}/(ch(h) + D(\beta,h)), \quad q = e^{h}/(ch(h) + D(\beta,h)),$$

$$\text{with}\quad D(\beta,h) = (e^{-4\beta} + sh^2(h))^{1/2}.$$

When $h = 0$ ($p = q$), this corresponds to a uniform marginal, and $\beta = 0$ ($p + q = 1$) corresponds to an independent field.

### *2.2.3. Besag's Auto-Models [14]*

Let $X$ be a Markovian field over $S = \{1, 2, \ldots, n\}$ such that $\mu(x) > 0$ for all $x \in \Omega$ and with cliques of, at most, two points:

$$\mu(x) = \mu(0)\exp(H(x)), \text{ with } H(x) = \sum_S \phi_i(x_i) + \sum_{\{i,j\}} \phi_{ij}(x_i, x_j)$$

s.t. for all $A$, $\phi_A(x^i) = 0$ if $i \in A$. As usual we consider 0 as a reference state in $E$.

**(2.2.2) Theorem.** *Assume additionally that for each $i \in S$, $\mu_i(x_i|\cdot)$ belongs to an exponential family:*

$$\log \mu_i(x_i|\cdot) = A_i(\cdot)B_i(x_i) + C_i(x_i) + D_i(\cdot), \quad B_i(0) = C_i(0) = 0. \tag{2.23}$$

*Then, (a) there exists $\alpha_i$, $\beta_{ij} = \beta_{ji}$, $i$,$j \in S$ s.t.*

$$A_i(\cdot) = \alpha_i + \sum_{j \neq i} \beta_{ij} B_j(x_j), \tag{2.24}$$

$$\phi_i(x_i) = \alpha_i B_i(x_i) + C_i(x_i), \quad \phi_{ij}(x_i, x_j) = \beta_{ij} B_i(x_i) B_j(x_j). \tag{2.25}$$

*(b) Conversely, conditional distributions $\mu_i(x_i|\cdot)$ that satisfy* (2.23) *and* (2.24) *are the conditional distributions of a Markov field whose potential is given by* (2.25).

REMARK: Condition (2.24) shows that in order for the conditional distributions $\mu_i(x_i|\cdot)$ to "reconstruct" a joint distribution, they must satisfy certain restrictions. In our case, conditions (2.24) are precisely those.

*Proof:*

*(a)* $H(x) - H(x^i) = \log(\mu_i(x_i|\cdot)/\mu_i(0|\cdot))$. Choosing $j \neq i$, and $x$ *s.t.* $x_j = 0$, we get $\phi_i(x_i) = A_i(0)B_i(x_i) + C_i(x_i)$. Fix $i = 1$ and $j = 2$ and pick $x$ with $x_k = 0$ for all $k \neq 1, 2$. Studying $H(x) - H(x^1)$ and $H(x) - H(x^2)$ for this $x$ yields

$$\phi_1(x_1) + \phi_{12}(x_1, x_2) = A_1(0, x_2, 0, \ldots, 0)B_1(x_1),$$

$$\phi_2(x_2) + \phi_{12}(x_1, x_2) = A_2(x_1, 0, 0, \ldots, 0)B_2(x_2),$$

$$\phi_{12}(x_1, x_2) = [A_1(0, x_2, 0, \ldots, 0) - A_1(0)]B_1(x_1)$$

$$= [A_2(x_1, 0, \ldots, 0) - A_2(0)]B_2(x_2).$$

Assume $B_2(x_2) \neq 0$:

$$A_2(x_1, 0, \ldots, 0) - A_2(0) = \frac{A_1(0, x_2, 0, \ldots, 0) - A_1(0)}{B_2(x_2)} \cdot B_1(x_1).$$

Thus, the quotient that appears on the right hand side is a constant, equal to $\beta_{21}$. We define $\beta_{12}$ analogously. It can be checked that

$$\beta_{12} = \beta_{21}, \quad \phi_{12}(x_1, x_2) = \beta_{12} B_1(x_1) B_2(x_2).$$

Identifying the conditional distributions and writing the potentials as we did above,

$$H(x) - H(x^i) = A_i(\cdot)B_i(x_i) + C_i(x_i) = \phi_i(x_i) + \sum_{j:j\neq i} \phi_{ij}(x_i, x_j),$$

which gives (2.24) with $\alpha_i = A_i(0)$.

(*b*) Conversely, if the $\mu_i(x_i|\cdot)$ satisfy (2.23) and (2.24), then for the potentials $\phi_i$ and $\phi_{ij}$ defined by (2.25) and $H$ the associated energy, we have

$$H(x)-H(x^i) = \Big[\alpha_i + \sum_{j:j\neq i} \beta_{ij} B_j(x_j)\Big] B_i(x_i) + C_i(x_i) = \log\{(\mu_i(x_i|)/(\mu_i(0|))\}$$

The $\mu_i(\cdot|\cdot)$, $i \in S$, are the conditional distributions of a Markovian field with energy $H$. □

**(2.2.2) Definition.** Auto-models. *A real valued field $X$ is an auto-model if its distribution $\mu$ can be written as*

$$\mu(x) = \mu(0)\exp H(x) \quad \textit{with} \quad H(x) = \sum_S \phi_i(x_i) + \sum_{i\neq j} \beta_{ij} x_i x_j, \quad \beta_{ij} = \beta_{ji}.$$

Thus, an auto-model is a Markovian field with cliques of, at most, 2 points. The distributions $\mu_i(\cdot|\cdot)$ can be written as

$$\log \mu_i(x_i|\cdot) = (\alpha_i + \sum_{j:j\neq i} \beta_{ij} x_j) x_i + C_i(x_i) + D_i(\cdot)\ .$$

**(2.2.4) Example.** Auto-Logistic Model, $E = \{0, 1\}$.

$$H(x) = \sum_i \alpha_i x_i + \sum_{i\neq j} \beta_{ij} x_i x_j, \quad \mu_i(x_i|\cdot) = \frac{\exp x_i\Big(\alpha_i + \sum_{j:j\neq i} \beta_{ij} x_j\Big)}{1 + \exp\Big(\alpha_i + \sum_{j:j\neq i} \beta_{ij} x_j\Big)}.$$

Notice that the map $x \longmapsto y = 2x - 1$ transforms $E$ into $E' = \{-1, +1\}$, which gives the Ising model with energy function

$$\widetilde{H}(y) = \sum_i \tilde{\alpha}_i y_i + \sum_{i\neq j} \tilde{\beta}_{ij} y_i y_j, \quad \text{with} \quad \tilde{\alpha}_i = \frac{1}{2}\Big(\alpha_i + \sum_{j:j\neq i} \beta_{ij}\Big), \quad \tilde{\beta}_{ij} = \frac{1}{4}\beta_{ij}.$$

**(2.2.5) Example.** Auto-Binomial Model.

$\mu_i(x_i|\cdot) \sim B(n, \theta_i(\cdot))$. Under the exponential form, we can identify

$$A_i(\cdot) = \log \frac{\theta_i(\cdot)}{1-\theta_i(\cdot)} = \alpha_i + \sum_{j:j\neq i} \beta_{ij} x_j; \quad B_i(x_i) = x_i$$

so that the parameter of the conditional binomial distribution is given by

$$\theta_i(\cdot) = \Big[1 + \exp -\Big(\alpha_i + \sum_{j:j\neq i} \beta_{ij} x_j\Big)\Big]^{-1}.$$

**(2.2.6) Example.** Auto-Poisson model. $E = \mathbb{N}$, $\mu_i(x_i|\cdot) \sim \mathcal{P}(\lambda_i(\cdot))$, with

$$\log \lambda_i(\cdot) = A_i(\cdot) = \alpha_i + \sum_{j:j\neq i} \beta_{ij} x_j, \quad B_i(x_i) = x_i.$$

Thus, $H(x) = \sum_i (\alpha_i x_i - \log(x_i!)) + \sum_{j:j\neq i} \beta_{ij} x_i x_j$. Summability on $j$ must be satisfied for all $x$; for example, for $x = (x_1, x_2, 0, 0, \ldots)$,

$$\exp H(x_1, x_2, 0, \ldots, 0) = \frac{\exp(\alpha_1 x_1 + \alpha_2 x_2 + \beta_{12} x_1 x_2)}{x_1! x_2!}.$$

If $\beta_{12} \leq 0$, the summability is assured. This is not true, however, if $\beta_{12} > 0$ since

$$\exp H(x_1, x_2, 0, \ldots, 0) \geq \frac{\exp(\frac{1}{2}\beta_{12} x_1 x_2)}{x_1! x_2!}$$

for big enough $x_1$ and $x_2$, and if we fix $x_2$, the series whose term is the right hand side (r.h.s.) diverges. Thus a necessary and sufficient condition is, if $i \neq j$, $\beta_{ij} \leq 0$: the interactions between $i$ and $j$ are competitive.

**(2.2.7) Example.** Auto-Normal models (see Section 2.3), $E = \mathbb{R}$.

$\mu_i(x_i|\cdot) \sim \mathcal{N}(\mu_i(\cdot), \sigma_i^2(\cdot))$. This family is an auto-model if $\sigma_i^2(\cdot) = \sigma_i^2$, $\mu_i(\cdot) = \nu_i + \sum_{j:j\neq i} a_{ij} x_j$. The conditional distributions $\mu_i(x_i|\cdot)$ belong to a Gaussian Gibbs field that satisfies

$$E(X_i|\cdot) = \nu_i + \sum_{j\neq i} a_{ij} X_j, \quad Var(X_i|\cdot) = \sigma_i^2,$$

$$\log \mu_i(x_i|\cdot) = \frac{x_i^2 - 2\nu_i x_i}{2\sigma_i^2} - \frac{1}{2\sigma_i^2} x_i \Big(\sum_{j\neq i} a_{ij} x_j\Big) - \frac{1}{2}\log \sigma_i^2 + C.$$

This gives $a_{ij}\sigma_j^2 = a_{ji}\sigma_i^2$. The global energy over a finite subset $S$ is

$$H(x) = -\frac{1}{2}\sum_S \frac{x_i^2 - 2\nu_i x_i}{\sigma_i^2} - \frac{1}{2}\sum_{i\in S,\ j\neq i} \frac{a_{ij}}{\sigma_i^2} x_i x_j.$$

## *2.2.4. Applications to Spatial Texture Modeling*

A nice example of the possibilities of modeling by Markovian fields appears in Cross and Jain [40]: the idea is to simulate a homogeneous texture in order to identify and estimate real textures. Pattern analysis is another field where Markovian tools have proved to be very useful (cf. § 2.4 and § 6.8).

Simulated textures are those of an *auto-binomial model*, with range, at most, four, based on the neighborhoods defined by the following figures:

| | | | | |
|---|---|---|---|---|
| | $o_1$ | $m$ | $q_1$ | |
| $o_2$ | $v$ | $u$ | $z$ | $q_2$ |
| $l$ | $t$ | $\boxed{\text{x}}$ | $t'$ | $\ell'$ |
| $q_2$ | $z'$ | $u'$ | $v'$ | $o_2'$ |
| | $q_1'$ | $m'$ | $o_1$ | |

*Neighbors of central pixel $x$*

range $1 : u, u', t, t'$

range $2 : v, v', z, z'$

range $3 : \ell, \ell', m, m'$

range $4 : o_i, o_i', q_i, q_i', i = 1, 2$

*Supports*

The auto-binomial conditional distributions (which determine the joint distribution) are

$$\mu_i(x_i|\cdot) \sim B(n, \theta(\cdot)), \quad \theta(\cdot) = \frac{\exp A(\cdot)}{1 + \exp A(\cdot)}, \quad \text{with}$$

$$A(\cdot) = a + b(1,1)(t + t') + b(1,2)(u + u') + b(2,1)(z + z') + b(2,2)(v + v') + b(3,1)(\ell + \ell') + b(3,2)(m + m') + b(4,1)(o_1 + o_1' + o_2 + o_2') + b(4,2)(q_1 + q_1' + q_2 + q_2').$$

Recall that if the model has $n$ gray levels ($n = 2$ for the auto-logistic model), the dimensions of the general stationary models with cliques of, at most, two points, and those of the auto-binomial model are given by:

| Support | General | Auto – Binomial |
|---|---|---|
| 1 | $(n-1) + 2(n-1)^2$ | 3 |
| 2 | $(n-1) + 4(n-1)^2$ | 5 |
| 3 | $(n-1) + 6(n-1)^2$ | 7 |
| 4 | $(n-1) + 8(n-1)^2$ | 9 |

with isotropy at $(o_1, o_2)$ and $(q_1, q_2)$. In particular, if $n = 2$, both columns coincide so that auto-logistic mode is the most general among binary homogeneous models with cliques of, at most, two points. In between general textures and auto-binomial textures there exists the class of *reversible* textures that satisfy for any two point clique: $\phi(x,y) = \phi(y,x)$. For these textures, it is enough to change $(n-1)^2$ for $\frac{n(n-1)}{2}$ in order to find the parametric dimension of the reversible model. Binary models are automatically reversible.

Changing the Markovian support and the values of parameters $a$ and $b$ of the auto-binomial models, Cross and Jain simulated different types of textures. The simulation algorithm, described in § 6.2.2., uses the Metropolis dynamic of spin exchange.

(I) *Binary Isotropic First-order Textures.* According to whether the state space is $\{0,1\}$ or $\{-1,1\}$ (see Example 1 above), they are given by the model

$$x_i \in \{0,1\} \quad P(x) = Z^{-1}(a,b)\exp\Big(a\sum x_i + b\sum_{<i,j>} x_i x_j\Big),$$

$$y_i \in \{-1,1\} \quad Q(y) = \Delta^{-1}(\alpha,\beta)\exp\Big(\alpha\sum y_i + \beta\sum_{<i,j>} y_i y_j\Big)$$

with, up to border effects, the following parametric correspondence for $y_i = 2x_i - 1$:

$$a = 2\alpha - 8\beta, \quad b = 4\beta \quad \text{or} \quad \alpha = \frac{1}{2}(a+2b), \beta = \frac{1}{4}b.$$

It is easy to see that the $y$-valued model has an equiprobable marginal over $\{-1,1\}$ if $\alpha = 0$, that is, the $\{0,1\}$ has an equiprobable marginal if $a = -2b$ (with $b = b(1,1) = b(1,2)$). The three simulations correspond to the choices $b = 0$ (white noise), $b = 0.75$ (beginning of heap formation), and $b = 3$ (geometrically regular 0 and 1 bands).

(II) *Binary Anisotropic Line Textures (with support of range* 2*).* The important difference among coefficients $b(1,1)$ and $b(1,2)$ in the first order models (the first two simulations), and $b(2,1)$ and $b(2,2)$ for the second order models (third simulation) creates the line anisotropies.

(III) *Binary Textures with Inhibition and Organization.* Diagonal inhibition (attraction for first-order and repulsion for second-order) creates a "maze"-like texture. Isotropic inhibition (attraction for orders 1 and 2, repulsion for orders 3 and 4) creates an "isotropic maze." The last texture with strong repulsion at order 1 is close to the alternate 0, 1 configuration (chessboard effect). Other examples of anisotropy with more gray levels can also be found in ([40]).

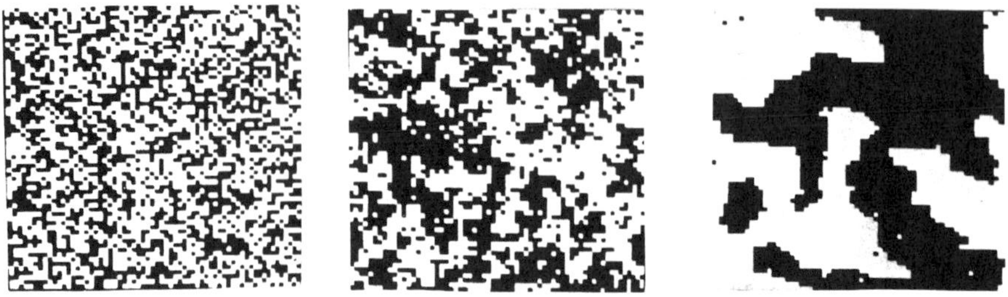

(I) *Simulation of three binary isotropic textures with four nearest neighbors* $(a = -2b) : b = 0,\ 0.75, 3$ *and formation of regular bands*

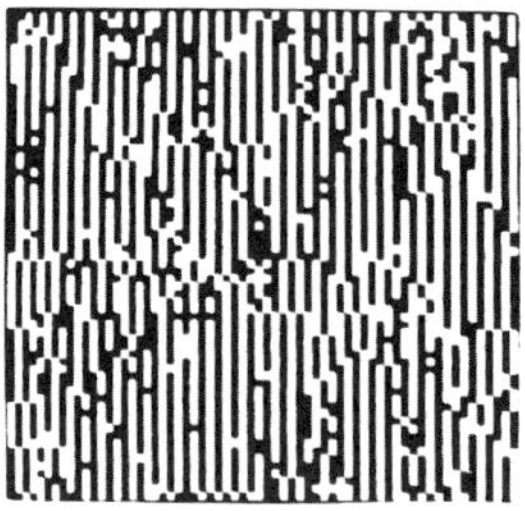

0.13 2.1 0.015
[-0.26] -2

0.02 0.16 0.02
[-2.4] 1.93

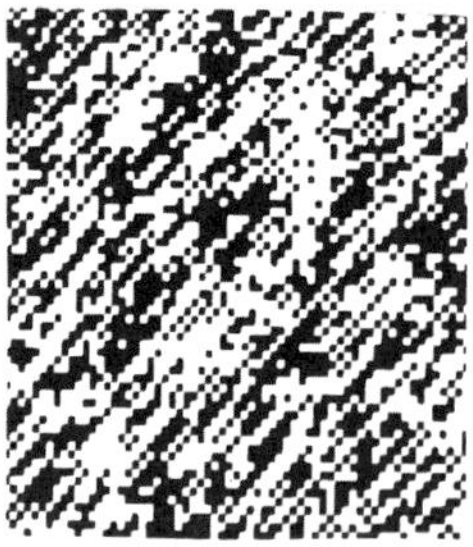

1.9 0.1 -0.075
[-1.9] -0.1

(II) *Anisotropic line textures: description of coefficients* $a$ *and* $b$ *for orders 1 and 2; the boxes stand for the central pixel*

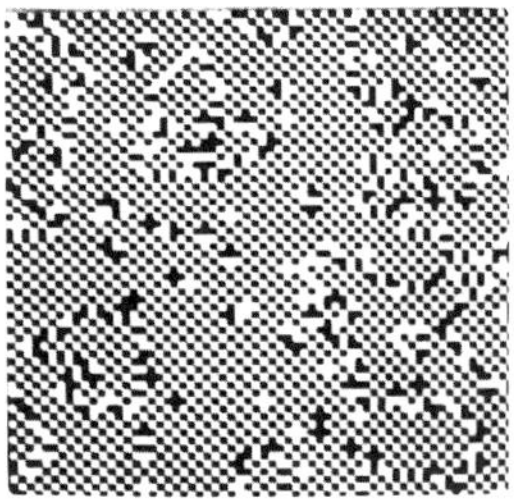

-2.03 2.05 -2.10
[0.16] 2.06

III-1 *Diagonal Inhibition (order 2)*

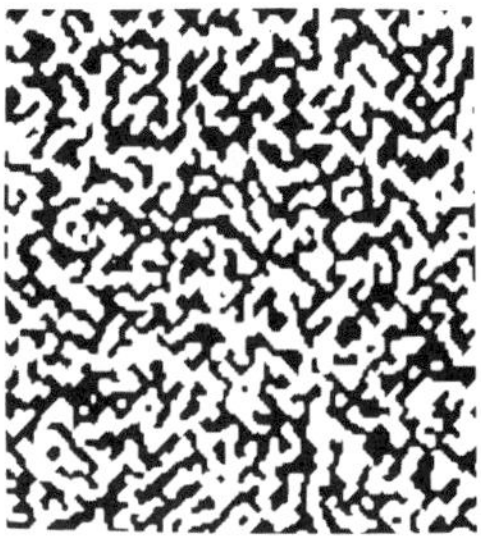

-0.85 -0.78 -0.85
-0.85 2.17 2.62 2.17 -0.85
[-4.6] 2.62 -0.78

III-2 *Isotropic Inhibition (order 4)*

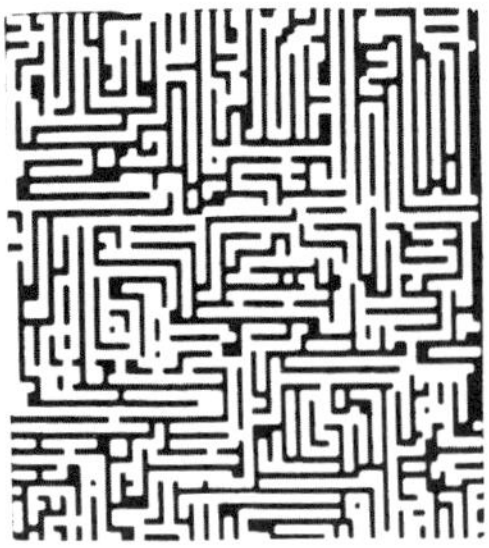

-2.16
[5.09] -2.25

III-3 *Organized Patterns (order 1)*

*Source: Cross and Jain, IEEE [40]*

*Real Texture Modeling by Markovian Fields.* Given a real homogeneous texture of gray levels, the problem is to identify and estimate a Markovian texture which adjusts well to the real texture. If it does, the simulated texture should look like the original image. The identification shall be done by a penalized pseudo-likelihood method (Akaike Information Criterium (AIC) or Bayesian Information Criterium (BIC), see § 3.5 et § 5.1). The real textures are from the Textures Album of Brodatz [24].

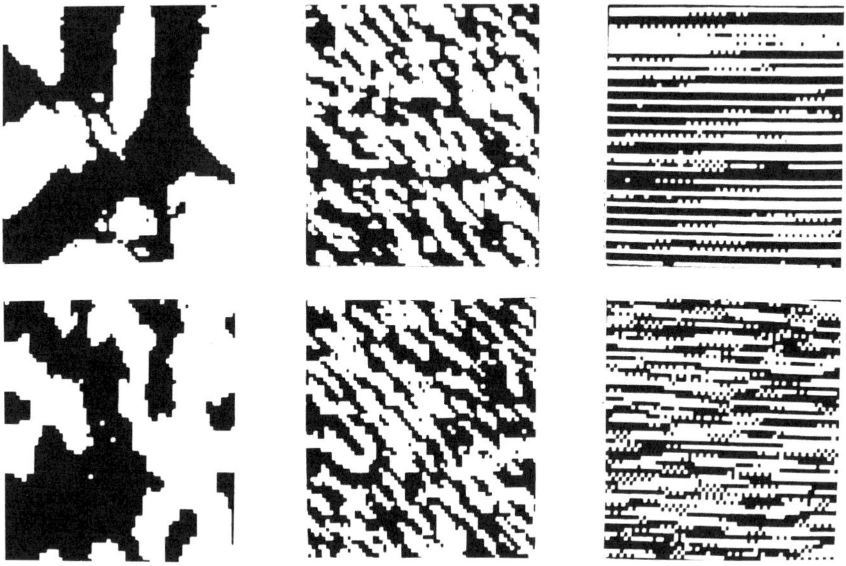

(a) (b) (c)

*Real textures $(x)$: fitting and simulation $(x_1)$*
*(a) stone; (b) cork; (c) curtain*
*Source: Cross and Jain, IEEE [40]*

## 2.2.5. Causal Markovian Fields and Bilateral Representation

For simplicity's sake, consider $\mathbb{Z}^2$ with the lexicographic order. Let $A$ be a finite subset of the negative half plane: $A \subseteq \mathbb{Z}^2_- = \{(i,j) < (0,0)\}$. Define $A_s = A + s$ for $s \in \mathbb{Z}^2$. Also, if $S$ is a finite subset of $\mathbb{Z}^2$, $S$'s past (relative to $A$) is defined as $S_- = \left\{\bigcup_{s \in S} A_s\right\} \backslash S$. Conditional to $S_-$, a causal Markovian field is defined by a family of transition probabilities $q_s(x_s \mid x_{A_s})$, and its distribution over $S$, conditional to $S_-$, is given by

$$P(x(S) \mid S_-) = \prod_S q_s(x_s \mid x_{A_s}).$$

This recursive calculation of the joint distribution, impossible in a bilateral model, allows us to use the likelihood and simulate $X$ recursively and quickly: the analogy with Markov chains over $\mathbb{Z}$ is complete. A class of unilateral efficient models was introduced by Pickard [127]. These models, which permit fast numerical processing, have been developed for image applications (cf., for example, Devijver [45]). However, if we are considering spatial phenomena that does not specifically favor any given orientation, there is no direct interpretation of these models. Nonetheless, they remain numerically performant tools which very much cover bilateral models by their density properties. The causal texture simulation carried out in [60] confirms the interest in these models. If $q_s(\cdot|\cdot)$ is constant on $s$ (homogeneity) and if the state space has $G$ points, the field has parametric dimension $(G-1)^{G^{|A|}}$.

*Bilateral Representation.* Markovian bilateral dependence is determined by the dependence on $x_t$, $t \neq s$ of the expression

$$\frac{p_s(x_s|\cdot)}{p_s(x'_s|\cdot)} = \frac{P(x_s, x_{S\setminus\{s\}})}{P(x'_s, x_{S\setminus\{s\}})}.$$

If $s \notin A_t$, the transition probabilities $q_t(x_t \mid x_{A_t})$ simplify in the above expression so that

$$\frac{p_s(x_s|\cdot)}{p_s(x'_s|\cdot)} = \prod_{t:s\in A_t} \frac{q_t(x)}{q_t(x')},$$

where $x = (x_s, x_{S\setminus\{s\}})$ and $x' = (x'_s, x_{S\setminus\{s\}})$. Thus, we can deduce

$$p_s(x_s \mid x_t, t \neq s) = p_s(x_s \mid x_t,\ t \in L+s), \quad L = (\widetilde{A}-\widetilde{A})\setminus\{0\}, \quad \widetilde{A} = A\cup\{0\}.$$

We have found the same kind of bilateral dependence as that of a conditional auto-regressive model based on a unilateral $AR$ model (cf. Theorem 1.3.3): the causal $A$ Markovian field is thus also a bilateral Markovian $L$-field.

**(2.2.8) Example.** $A=\{(-1,0),(0,-1)\}$; thus,

$$L=\{(-1,0),(1,0),(0,-1),(0,1),(-1,1),(1,-1)\}.$$

There are six kinds of cliques for an $L$-field: 1 singleton, 3 with 2 points ($\{(0,0),(1,0)\}$, $\{(0,0),(0,1)\}$, and $\{(0,0),(-1,1)\}$) and 2 with 3 points ($\{(0,0),(-1,0),(0,-1)\}$ and $\{(0,0),(1,0),(0,1)\}$). If $X$ has $G$ states, the space of $L$-fields has parametric dimension $(G-1)+3(G-1)^2+2(G-1)^3 > (G-1)G^2$; there are $L$-fields that are not causal fields.

The potentials in a bilateral Markovian representation are defined by the transition probabilities. For example, for a binary, causal, and homogeneous field with transition probabilities

$$a_{xy} = P(X_{00} = 1 \mid X_{-1,0} = x,\ X_{0,-1} = y).$$

The associated potentials are:

$\phi_0(x) = hx;$

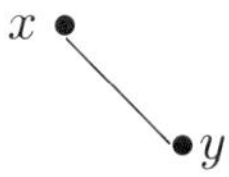

$\phi_3(x, y) = b_3xy;$

$\phi_1(x, y) = b_1xy;$

$\phi_2(x, y) = b_2xy;$

$\phi_4(x, y, z) = cxyz;$

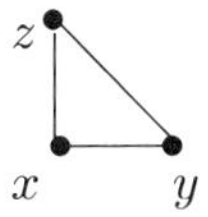

$\phi_5(x, y, z) \equiv 0;$

with $h = \alpha_{00} + \beta_{01} + \beta_{10} - 2\beta_{00}$, $c = \theta_{11} - \theta_{10} - \theta_{01} + \theta_{01}$, $\quad \theta = \alpha - \beta$,

$$b_1 = \alpha_{10} - \alpha_{00},\ b_2 = \alpha_{01} - \alpha_{00},\ b_3 = \beta_{11} - \beta_{10} - \beta_{01} + \beta_{00},$$

$$\text{and} \quad \alpha_{xy} = \log a_{xy}, \quad \beta_{xy} = \log(1 - a_{xy}).$$

*Case* $S = \mathbb{Z}$. Over $\mathbb{Z}$, the class of homogeneous Markov chains coincides with that of the homogeneous Markovian fields with nearest neighbors. A priori we know that the class of chains is smaller or equal. A parametric dimension argument helps understand why they are the same: the family of chains has dimension $G(G-1)$ (there are $G$ states), which is also the dimension of the class of fields: $(G-1) + (G-1)^2$. The correspondence is shown explicitly by Spitzer [151]:

$$p(y \mid x, z) = \frac{q(x, y)q(y, z)}{q^2(x, z)},$$

where $q^2$ is the square of matrix $q$. Conversely, let 0 be a reference element of $E$. Define the positive matrix

$$Q(x, y) = \frac{p(x \mid 0, y)}{p(0 \mid 0, y)}.$$

If $\lambda$ is the biggest eigenvalue (in absolute value) of $Q$, and $r$ is an associated positive eigenvector, then

$$q(x, y) = \frac{1}{\lambda} \frac{Q(x, y)r(y)}{r(x)}.$$

For $E = \{-1, 1\}$, we can check that this correspondence, given in example 2, § 2.2.2, is one to one.

## 2.2.6. *Dynamics for a Markovian Field: Reversibility*

Let $S$ be a finite subset of sites, $S = \{1, 2, \ldots, n\}$, and let $X(t) = \{X_i(t), i \in S\}$ be a configuration at time $t$, $X(t) \in \Omega = E^S$. A Markovian evolution in time of $X = \{X(0), X(1), \ldots\}$ is given by

(1) an initial distribution $\pi_0(x)$ for $X(0)$, with energy $u_0(x)$; and
(2) for each $t$, a transition $P_t(x(t) \mid x(t-1))$ with energy $U_t(x \mid y)$.

We have a *Markovian Dynamic of a Markovian field* if, for a fixed Markov graph over $S$, $u_0(\cdot)$ and $U_t(\cdot|y)$ are Markovian energies for all $t$ and $y \in \Omega$. We have a *homogeneous dynamic* if $U_t$ is independent of $t$.

The general form of $U_t(\cdot \mid y)$ is $U_t(x \mid y) = \sum_{C \in \mathcal{C}} \phi_{C,t}(x \mid y)$, defined by the potentials on the cliques. In order for the model to be interpretable and for the parametric dimension to be reasonable, the dependence of the potentials on $y$ must be more clearly specified.

**(2.2.9) Example.** Modelling of an occupation/nonoccupation phenomena which evolves stationarily in time. $E = \{0, 1\}$, and we shall assume that the cliques have, at most, two points. Then

$$\phi_i(x(t) \mid x(t-1)) = \alpha_i(x(t-1))x_i(t),$$

$$\phi_{i,j}(x(t) \mid x(t-1)) = \beta_{ij}(x(t-1))x_i(t)x_j(t), \ j \in V_i, \ j \neq i,$$

where $V_i$ stands for the set of neighbors of $i$. Since $\Omega$ has $2^n$ points, $\alpha_i(\cdot)$ as well as $\beta_{ij}(\cdot)$ must have restrictions imposed on them; for example,

$$\alpha_i(x(t-1)) = \sum_{j \in V_i \cup \{i\}} \alpha_{ij} x_j(t-1), \quad \beta_{ij}(x(t-1)) = \beta_{ij}.$$

The first condition translates into a local linear dependence of $\alpha_i$ on $x(t-1)$, and the second translates into a temporal independence of the spatial dependence factor at time $t$.

For simulation or estimation by conditional pseudo-likelihood methods, we shall use the conditional density at each site. In this particular example, this can be written as

$$P_i(x_i(t) \mid x_j(t), \ j \neq i, \ x_\ell(t-1), \ell \in S) = \frac{\exp\left\{x_i(t)\left[\alpha_i(x(t-1)) + \sum_{j:\langle i,j\rangle} \beta_{ij} x_j(t)\right]\right\}}{1 + \exp\left\{\alpha_i(x(t-1)) + \sum_{j:\langle i,j\rangle} \beta_{ij} x_j(t)\right\}}.$$

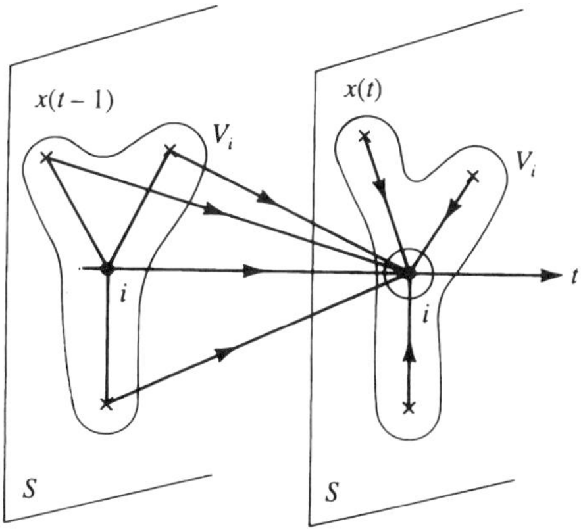


*Sketch of the spatial-temporal dependence*

**2.2.6.1. Time Reversibility and Reversibility of the Dynamic.**

Assume $E$ is finite and consider the homogeneous transition $P_t(x,y) = P(X_1 = y \mid X_0 = x)$. Assume that transition $P$ is associated with potentials $\{\phi_{w_1,w_2}(y,x),\ w_1 \in \mathcal{S}, w_2 \in \mathcal{S}_0\}$, where $\mathcal{S}$ is the set of nonempty subsets of $S$ and $\mathcal{S}_0 = \mathcal{S} \cup \{\emptyset\}$:

$$P(x,y) = Z_x^{-1} \exp\{ \sum_{w_1 \in \mathcal{S}, w_2 \in \mathcal{S}_0} \phi_{w_1,w_2}(x,y)\}.$$

If

$$\phi_{w_1,w_2}(y^a, x) = \phi_{w_1,w_2}(y, x^b) = 0,\ \ a \in w_1, b \in w_2,$$

the potentials can be uniquely determined ($z^c$ is the configuration equal to $z$ everywhere except at site $c$, where it is equal to 0; 0 is a reference state of $E$). The transition kernel over $V \in \mathcal{S}$ of the Markovian field, conditional to $x$, is, if $y = (u,z)$ where $u$ is the configuration over $V$,

$$\begin{aligned}\pi_x^V(u \mid z) &= P(X_{t-1}(V) = u \mid X_{t+1}(S \backslash V) = y, X_t = x) \\ &= Z_x^V(z)\ \exp \sum_{w_1 \cap V \neq \emptyset, w_2 \in \mathcal{S}_0} \phi_{w_1,w_2}((u,z),x).\end{aligned}$$

We are interested in the distribution of $(X_0, X_1)$ over $\Omega^{\{0,1\}}$. Let $\mu$ be the distribution of $X_0$, $\mu(x) > 0$. $\mu$ can be represented in terms of potentials $\{\psi_w, w \in \mathcal{S}\}$. The joint distribution $\nu$ of $(X_0, X_1)$ is $\nu(x,y) = \mu(x)P(x,y)$ and the *time reversed transition* is

$$\hat{P}(y,x) = P(X_0 = x \mid X_1 = y) = \frac{\nu(x,y)}{\nu_1(y)},$$

where $\nu_1$ is the marginal distribution (in $y$) of $\nu$. Let $\{\hat{\pi}_y^V(\cdot \mid \cdot), V \in \mathcal{S}\}$ be the associated specification.

**(2.2.3) Theorem.** (Künsch [108])

*(a) $\hat{P}(\hat{\pi})$ has associated potentials*

$$\hat{\phi}_{w_1,w_2}(x,y) = \phi_{w_2,w_1}(y,x), \quad w_1, w_2 \in \mathcal{S}.$$

*Potentials $\hat{\phi}_{w_1,\phi}$ satisfy*

$$\exp \sum_{w \ni a} \psi_w(x) = \frac{Z_x}{Z_{x^a}} \exp \sum_{w_1 \ni a} \hat{\phi}_{w_1,\phi}(x), \quad a \in \mathcal{S}. \tag{2.25.1}$$

*(b) $P(x, \cdot\,)$ is reversible with stationary measure $\mu$ if and only if*

$$\phi_{w_1,w_2}(y,x) = \phi_{w_2,w_1}(x,y), \quad w_1 \in \mathcal{S}, w_2 \in \mathcal{S}. \tag{2.25.2}$$

*$\mu$'s associated potentials then satisfy Equation (2.25.1), replacing $\hat{\phi}_{w_1,\phi}$ by $\phi_{w_1,\phi}$.*

*Proof:*

(a) Following the definition of a conditional probability, for each $a$ in $S$ we find

$$\frac{\mu^a(x_a \mid x_{S\backslash a})}{\mu^a(0 \mid x_{S\backslash a})} = \frac{P(x^a, y)}{P(x,y)} \frac{\hat{\pi}^a_y(x_a \mid x_{S\backslash a})}{\hat{\pi}^a_y(0 \mid x_{S\backslash a})}. \tag{2.25.3}$$

As $\phi_{w_1,w_2}(y, x^a) = 0$ if $a \in w_2$ and $\phi_{w_1,w_2}(y,x) = \phi_{w_1,w_2}(y,x^a)$ if $a \notin w_2$, we have

$$\frac{P(x^a, y)}{P(x,y)} = \frac{Z_x}{Z_{x^a}} \exp(- \sum_{w_1 \in \mathcal{S}, w_2 \ni a} \phi_{w_1,w_2}(y,x)).$$

On the other hand, a straightforward calculation shows that

$$\frac{\hat{\pi}^a_y(x_a \mid x_{S\backslash a})}{\hat{\pi}^a_y(0 \mid x_{S\backslash a})} = \exp \sum_{w_1 \ni a, w_2 \in \mathcal{S}_0} \hat{\phi}_{w_1,w_2}(x,y).$$

Using both equations, and observing that the left hand side of (2.25.3) does not depend on $y$, we find that

$$\sum_{w_1 \in \mathcal{S}, w_2 \ni a} \phi_{w_1,w_2}(y,x) - \sum_{w_1 \ni a, w_2 \in \mathcal{S}_0} \hat{\phi}_{w_1,w_2}(x,y)$$

does not depend on $y$. This yields (a).

(b) • Assume we have reversibility. Then $\hat{\phi}_{w_1,w_2}(x,y) = \phi_{w_1,w_2}(y,x)$ for $w_1 \in \mathcal{S}, w_2 \in \mathcal{S}_0$, which yields condition (2.25.2). $\mu$ satisfies (2.25.2) with $\hat{\phi}_{w_1,\phi}(z) = \phi_{w_1,\phi}(z)$.

•• Conversely, assume condition (2.25.2), and define

$$\phi_{\phi,w_2} = \phi_{w_2,\phi}. \tag{2.25.4}$$

Let $\nu$ be a field over $S \times \{0,1\}$ with potentials $\{\phi_{w_1,w_2}, \quad w_1, w_2 \in \mathcal{S}_0\}$. By construction, $\nu$ is a symmetric distribution ($\nu(x,y) = \nu(y,x)$), with transitions $P = \hat{P}$, and which admits $\mu = \nu_0$, the marginal of $\nu$ at time 0, as an invariant measure. □

REMARKS:

(1) If $P(x,\cdot)$ has a stationary, reversible measure $\mu$, then we have the following detailed balance equation:

$$\nu(x,y) = \mu(x)P(x,y) = \mu(y)P(y,x) = \nu(y,x), \quad x,y \in \omega.$$

In this case, it is easy to find a closed form for $\mu$ in terms of $P$ (cf., for example, [163]).

(2) Let $\nu$ be the measure defined in b(••): $\nu$ is the invariant measure for transition over $\Omega^{\{0,1\}}$

$$P^*((x,y),(z,t)) = \mathbb{1}(y=z)\ P(z,t).$$

In general, it isn't easy to find an analytical expression for the invariant distribution for a given transition $P$.

(3) *Synchronous transitions* (cf. [102],[176] and Ex. 17, Chapter 6). $P(x,\cdot)$ is said to be *synchronous* if, for all $x, y$, it can be written as

$$P(x,y) = \prod_{s\in S} p_s(x,y_s), \quad p_s(x,y_s) > 0, \quad \sum_{u\in E} p_s(x,u) = 1, \quad s \in \mathcal{S}.$$

The above expression says that the values at sites $s$, $s \in S$ are relaxed independently and at the same time with distributions $p_s(x,\cdot)$ over $E$. On the other extreme we have *asynchronous dynamics*, where the values $y_s$, $s = 1,2,\ldots,n$ are relaxed sequentially:

$$P(x,y) = \prod_{s=1}^{n} p_s(y_1,\ldots,y_{s-1},x_s,\ldots,x_n \mid y_s).$$

A fundamental example of an asynchronous dynamic is that of the Gibbs sampler (c.f. §6.2.2).

Suppose that $P$ is a synchronous dynamic with potentials $\phi_{w_1,w_2}$ such that $\phi_{w_1,w_2} \equiv 0$ if $|w_1| > 1$. From (2.25.2) we get that if the dynamic is reversible, $\phi_{w_1,w_2} \equiv 0$ if either $|w_1|$ or $|w_2| > 1$, which only permits symmetric pairwise potentials

$$\phi_{\{i\},\{j\}}(x_i,y_j) = \phi_{\{j\},\{i\}}(x_j,y_i).$$

This property as well as the use of associated synchronous kernels, is studied in [176] for a Markovian transition with memory $q > 1$.

### *2.2.7. Modeling and Identification in the Context of "Pattern Theory"*

Let $\mathcal{C}^*$ be the subfamily of maximal cliques. By defining a way to appoint each nonmaximal clique to a maximal clique that contains it, we can write the energy of configuration $x$ as $H(x) = \sum_{A\in\mathcal{C}^*} \psi_A(x_A)$.

What constraints permit this $\psi$ representation to be identifiable? This problem is considered in [75] and [76] in the case of cliques with, at most, two points.

Let $\sigma$ be a set of (ordered) pairs $(i,j)$, $i,j \in S$, $i \neq j$ *s.t.* if $(i,j) \in \sigma$, $(j,i) \notin \sigma$. For example, $\sigma$ can be obtained from a Markovian graph by arbitrarily choosing a direction for each edge. Then we shall consider an energy function:

$$H(x) = \sum_{s \in \sigma} a_s(x_i, x_j), \quad \text{if } s = (i,j).$$

In the context of "Pattern Theory," $\sigma$ is called the *graph connector* and $a_s$ the *acception functions.*

(1) *Nonparametric case.* No restrictions (except for eventually the identification ones) are placed over the $a_s$. The identification conditions can be expressed with the help of the *cut graph* $\overline{\sigma}$. For a given pair $(i,j)$, (also called the *arrow* $i \longrightarrow j$), there are four possibilities:

close cut (to $i$) — $(CC)$

far cut (to $i$) — $(FC)$

double cut — $(DC)$

no cut — $(NC)$

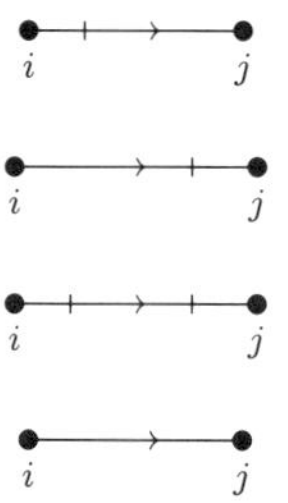


If for each $(i,j)$ of $\sigma$ we chose one of the above, we get $\overline{\sigma}$, a cut graph. Now we can write the following constraints in terms of $\overline{\sigma}$:

(C)

| $s = (i,j)$ | constraints |
|---|---|
| Close cut $(CC)$ | $a_s(\cdot, 0) = 0$ |
| Far cut $(FC)$ | $a_s(0, \cdot) = 0$ |
| Double cut $(DC)$ | $a_s(0, \cdot) = a_s(\cdot, 0) = 0$ |
| No cut $(NC)$ | $a_s(0,0) = 0$ |

**(2.2.4) Theorem.** *Assume $\overline{\sigma}$ is a cut graph such that for each site $i$ of $S$ there exists one and only one arrow without a close cut (to $i$)* (*)*. Then if $a$ satisfies constraints $(C)$ relative to $\overline{\sigma}$, $a$ is identifiable.*

---

(*) Such a $\overline{\sigma}$ always exists; cf. [75] for a description of an algorithm that allows its construction.

*Proof:* Because of Theorem 2.6, there exists a unique representation of $H$ which can be written in terms of potentials of first and second-order, with $C_i$ and $B_{ij}$, *s.t.*:

$$\begin{cases} H(x) = \sum_s a_s(x_i, x_j) = \sum_i C_i(x_i) + \sum_s B_s(x_i, x_j), \\ C_i(0) = B_{ij}(0, \cdot) = B_{ij}(\cdot, 0) = 0, \quad i, j \in S. \end{cases}$$

Under the Theorem's assumptions, we have the following relationships between $a_s$, $B_s$, $C_i$ and $C_j$ with $s = (i, j)$:

| Arrow Type | $B_s,\ s = (i \longrightarrow j)$ | $C_i$ | $C_j$ |
|---|---|---|---|
| $CC$ | $a_s(g_i, g_j) - a_s(0, g_j)$ | / | $a_s(0, g_j)$ |
| $FC$ | $a_s(g_i, g_j) - a_s(g_i, 0)$ | $a_s(g_i, 0)$ | / |
| $DC$ | $a_s(g_i, g_j)$ | / | / |
| $NC$ | $a_s(g_i, g_j) - a_s(g_i, 0) - a_s(0, g_j)$ | $a_s(g_i, 0)$ | $a_s(0, g_j)$ |

*Decomposition of $a_s$ according to the kind of arrow*

$B_s$ is defined for all $s$ and $C_i$ is defined for all $i$ because of the properties of $\overline{\sigma}$. We can thus deduce the following necessary unique representation of the $a_s$:

| Type | $CC$ | $FC$ | $DC$ | $NC$ |
|---|---|---|---|---|
| $a_s$ | $B_s + C_j$ | $B_s + C_i$ | $B_s$ | $B_s + C_i + C_j$ |

□

**(2.2.10) Example.**

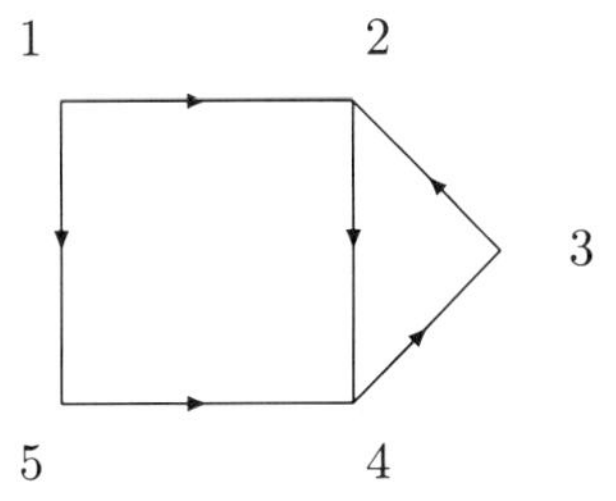


*$\sigma$: Directed graph*

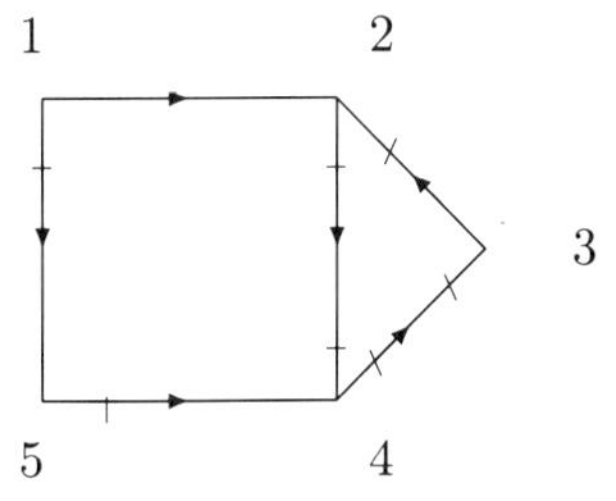


*A choice of $\overline{\sigma}$: Cut graph*

*Identification constraints*:

$$(CC) \quad a_{1,5}(\cdot,0) = a_{5,4}(\cdot,0) = 0, \qquad (FC) \quad a_{3,2}(0,\cdot) = 0;$$

$$(DC) \begin{cases} a_{2,4}(0,\cdot) = a_{2,4}(\cdot,0) = 0 \\ a_{4,3}(0,\cdot) = a_{4,3}(\cdot,0) = 0 \end{cases}, \qquad (NC) \quad a_{1,2}(0,0) = 0.$$

*Parametric dimension.* Assume that the state space has $K$ states: for $a_{1,5}$, $a_{5,4}$ and $a_{3,2}$, the dimension is $K(K-1)$; for $a_{2,4}$ and $a_{4,3}$ the dimension is $(K-1)^2$; and for $a_{1,2}$, the dimension is $K^2-1$. So that

$$\text{dimension} = 3K(K-1) + 2(K-1)^2 + K^2 - 1.$$

In the nonordered Markovian approach, this dimension can be expressed as $n(K-1) + m(K-1)^2$ if $|S| = n$ and $m = |\sigma|$. In our case,

$$\text{dimension} = 5(K-1) + 6(K-1)^2.$$

(2) *Homogeneous case.* $a_s \equiv a$ is a fixed function. Define for each $i \in S$

$$\begin{aligned} \omega_i^+ &= \text{number of arrows } (k,i) \text{ of } \sigma \text{ (number of arrivals at } i), \\ \omega_i^- &= \text{number of arrows } (i,\ell) \text{ of } \sigma \text{ (number of departures of } i), \text{ and} \\ \omega_i &= \omega_i^+ + \omega_i^- = \text{number of neighbors of } i \text{ different from } i. \end{aligned}$$

**(2.2.5) Theorem.** *Assume $a_s \equiv a$ for all $s$ of $\sigma$. Then $a$ shall be identifiable if one of the following conditions is true*:

(2.1) *There exist two sites $i$ and $k$ of $S$ such that*

$$\omega_i^+\omega_k^- - \omega_i^- \ \omega_k^+ \neq 0 \quad \text{and} \quad a(0,0) = 0.$$

(2.2) *For all $i$,$k \in S$, $\omega_i^+ = \omega_k^+$ and $\omega_i^- = \omega_k^-$ and $a(\,\cdot\,,0) = 0$.*

*Proof:* The decomposition of $a(g_i, g_j)$,

$$a(g_i,g_j) = (a(g_i,g_j) - a(g_i,0) - a(0,g_j)) + a(g_i,0) + a(0,g_j),$$

permits us to write potentials $B_s$ and $C_i$ as

$$B_s(g_i,g_j) = B(g_i,g_j) = a(g_i,g_j) - a(g_i,0) - a(0,g_j),$$

$$C_i(g_i) = \sum_{s=(i,\cdot)\in\sigma} a_s(g_i,0) + \sum_{s=(\cdot,i)\in\sigma} a_s(0,g_i) = \omega_i^- a(g_i,0) + \omega_i^+ a(0,g_i).$$

Case (2.1): For $i$, $k$, and a given argument $g$, we have

$$C_i(g) = \omega_i^- a(g,0) + \omega_i^+ a(0,g), \ C_k(g) = \omega_k^- a(g,0) + \omega_k^+ a(0,g)$$

so that we can write $a(g,0)$ and $a(0,g)$ in terms of $C_i$ and $C_k$, and then $a(g,h)$ in terms of $B$, $C_i$ and $C_k$ as

$$\begin{cases} a(g,0) = \dfrac{\omega_k^+ C_i(g) - \omega_i^+ C_k(g)}{\omega_i^- \omega_k^+ - \omega_k^- \omega_i^+}, \; a(0,g) = \dfrac{\omega_k^- C_i(g) - \omega_i^- C_k(g)}{\omega_k^- \omega_i^+ - \omega_i^- \omega_k^+}, \\ a(g,h) = B(g,h) - a(g,0) - a(0,h) \end{cases}$$

so that $a$ is identifiable.

Case (2.2): Let $m$ be the number of arrows of $\sigma$, and let $n$ be the number of sites. Then

$$m = \sum_{i \in S} \omega_i^+ = \sum_{i \in S} \omega_i^- = n\omega^+ = n\omega^-, \quad \omega^+ = \omega_i^+, \quad \omega^- = \omega_i^-$$

so that $\omega_i^+ = \omega_i^- = \omega^+ = \omega^- = \frac{\omega}{2}$.

The connection between $a$ and potentials $C_i$ and $B_s$ can be obtained as follows:

$$\begin{aligned} \sum_s a_s(g_i, g_j) &= \sum_s a_s(g_i, 0) + \sum_s (a_s(g_i, g_j) - a_s(g_i, 0)) \\ &= \frac{\omega}{2} \sum_i a(g_i, 0) + \sum_s (a(g_i, g_j) - a(g_i, 0)), \end{aligned}$$

$$C_i(g_i) = C(g_i) = \frac{\omega}{2} a(g_i, 0), \quad B_s(g_i, g_j) = B(g_j, g_j) = a(g_i, g_j) - a(g_i, 0)$$

so that $a(g,h) = \frac{2}{\omega} C(g) + B(g,h)$. □

**(2.2.11) Example.**

*Conditions* (2.1)

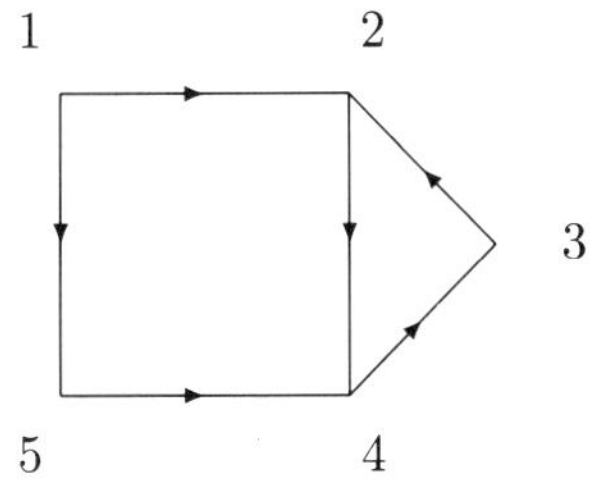


Sites 1 and 2 satisfy conditions (2.1). The identification condition is given by $a(0,0) = 0$. If $E$ has $K$ states, the dimension is $K^2 - 1$.

*Conditions* (2.2)

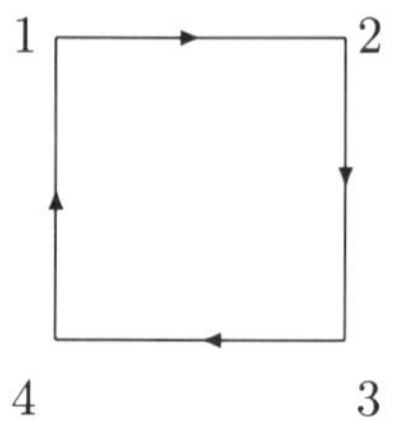


Here, $\omega_i^+ = \omega_i^- = 1$ if $i \in S$; $a(\cdot, 0) = 0$; the dimension is equal to $K(K-1)$. The Markovian approach yields $(K-1) + (K-1)^2$.

## 2.3. Gaussian Specifications, Gaussian Fields, and Gibbs Fields

### *2.3.1. Gaussian Fields as Gibbs Measures*

Here, $E = \mathbb{R}$, $\mathcal{E}$ is the Borel $\sigma$-algebra, and $\lambda$ is the Lebesque measure. $X = \{X_i, i \in S\}$, with distribution $\mu$, is Gaussian if and only if its finite-dimensional marginal distributions $X_\Lambda, \Lambda \in \mathcal{S}$ are Gaussian. $X$ is characterized by its mean $m = \{m_i, i \in S\}$ and its covariance $\sum = (\sum(i,j), i, j \in S)$.

We shall assume that $X$ is *Markovian at each* site, that is to say, for all $i \in S$, there exist finite subsets $\partial i$, $i \notin \partial i$ and constants $c_{i,j}$ and $c_{i,j} = 0$ if $j \notin \partial i$, $v_i \geq 0$ *s.t.*:

$$(2.26)\qquad \begin{cases} \mathcal{L}(X_i \mid X_j, j \neq i) = \mathcal{L}(X_i \mid X_j, j \in \partial i) & (X\ \partial i\text{–Markovian}), \\ \qquad = \mathcal{N}(m_i + \sum_{j \in \partial i} c_{i,j}(X_j - m_j), v_i) & (X\text{–Gaussian}). \end{cases}$$

The $v_i$ are just the variances of the conditional residuals: $e_i = (X_i - m_i) - \sum_{j \in \partial i} c_{i,j}(X_j - m_j)$ is also called the prediction mean square error. The $e_i$ are characterized by

$$(2.27)\qquad e_i \perp X_j, \quad \text{if} \quad i \neq j.$$

We shall assume in what follows that $v_i > 0$, $i \in S$. Calculating $Cov(e_i, e_j)$ in two different ways, we get

$$(2.28)\qquad c_{ij} v_j = c_{ji} v_i, \quad i, j \in S,$$

and the Yule–Walker equations can be written as

$$(2.29)\qquad \Sigma = C\Sigma + V,$$

where $C = (c_{ij}, i, j \in S)$, and $V = Diag(v_i, i \in S)$. (2.29) can be written as

$$J \cdot \Sigma = I \quad \text{with} \quad J = V^{-1}(I - C). \tag{2.30}$$

Because of (2.28), $J$ is a symmetrical matrix:

$$J(i,i) = v_i^{-1}, \quad J(i,j) = -\frac{c_{i,j}}{v_i}, \quad i \neq j, \quad i,j \in S.$$

Define

$$h_i = -J(i,i)\nu_i, \quad i \in S \tag{2.31}$$

with $\nu_i = \left(m_i - \sum_{j \in \partial i} c_{i,j} m_j\right)$. The conditional distribution in $i$ is

$$\begin{aligned}\pi_{\{i\}}(x_i|\cdot) &= (2\pi v_i)^{-1/2} \exp -\frac{1}{2v_i}\left(x_i - \nu_i - \sum_{j \in \partial i} c_{i,j} x_j\right)^2 \\ &= \sqrt{\frac{J(i,i)}{2\pi}} \exp -\Big[\frac{1}{2} J(i,i) x_i^2 + h_i x_i \\ &+ \sum_{j \in \partial i} J(i,j) x_i x_j + \frac{1}{2v_i}(v_i - \sum_{j \in \partial_i} c_{i,j} x_j)^2\Big].\end{aligned}$$

The choice for the Gibbs potential is thus

$$-\phi_A = \begin{cases} \frac{1}{2} J(i,i) x_i^2 + h_i x_i & \text{if } A = \{i\}, \\ J(i,j) x_i x_j & \text{if } A = \{i,j\},\ i \neq j. \end{cases} \tag{2.32}$$

$(J, h)$ is the potential in the parametrization of (2.26). Given the conditional residuals process $e = \{e_i, i \in S\}$ and

$$\Gamma(i,j) = Cov(e_i, e_j), \quad i,j \in S, \tag{2.33}$$

we have the following:

**(2.3.1) Theorem.** *Let $X$ be a Gaussian field Markovian at each site such that $v_i = \Gamma(i,i) > 0$ for each $i$. Then*

(a) *$J(i,j) = \Gamma(i,j)/(\Gamma(i,i)\Gamma(j,j))$, $h_i = -\sum_{j \in \partial i \cup \{i\}} J(i,j) m_j$, and*

(b) *$J$ is positive definite.*

*Proof:*

$(a)$ The second identity is just (2.31). On the other hand, the identity involving $J$ and $\Gamma$ is true if $j \notin \partial i$. Now consider $j \in \partial i$; then

$$\Gamma(i,j) = E\Big(e_j \cdot \Big(X_i - \nu_i - \sum_{k\in\partial i} c_{i,k}X_k\Big)\Big) = -c_{ij}E(e_jX_j)$$

$$= -c_{ij}\Gamma(j,j) = J(i,j)\Gamma(i,i)\Gamma(j,j).$$

$(b)$ Because of $(a)$, $J$ is semi definite positive. It is thus enough to show that for all $\Lambda \in \mathcal{S}$, $J_\Lambda = (J(i,j), i,j \in \Lambda)$ is regular to conclude that $J$ is positive definite. In order to do this, we show that if we call $\Gamma_\Lambda = (\Gamma_\Lambda(i,j), i,j \in \Lambda)$ the matrix of covariances conditional to $\Lambda^c$:

$$\Gamma_\Lambda(i,j) = E((X_i - X_i^\Lambda)(X_j - X_j^\Lambda)), \quad X_i^\Lambda = E(X_i|X_k, k \notin \Lambda).$$

Then, if $i \neq j$,

$$E((X_i - E(X_i|\cdot))(X_j - X_j^\Lambda)) = E[E(X_i - E(X_i|\cdot)|X_k, k \notin \Lambda)(X_j - X_j^\Lambda)] = 0.$$

Now for $i,j \in \Lambda$:

$$\sum_{k\in\Lambda} J(i,k)\Gamma_\Lambda(k,j)$$

$$= \sum_{k\in\Lambda} J(i,k)E(X_k(X_j - X_j^\Lambda)) = \sum_{k\in S} J(i,k)E(X_k(X_j - X_j^\Lambda))$$

$$= J(i,i)E((X_i - E(X_i|\cdot))(X_j - X_j^\Lambda))$$

$$= \begin{cases} 0 & \text{if } i \neq j, \\ J(i,i)E((X_i - E(X_i|\cdot))X_i) = J(i,i)\Gamma(i,i) = 1, & \text{if not.} \end{cases}$$

□

**(2.3.2) Theorem.** Gaussian Markovian Specifications. *Consider $h \in \mathbb{R}^S$ and $J: S \times S \to \mathbb{R}$ a symmetric, bounded support positive definite function. Then $\phi = \phi(J,h)$, the potential defined by* (2.32), *is admissible and the conditional distribution $\pi_V(J,h)$, $V \in \mathcal{S}$ is the Gaussian distribution $\mathcal{N}_\Lambda(m_\Lambda, J_\Lambda^{-1})$ with*

$$\begin{cases} m_\Lambda = -J_\Lambda^{-1}[h_\Lambda + J_{\Lambda,S\setminus\Lambda}x(X\setminus\Lambda)], \\ J_{\Lambda,\Delta} = (J(i,j), i \in \Lambda, j \in \Delta), J_\Lambda = J_{\Lambda,\Lambda}. \end{cases}$$

*Proof:* Setting $x = x_\Lambda$, $y = x_{S\setminus\Lambda}$, we have

$$-H_\Lambda^{J,h}(x,y) = \frac{1}{2}(x - m_\Lambda)'J_\Lambda(x - m_\Lambda) - \frac{1}{2}m_\Lambda' J_\Lambda m_\Lambda,$$

where $J_\Lambda$ and $m_\Lambda$ are those defined by the Theorem. This ends the proof.

□

**(2.3.3) Theorem.** *Let $X$ be a Gaussian field, Markovian at each site, defined by (2.26) such that $X_i \neq E(X_i|\cdot)$ a.s. and its associated $(h,J)$. Then $\pi(J,h)$ defined by the potential $\phi(J,h)$ is well defined and the distribution of $X$ belongs to $\mathcal{G}(J,h)$.*

*Proof:* Combine Theorem 2.3.1 ($J$ positive definite), and 2.3.2 (which assures that the potentials are well defined) and the result which assures that $\mu \in \mathcal{G}(\pi) \Longleftrightarrow \mu\pi_{\{i\}} = \mu$ for each $i$ (cf. Appendix 1). □

## *2.3.2. Description of Gibbs Measures for a Given Gaussian Specification*

Let $\phi = \phi(J,h)$ be a Gaussian potential of *bounded support* ($J$ positive definite and of bounded support). Define

$$M_{J,h} = \{m \in \Omega = \mathbb{R}^S : h_i + \sum_{j \in S} J(i,j)m_j = 0, \quad i \in S\}.$$

It is important to know for modeling purposes under which conditions the specification $(J,h)$ (or the model $(C,V,m)$ in terms of (2.26)) is not "empty."

**(2.3.4) Theorem.** *Assume that (a) $\sup_{\Lambda} J_{\Lambda}^{-1}(i',i) < \infty$; (b) $M_{J,h} \neq \emptyset$; then $g(J,h) \neq \emptyset$. More precisely, for all $i,j \in S$, $\lim_{\Lambda} J_{\Lambda}^{-1}(i,j) = \sum(i,j)$ exists. In particular, if $m \in M_{J,h}$, the Gaussian field $(m,\sum)$ belongs to $\mathcal{G}(J,h)$.*

*Proof:* Let $\emptyset \neq \Lambda \subset \Delta \in \mathcal{S}$, $t = (t_i, i \in \Lambda)$. We will show that

$$a = \sum_{i,j\in\Lambda} J_{\Lambda}^{-1}(i,j)t_it_j \leq \sum_{i,j\in\Lambda} J_{\Delta}^{-1}(i,j)t_it_j = b. \tag{2.34}$$

$J_{\Lambda}^{-1}$ is the covariance of the Gaussian specification $\pi_{\Lambda}(J,h)$, and thus,

$$a = E_{\pi_\Lambda}[(Y - E_{\pi_\Lambda}Y)^2], \quad \text{with } Y = \sum_{i\in\Lambda} t_iX_i.$$

As this term does not depend on $x(S\backslash\Lambda)$, we have

$$\begin{aligned} a &= E_{\pi_\Delta}(E_{\pi_\Lambda}[Y - E_{\pi_\Lambda}Y]^2) = E_{\pi_\Delta}(E_{\pi_\Lambda}(Y^2) - (E_{\pi_\Lambda}Y)^2) \\ &= E_{\pi_\Delta}(Y^2) - E_{\pi_\Delta}((E_{\pi_\Lambda}Y)^2) \leq E_{\pi_\Delta}(Y^2) - (E_{\pi_\Delta}Y)^2 = b \end{aligned}$$

because of Jensen's inequality. $J_{\Lambda}^{-1}(i,i)$ is thus increasing over $\Lambda$ and bounded, so $\Sigma(i,i) = \lim_{\Lambda} J_{\Lambda}^{-1}(i,i)$ exists. Now let $i \neq j$; for $\Delta = \{i,j\}$, (2.34) yields that

$$J_\Lambda^{-1}(i,i) + J_\Lambda^{-1}(j,j) + 2J_\Lambda^{-1}(i,j) \tag{2.35}$$

is increasing over $\Lambda$. However, since $J_\Lambda^{-1}$ is a covariance,

$$J_\Lambda^{-1}(i,j)^2 \le J_\Lambda^{-1}(i,i)\ J_\Lambda^{-1}(j,j)$$

so that $\lim_\Lambda J_\Lambda^{-1}(i,j) = \Sigma(i,j)$ exists. $\Sigma$ is defined, nonnegative, and since $J$ is of bounded support, $\Sigma$ satisfies:

$$\sum_{j\in S} J(i,j)\Sigma(j,k) = \lim_\Lambda \sum_{j\in\Lambda} J(i,j)J_\Lambda^{-1}(i,k) = \delta_{i,k},$$

which means $\Sigma$ is positive definite. If $M_{J,h} \ne \emptyset$, and if $m \in M_{J,h}$, then the Gaussian field $(m,\Sigma)$ belongs to $\mathcal{G}(J,h)$. □

OBSERVATIONS:

(1) Conditions $(a)$ and $(b)$ are necessary even if $J$ isn't of bounded support.

(2) Assume that the set of solutions $a$ of

$$\sum_{j\in S} J(i,j)a_j = 0, \quad i \in S$$

(called *J-harmonic* functions) is different from 0. Then if $m \in M_{J,h}$, $m+a \in M_{J,h}$ and there is phase transition, $\mu \sim \mathcal{N}(m,\Sigma)$ and $\mu' \sim \mathcal{N}(m+a,\Sigma)$ belongs to $\mathcal{G}(J,h)$.

Over $\mathbb{Z}$, if $J(i,j) = -\dfrac{\alpha}{1+\alpha^2}$ if $|i-j|=1$, 1 if $i=j$, 0 if not, the space of $J$-harmonic functions has dimension 2 and is generated by the sequences $(\alpha^j)$ and $(\alpha^{-j})$. Over $\mathbb{Z}^2$, defining $J(i,j) = 1$ if $i=j$, $\alpha$ if $|i-j|_\infty = 1$, 0 if not, and if $\alpha \ne 0$, the space of harmonic functions has infinite dimension.

(3) *Description* of $\mathcal{G}(J,h) \ne \emptyset$:

• The extremal measures of $\mathcal{G}(J,h)$ are precisely the Gaussian measures $\mathcal{N}(m,\Sigma)$ with $m \in M_{J,h}$.

•• If we define the convolution product

$$\mu_1 * \mu_2(f) = \int \mu_1(dx) \int \mu_2(dy) f(x+y),$$

then $\mathcal{G}(J,h) = \{\mu_C * \nu : \nu \in \mathcal{P}(\Omega,\mathcal{F})\ s.t.\ \nu(M_{J,h}) = 1\}$ with $\mu_C$ the Gaussian centered measure of $\mathcal{G}(J,0)$. Set

$$\alpha = \sup_i \sum_{j\ne i} |c_{i,j}| < 1 \tag{2.36}$$

with $c_{i,j} = -J(i,j)J(i,i)^{-1}$ for $i \ne j$, $c_{i,i} = 0$ if not. If $c_{ij}^n$ is the $(i,j)$-*th* coefficient of matrix $C^n$, we define coefficients $(a_{ij})$ and $(b_{ij})$ associated respectively to the series developments of $(I-C)^{1/2}$ and $(I-C)^{-1/2}$,

(2.37)
$$\begin{cases} a_{i,j} = \delta_{i,j} - \frac{1}{2}c_{ij} - \frac{1}{2.4}c_{ij}^2 \cdots - \frac{1.3\cdots(2n-3)}{2.4\cdots 2n}c_{ij}^n - \cdots, \\ b_{i,j} = \delta_{i,j} + \frac{1}{2}c_{ij} + \frac{1.3}{2.4}c_{ij}^2 \cdots + \frac{1.3\cdots(2n-1)}{2.4\cdots 2n}c_{ij}^n + \cdots. \end{cases}$$

**(2.3.5) Theorem.** *Consider* $(J,h)$ *a Gaussian potential of bounded support which satisfies* (2.36). *Then*

(1) $\mathcal{G}(J,h) \neq \emptyset$ *if* $M_{J,h} \neq \emptyset$.

(2) *Let* $m \in M_{J,h}$ *and let* $X$ *be the Gaussian field* $\mathcal{G}(J,h)$ *with mean* $m$ *and covariance matrix* $\Sigma$.

(a) $\Sigma(i,j) = \left(\sum_{n\geq 0} c_{ij}^n\right) v_j \quad (v_j = J(j,j)^{-1})$.

(b) *X can be written as the MA and as the AR*

$$X_i = \sum_{j\in S} b_{i,j}\varepsilon_j, \quad \varepsilon_j \sim \mathcal{N}(0,v_j) \text{ independent}; \quad \varepsilon_i = \sum_{j\in S} a_{i,j}X_j.$$

(c) $\alpha_{a,b}^X(k) \leq C\cdot b\gamma^k$ *for a certain* $\gamma < 1$.

*Proof:*

(1) For $C = (c_{i,j})$, $V = \mathrm{diag}(v_i)$, $J = V^{-1}(I - C)$ and $J_\Lambda = V_\Lambda^{-1}(I - C_\Lambda)$. So that because of (2.36), $J_\Lambda^{-1} = \left(\sum_{n\geq 0} C_\Lambda^n\right)V_\Lambda$, and thus $|J_\Lambda^{-1}(i,i)| \leq J(i,i)(1-\alpha)^{-1}$.

$(2-a)$ follows from the above decomposition of $J_\Lambda^{-1}$ and the fact that $\Sigma = \lim J_\Lambda^{-1}$.

$(2-b)$ The coefficients $b_{i,j}$ are associated with the series development of $(I-C)^{-1/2}$. We shall check that $b_{ik}v_k = b_{ki}v_i$. In order to do this, it is enough to check the same identity with $c^n$ instead of $b$. The latter is true for $n = 1$, and by recurrence

$$c_{ik}^n v_k = \sum_\ell c_{i\ell}^{n-1} c_{\ell l}\, v_k = \sum_\ell c_{i\ell}^{n-1} v_\ell\, c_{k\ell} = \sum_\ell c_{\ell i}^{n-1} v_i\, c_{k\ell} = c_{ki}^n\, v_i.$$

If the variables $(\varepsilon_j)$ are independent $\mathcal{N}(0,v_j)$ and if $X$ is defined by the $MA$ representation, we have

$$E(X_iX_j) = \sum_k b_{ik}b_{jk}v_k = \left(\sum_k b_{ik}b_{kj}\right)v_j = \left(\sum_{n\geq 0} c_{ij}^n\right)v_j = \Sigma(i,j).$$

By similar arguments and using the fact that the coefficients of $a$ are associated to the series development of $(I-C)^{1/2}$, it can be shown that

$$Cov\Big(\sum_j a_{ij}X_j, \sum_k a_{mk}X_k\Big) = \delta_{im}v_i.$$

Consider $\ell$ the range of $J$: $J(i,j) = 0$ if $|i-j| > \ell$. Let $k \in \mathbb{N}^*$ and assume that $d(i,j) > k\ell$. Then if $n \le k$, $c_{ij}^n = 0$. Indeed,

$$c_{ij}^n = \sum_{i_1, i_2, \ldots, i_{n-1}} c_{ii_1} \cdot c_{i_1 i_2} \cdots c_{i_{n-1}j} = 0$$

since there isn't any sequence $i_1, \ldots, i_{n-1}$ with $|i_s - i_{s+1}| \le \ell$ for $s = 0,n-1$ (here, $i_0 = i$, $i_n = j$). So that

$$|b_{i,j}| = \Big|\sum_{k+1}^{\infty} c_{ij}^n\Big| \le \alpha^k (1-\alpha)^{-1}$$

because $|c_{ij}^n| \le \alpha^n$. In fact, we have the following stronger result:

$$\sum_j |c_{ij}^n| \le \alpha^n$$

This gives the exponential decay of the coefficients. Estimation of the mixing coefficients is then a consequence of Corollary 1.7.3. □

REMARK: Condition (2.36) isn't necessary, as shall be proven in Theorem 2.3.5, which assures that $\mathcal{G}(J,0) \neq \emptyset$ for the stationary following examples:

$d=1$: $J(i,j) = J(|i-j|) = 1$ if $i=j$, $\frac{1}{2}$ if $|i-j| = 1$, $\frac{1}{4}$ if $|i-j| = 2, 0$ if not.

$d \ge 2$: cf. Example 2, § 4.2 (Whittle estimation of a Markovian Gaussian field).

### *2.3.3. Homogeneous Case*

Assume $S = \mathbb{Z}^d$, $h \in \mathbb{R}^S$ and consider $J$ the positive defined, invariant under translation, bounded support potential

$$J(i,j) = J(|i-j|), \quad J(i,j) = 0 \text{ if } d(i,j) > \ell.$$

Let $\widehat{J}$ be the Fourier transform of $J$, with $z = \exp i < \lambda, \mathbb{1} >$, $\lambda \in [0, 2\pi[^d$

$$\widehat{J}(z) = \sum_{j \in S} J(j) z^j = J(0) + \frac{1}{2} \sum_{j \neq 0} J(j) \cos < \lambda, j > .$$

**(2.3.6) Theorem.**

(1) *$\mathcal{G}(J,h) \neq \emptyset$ if $M(J,h) \neq \emptyset$ and if $\widehat{J}^{-1}(\lambda)$ is integrable.*

(2) *The Gaussian measures of $\mathcal{G}_s(J,0)$ are characterized by the spectral measures:*

$$F(d\lambda) = \widehat{J}^{-1}(\lambda)d\lambda + \alpha(d\lambda)$$

*with $\alpha$ supported over $\{\lambda{:}\widehat{J}(\lambda) = 0\}$. If $d \leq 2$, there is only one centered Gaussian measure in $\mathcal{G}_s(J,0)$.*

*Proof:* (1) Assume $\widehat{J}^{-1}$ is integrable, then

$$\Sigma(j,k) = \Sigma(|j-k|) = \int_{T^d} e^{i<j-k,\lambda>}\widehat{J}^{-1}(\lambda)d\lambda$$

is symmetric positive definite and the Gaussian field with mean $m$ and covariance $\sum$ belongs to $\mathcal{G}(J,h)$.

That $\widehat{J}^{-1}(\lambda)$ is integrable is also a necessary condition. For (2), cf. [71]. □

## 2.4. Gibbs Modeling in Image Analysis: Some Problems—Segmentation, Deconvolution, Edge Detection, and Motion Detection

Consider $x = \{x_s, s \in S\}$, $S$ is finite, and $x_s \in E$ ($E$ discrete or continuous) an image for which we can only observe the degraded image $y$:

$$y = f(H(x), b) \quad y = \{y_t, t \in T\}, \quad T \text{ finite}, \quad y_t \in F. \tag{2.38}$$

The noise and image formation models: $b$, $f$ and $H$ are known and we assume that the transformation $(x, b) \longrightarrow (x, y)$ is invertible; that is, $b$ can be univocally expressed in terms of $(H(x), y)$

$$b = g(H(x), y). \tag{2.39}$$

To *restore* $x$ (to *filter* in signal processing terminology) is to propose an estimator $\widehat{x}$ of $x$ on the basis of $y$. If the cardinal of $T$ is much smaller than that of $S$, the inverse problem is strongly underdetermined or *mal posé*. This is the case, for example, if $T$ is a (coarser) sublattice of $S$, or in tomography, when the number of tomographic directions is reduced. *Bayesian image restoration methods* are based on the following principles:

(1) We translate any prior information we have of the object $x$ we want to reconstruct into a global energy prior $U(x)$: the weaker $U(\widehat{x})$ is, the closer the reconstruction is to this prior knowledge. In this setting, $x$ appears as a realization of a field with energy $U$ and density $P(x) = Z_1^{-1}\exp\{U(x)\}$.

(2) We shall assume that the noise is independent of $x$, with energy $V(b)$. So that because of (2.39), the joint density of $(x, y)$ is given by

$$P(x,y) = Z_2^{-1} \exp\{U(x) + V(g(H(x), y))\}. \tag{2.40}$$

(3) $\widehat{x}$ will be chosen following the posterior distribution given by

$$P(x|y) = Z^{-1}(y) \exp\{U(x) + V(g(H(x), y))\}.$$

Indeed, up to a constant (which depends on $y$), $P(x,y)$ and $P(x|y) = \frac{P(x,y)}{\int P(x,z)dz}$ have the same energy.

Two reconstruction criteria are classically considered (cf. § 6.8):

- $MAP : \widehat{x} = \underset{x}{\text{ArgMax}}\, P(x|y)$.
- •• $MPM : \widehat{x} = \{\widehat{x}_s = \underset{x_s}{\text{Arg Max}}\, P_s(x_s|y), s \in S\}$.

Here, $P_s(x_s|y)$ is the marginal distribution of $x_s$, conditional to $y$. These choices of $\widehat{x}$ respond to the compromise of both satisfying the prior distribution and being faithful to observations $y$. The problem of which model to choose, which parameters to consider, which criteria to accept, and which reconstruction algorithm to consider form part of the *knowhow* of the image analyzer. We shall not consider this problem in depth, but we invite the reader to consult the books and articles listed in the bibliography.

*Description of Some Choices for the Conditional Distribution* $P(x \mid y)$. Assume that $P(x)$ and $P(b)$ are defined by potentials $\phi_C(x)$, $C \in \mathcal{C}_1$, $\psi_D(b)$, $D \in \mathcal{C}_2$, $\mathcal{C}_1$ (resp., $\mathcal{C}_2$) sets of cliques of $S$ (resp., of $T$). The conditional energy $U(x|y)$ is then:

$$U(x|y) = \sum_{\mathcal{C}_1} \phi_C(x) + \sum_{\mathcal{C}_2} \psi_D(g(H(x), y)). \tag{2.41}$$

If $H(x)$ does not depend on $x$ in a local sense (e.g., in tomography), $U(x|y)$'s dependence on $x$ shall not be local either. If we ask that in model (2.38) $S = T$ and $H(x) = (H_s(x(V_s)), s \in S)$, with $V_s$ a local window which contains $s$, then

$$U(x|y) = \sum_{\mathcal{C}_1} \phi_C(x) + \sum_{D \in \mathcal{C}_2} \overline{\psi}_{\overline{D}}(x|y) \tag{2.42}$$

with

$$\overline{\psi}_{\overline{D}}(x|y) = \psi_D[(g_s(H_s(x(V_s)), y_s), s \in D)] \quad \text{and} \quad \overline{D} = \bigcup_{s \in D} V_s.$$

The distribution $P(x|y)$ is Markovian for the family of cliques

$$\mathcal{C}^* = \mathcal{C}_1 \cup \overline{\mathcal{C}}_2 \quad \text{with} \quad \overline{\mathcal{C}}_2 = \{\overline{D} : D \in \mathcal{C}_2\}. \tag{2.43}$$

Expression (2.43) completely specifies the Markovian dependence neighborhoods for the posterior distribution.

*Particular case: b, an independent noise.* So $\mathcal{C}_2 = \{\{s\}, s \in S\}$. If we define

$$V^2(s) = \{t \in S : \ \exists i \in S \quad s.t.\ s \in V(i) \text{ and } t \in V(i)\}, \tag{2.44}$$

then the neighborhood $\mathcal{V}^*(s)$ of $s$ given $\mathcal{C}^*$ is

$$\mathcal{V}^*(s) = \mathcal{V}(s) \cup V^2(s),$$

with $\mathcal{V}(s)$ the neighborhood of $s$ under the prior distribution.

**(2.4.1) Example.** Image Segmentation.
Consider $x = \{x_s,\ s \in S\}$, $S = \{1, 2, \ldots, n\}^2$, and $x_s \in E = \{1, 2, \ldots, K\}$, an image with $K$ (qualitative) states observed through $y = \{y_s, s \in S\}$, an independent process with distribution

$$\mathcal{L}(y_s | x_s = k) = F_k(y_s), \quad s \in S. \tag{2.45}$$

Our prior information is that the $K$-colored regions $C_k$ associated to each possible realization of $x$

$$C_k = \{s \in S : x_s = k\}, \quad k = 1, K$$

are geometrically regular. For example, we can model this prior information by a Gibbs field over $x$ with pairwise interaction potentials and eight nearest neighbors ([154], [18]):

$$\begin{cases} \phi_s(x) = \alpha_k, & \text{if } x_s = k, \\ \phi_{\{s,t\}}(x) = \beta_{k\ell}, & \text{if } \|s - t\| \le 2,\ (x_s, x_t) = (k, \ell), s \neq t. \end{cases}$$

The energy of configuration $x$ will be

$$U(x) = \sum_k \alpha_k n_k + \sum_{k \neq \ell} \beta_{k\ell} n_{k,\ell}, \tag{2.46}$$

where $n_k$ is the number of sites of "color" $k$ and $n_{k,\ell}$ is the number of neighboring sites of "colors" $(k, \ell)$. $\alpha_k$ measures the marginal importance of state $k$ (which increases if $\alpha_k$ increases) and $\beta_{k\ell}$ measures the likelihood of neighboring configurations $(k, \ell)$ (if we want to prohibit neighboring states $k$ and $\ell$, we can choose $\beta_{k,\ell}$ big and negative). Notice that since $\sum_k n_k = n^2$, model (2.46) is not well parametrized: we could, for example, fix $\alpha_K \equiv 0$. If there is exchangeability between states, the energy can be written (up to a constant) as

$$U(x) = \beta n(x), \quad n(x) = \sum_{<s,t>} 1(x_s = x_t). \tag{2.47}$$

If we rewrite the model for $(y_s|x_s)$ as $F_k(y_s) = \exp\{u_k(y_s)\}$, the posterior energy, under the exchangeability assumption (2.47), becomes (up to a constant in $y$):

$$U(x|y) = \beta n(x) + \sum_{s \in S} u_{x_s}(y_s). \tag{2.48}$$

Since $H(x) = x$, that is, $V_s = \{s\}$ and the noise is independent, the posterior distribution is also an eight nearest neighbor distribution. Choosing $\widehat{x}$ by $MAP$ is a compromise between big $n(\widehat{x})$ (if $\beta > 0$), or in other words, that $\widehat{x}$ has geometrically regular zones $\widehat{C}_k$, and a faithful reconstruction $\widehat{x}_s$ based on observation $y_s$, $s \in S$. The simulated annealing algorithm for $MAP$ (cf. Chapter 6) requires an explicit form of the conditional distribution of $x$ at $s$, with conditional energy

$$U_s(x_s|x_t,\, t \neq s, y) = -\beta n_s(x) + u_{x_s}(y_s), \text{ with } n_s(x) = \sum_{t, t\neq s, \|t-s\| \leq 2} 1(x_t = x_s).$$

Consider the following two models for (2.45):

*Transmission with Channel Noise.* $y_s \in E = \{1, 2, \ldots, K\}$ and

$$F_k(\ell) = \begin{cases} p & \text{if } \ell = k, \\ \dfrac{1-p}{K-1} & \text{if not.} \end{cases}$$

Condition $Kp > 1$ assures that the most probable is a good transmission. The conditional distribution in $x$ and in $s$ is

$$P(x_s = k|x_t, t \neq s, y) = \pi_s(k|x_{\partial s}, y_s) = \begin{cases} \dfrac{p\, e^{\beta n_s(k)}}{D} & \text{if } k = y_s, \\ \dfrac{(1-p)e^{\beta n_s(k)}}{(1-K)D} & \text{if not.} \end{cases}$$

with

$$n_s(k) = \sum_{t \neq s, \|t-s\| \leq 2} 1(x_t = k), \quad D = pe^{\,n_s(y_s)} + \frac{1-p}{1-K} \sum_{\ell : \ell \neq y_s} e^{\beta n_s(\ell)}.$$

*Observation of a Multidimensional Gaussian Texture.*
$y_s \in \mathbb{R}^p$ and and $F_k \sim \mathcal{N}_p(\mu_k, \Sigma_k)$. Image $x$ is observed through a multidimensional Gaussian response for each particular label; this is a standard situation in teledetection.

$$U(x|y) = \beta n(x) - \frac{1}{2}\sum_S {}^t(y_s - \mu_{x_s})\Sigma_{x_s}^{-1}(y_s - \mu_{x_s}) - \frac{1}{2}\sum_k n_k(x)\log|\Sigma_k|$$

with $n_k(x)$ the number of sites with level $k$. If, for example, textures can be differentiated only by their mean level ($\Sigma_k \equiv \Sigma$ for $k = 1, K$), the conditional distribution of $x$ in $s$ becomes

$$\pi_s(k|x_{\partial s}, y_s) = \frac{\exp(n_s(k) - \frac{1}{2}{}^t(y_s - \mu_k)\Sigma^{-1}(y_s - \mu_k))}{\sum_{\ell=1,K} \exp(\beta n_s(\ell) - \frac{1}{2}{}^t(y_s - \mu_\ell)\Sigma^{-1}(y_s - \mu_\ell))}.$$

**(2.4.2) Example.** Deconvolution of a Gray Level Image.
Consider $x = \{x_s, s \in S\}$, where $S = \{1, \dots, n\}^2$ is a gray level field with regular variation, observed through

$$y_i = \sum_{j \in V_i} h_{i,j} x_j + \varepsilon_j \tag{2.49}$$

with $\varepsilon$ a Gaussian white noise $\mathcal{N}(0, \sigma_\varepsilon^2)$. One way to model the prior regularity in the gray level dynamic of $x$ is to consider a Markovian field.

*$\Phi$-Model for Gray Level Fields* ([70], [66]).

$$U(x) = -\sum_{i=1,L} \sum_{<s,t>_i} \theta_i \Phi(x_s - x_t). \tag{2.50}$$

There are $L$ types of neighborhoods $< s, t >_i$, each with weight $\theta_i$ and

$$\Phi(u) = \left(1 + \left(\frac{u}{\delta}\right)^2\right)^{-1}. \tag{2.51}$$

$\theta_i > 0$ favors $x_s \sim x_t$ for $< s, t >_i$; $\theta_i < 0$ does not favor $x_s \sim x_t$ for $< s, t >_i$.

*Gaussian Markovian L-field Models.* $x$ is a realization of a Gaussian Markovian $L$-field $X$

$$E(X_i|\cdot) = \sum_{j \in L} \theta_j X_{i+j}, \quad Var(X_i|\cdot) = \sigma_e^2 \tag{2.52}$$

with $\theta$ a smoothing parameter and $\sigma_e^2$ a roughness parameter.

$\phi$-models are often preferred to the Gaussian models since big differences of $|x_s - x_t|$ are penalized less.

The conditional energy for both of the above models is given by

$$U(x|y) = U(x) + \frac{1}{2\sigma^2}\sum_i (y_i - (h_i * x)_i)^2,$$

and the neighborhood of $s$ under the posterior distribution is $\mathcal{V}^*(s) = \{s + (L \cup \{0\})\} \cup V^2(s)$. Here, $L$ stands for the neighborhood of 0 under the prior distribution and $V^2(s)$ is defined by (2.44). If the prior model is Gaussian, the posterior potential is Gaussian:

$$-\phi_{\{i\}}(x|y) = \frac{x_i^2}{2}\left\{\frac{1}{\sigma_e^2} + \frac{1}{\sigma_\varepsilon^2}\Big(\sum_{k:i\in V_k} h_{k,i}^2\Big)\right\} - \frac{1}{\sigma_\varepsilon^2} x_i \Big(\sum_{k:i\in V_k} h_{k,i} y_k\Big), \quad i \in S$$

and if $i, j \in S$, $i \neq j$: $-\phi_{\{i,j\}}(x|y) = x_i x_j \left\{\frac{\theta_{i-j}}{\sigma_e^2} + \frac{1}{\sigma_\varepsilon^2}\Big(\sum_{k:i,j\in V_k} h_{ki} h_{kj}\Big)\right\}$.

If the prior distribution is a $\Phi$–model, posterior potentials are obtained by eliminating $\frac{x_i^2}{2\sigma_e^2}$ in $\phi_{\{i\}}$ and replacing $-\frac{\theta_{i,j}}{\sigma_e^2} x_i x_j$ by $\sum_{\ell=1,L} \theta_\ell \Phi(x_i - x_j)$ in $\phi_{\{i,j\}}$. If the convolution model (2.49) is stationary, we just define

$$V_i = (i + V_0) \cap S, \quad h_{i,j} = h_{i-j}.$$

REMARKS:

(1) If we think the image is formed by different heterogeneous gray level subimages, the prior model can be enriched by introducing a *label* process $\lambda = \{\lambda_s, s \in \Lambda\}$ for the $K$ types of subimages. Then $\lambda$ can be modeled as in Example 1 and $(x|\lambda)$ as we did above under the hypothesis of independence for two different areas. Here, we only observe $y$ given by (2.49) and both $x$ and $\lambda$ must be reconstructed. This kind of two level $(X, \Lambda)$ model is used for texture segmentation ([68], [8]), in deconvolution ([110], [111]), and in tomography ([47],[96]).

(2) The crucial problem of how to choose the parameters has several solutions. One way is to fix the prior values by experience, although this might mean choosing among different restorations for different parameter values. Another way is to determine these values in a more "objective" way: for example, choosing values for the prior model that are compatible with those we expect for the image ([136], [8]). Still another way is to consider estimation with incomplete observations $y$ (cf. § 5), or based on an initial estimator $\widehat{x}(0)$ of $x$. Also, nonparametric techniques like crossvalidation have been considered [48].

**(2.4.3) Example.** Edge Detection ([67], [29]).
One way to improve image segmentation is to consider the segmentation's edge process: segmentation will be better if we consider prior information on edge regularity together with the regularity of the different "color" areas. It must be noted, however, that this kind of procedure leads to other problems, such as the appropriate choice of new parameters whereas other aspects, such as excessive complication of the restoration algorithms and computer time, should also be taken into account.

We shall study the joint model (Gray Level × Edges) given in Chalmond [29] and first considered in Geman [67]. The set of sites $S = S_G \cup S_B$

is the union of two lattices, the *site lattice* $S_G = \{1, 2, \ldots, n\}^2$ of *Gray Levels* where the gray level process $y = \{y_i,\ i \in S_G\}$ is observed, and the dual lattice of *edge sites* $S_G$: if $i$ and $j$ are two neighboring pixels of $S_G$, $|i_1 - j_1| + |i_2 - j_2| = 1$, there is an edge site $k = k(i, j)$ located in the center of $[i, j]$. For each $k$, either there is an edge ($\xi_k = 1$) or there isn't ($\xi_k = 0$).

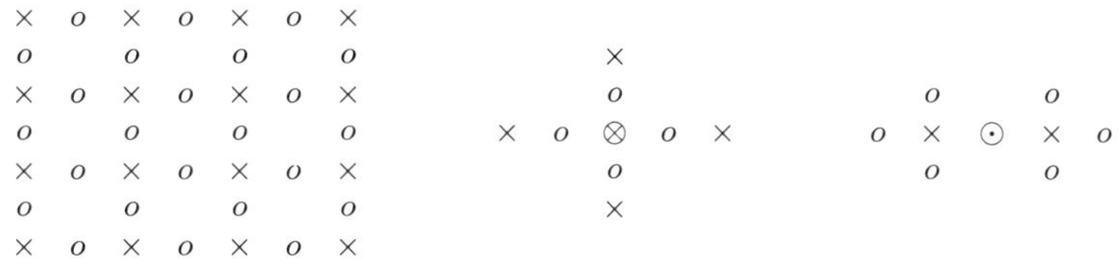

$\times$ stands for site of $S_G$ | $o$ stands for site of $S_B$ — Neighborhood of gray level site $\otimes$ — Neighborhood of edge site $\odot$

*Sites and neighbors for edge detection models*

The complete observation would be $x = (y, \xi)$, $\xi = \{\xi_k, k \in S_B\}$, but only the gray level $y = P_1(y, \xi)$ can be observed. Here, $P_1$ stands for the projection over the gray level space.

*Energy of* $x = (y, \xi)$. The energy $U(y, \xi) = U_G(y) + U_B(\xi) + U_{G,B}(y, \xi)$ shall be chosen so that $U_G$ smoothes the gray levels, $U_B$ organizes the contours, and $U_{G,B}$ preserves discontinuities.

(*a*) *Gray Level Energy* $U_G(y)$: The only cliques are the singletons and pairs $\{i, j\}$ *s.t.* $|i_1 - j_1| + |i_2 - j_2| = 1$. For $U_G$ we can choose, for example, the $\Phi$–energy: $U_G(y) = -\alpha \sum_{<i,j>} \phi(x_i - x_j)$.

(*b*) *Edge Process Energy* $U_B(\xi)$: Cliques for the edge process are the following:

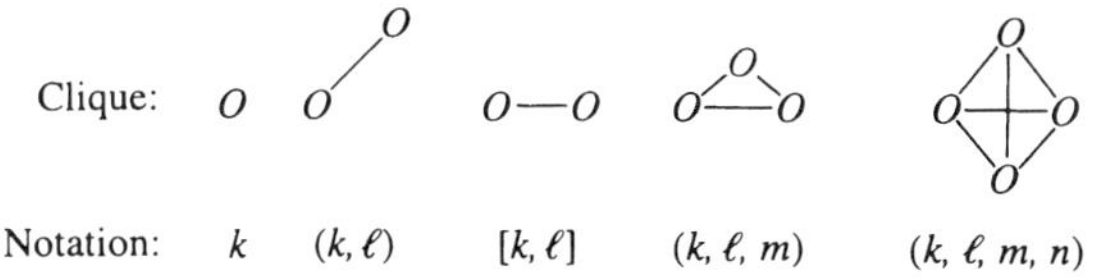


and those originated by rotation. Since $\xi$ is binary, under isotropy, the energy depends on five parameters:

$$U_B(\xi) = b_1 \sum_{S_B} \xi_k + b_2 \sum_{(k,\ell)} \xi_k \xi_\ell + b_3 \sum_{[k,\ell]} \xi_k \xi_\ell + b_4 \sum_{(k,\ell,m)} \xi_k \xi_\ell \xi_m + b_5 \sum \xi_k \xi_\ell \xi_m \xi_n,$$

and the last sum is over clique of order four. Considering the following five kinds of configurations for the fourth-order clique:

| | | | | | |
|---|---|---|---|---|---|
| Edge configuration | 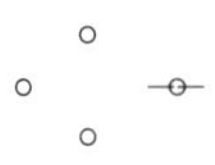 | 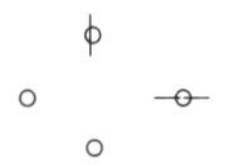 | 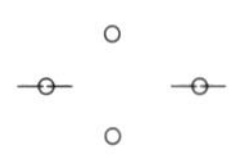 | 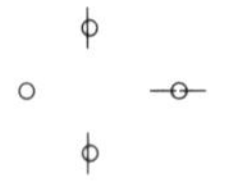 | 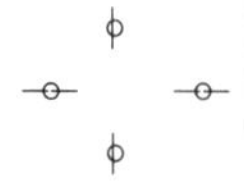 |
| Type n° | 1 | 2 | 3 | 4 | 5 |

and $n_i(\xi)$ the number of configurations of type n° $i$, $i = 1, 5$, we have the equivalent expression of $U_B(\xi)$: $U_B(\xi) = \sum_{i=1}^{5} \beta_i n_i(\xi)$, with $\beta_1 = \frac{1}{2} b_1$, $\beta_2 = b_1 + b_2$, $\beta_3 = b_1 + b_3$, $\beta_4 = \frac{3}{2} b_1 + 2b_2 + b_3 + b_4$, $\beta_5 = 2b_1 + 4b_2 + 2b_3 + 4b_4 + b_5$. Notice that the realization $(0, 0, 0, 0)$ doesn't have to be considered since we have the identification constraints. Choosing $b_3 > 0$ and letting it grow (just like $b_2$ to a lesser extent) is equivalent to favoring edge prolongation (with change of orientation). If, on the contrary, we want to disencourage endpoints of type 1 (or quadruple points of type 5), we should choose $b_1 < 0$ ($b_5 < 0$) with $|b_1|$ bigger or smaller according to our prior information.

(c) *Joint Energy* $U_{G,B}(y, \xi)$. We shall only consider the potential defined over the clique $\{i, k(i, j), j\}$ and

$$U_{G,B}(y, \xi) = \Delta^2 \sum_{<i,j>} (y_i - y_j)^2 \xi_{k(i,j)}.$$

If the gradient $|y_i - y_j|$ is big, we shall favor the appearance of an edge at $k(i, j)$, the more so, the bigger $\Delta^2$. The reconstruction of $\xi$ shall be done on the basis of the Markovian distribution $P(\xi|y)$, which admits

$$\begin{cases} \pi_k(\xi_k = 0 | \xi_\ell,\ \ell \neq k, y) = [1 + e^{a(\xi_{\partial k}, y_i, y_j)}]^{-1}, \quad \text{with} \\ a(\xi_{\partial k}; y_i, y_j) = b_1 + \sum_2^5 b_\ell \ v_\ell(\xi_{\partial k}) + \Delta^2 (y_i - y_j)^2 \end{cases}$$

as its conditional distribution at site $k = k(i, j)$. Here $v_\ell$, $\ell = 2, 5$ are associated to the four kinds of cliques other than the singleton. For example, for $l = 2$:

$$v_2(\xi_{\partial k}) = \sum_{\ell:(\ell,k)} \xi_\ell.$$

**(2.4.4) Example.** Detection of Movement [57].
We want to detect moving objects in a scene observed by a camera which is also in movement. Observations are the intensity $I$ over $S = \{1, 2, \dots, n\}^2 \times \{1, 2, \dots, T\}$ with time–space derivatives

$$y_s = \{(\nabla I(s), I_t(s)); s \in S\}, \quad s = (x, y; t).$$

$\nabla I = (I_x, I_y)$ and $I_u$ are the discretized gradient in space and time. We must choose the appropriate labels $\lambda_s \in \{1, 2, \dots, R\}$, $R \geq 2$ corresponding to each of the $R$ objects whose movement we are considering ($\lambda_s = 1$, e.g., corresponds to the camera's movement, that is, to the part of the scene which remains still). We observe only $y$ and seek to reconstruct $\lambda$.

The proposed energy model, $U(y, \lambda) = U_1(\lambda) + U_2(y, \lambda)$, consists of a regularization term $U_1(\lambda)$ (cf. Example 2.4.1) and of a data fitting term: for the *r-th* figure in movement, we adopt a linear velocity field, parametrized by $\theta_r = (a, b, \alpha, \beta, \gamma, \delta)$. Velocity at point $(\Delta x, \Delta y)$ (with respect to the center of the object ) is given by

$$\mathbf{v}_{\theta_r}(x, y) = \begin{pmatrix} a & + & \alpha \Delta x & + & \beta \Delta y \\ b & + & \gamma \Delta x & + & \delta \Delta y \end{pmatrix}.$$

Here, $R$ and $\theta = (\theta_r, r = 1, R)$ are additional parameters whose values must be updated at each step of the movement detection algorithm. If the true speed is $v(x, y, t)$, the chosen global fit measure is given by:

$$U_2(y, \lambda) = \frac{1}{2\sigma^2} \sum_s \varepsilon_s^2, \quad \text{with} \quad \varepsilon_s =< \nabla I(s), (v_{\theta_{\lambda(s)}}(x, y) - v(s)) > .$$

Because of the optic flow constraints $\nabla I \cdot \mathbf{v} + I_t = 0$, $U_2$ can only be expressed in terms of $(y, \lambda)$ since $\varepsilon_s = \nabla I(s) \cdot v_{\theta_{\lambda(s)}}(s) + I_t(s)$.

## 2.5. Appendix

### Appendix 1: Reconstruction of a Specification $\pi$ Given By Its Specifications $\pi_{\{i\}}$ For Each Site $i \in S$

We shall examine the case of a joint distribution $\pi$ over finite $\Omega = \prod_S E_i$.

**(2.5.1) Theorem.**

(*a*) *Assume that for all* $x \in \Omega$, $\pi(x) > 0$. *Then* $\pi$ *is only determined by its conditional distributions* $\pi_i$, $i \in S$.

(*b*) *A family of conditional distributions* $\{\pi_i, i \in S\}$ *in general does not induce a joint distribution.*

*Proof:*

($a$) Let $S = \{1, 2, \ldots, n\}$ be an arbirtary enumeration of $S$. If we define $x^i$ to be the configuration equal to $x$ everywhere except at $i$, where it is 0 (0 is the reference element of di $E_i$), then

$$\frac{\pi(x)}{\pi(x^i)} = \frac{\pi_i(x_i|x_1, x_2, \ldots, x_{i-1}, x_{i+1}, \ldots, x_n)}{\pi_i(0|x_1, x_2, \ldots, x_{i-1}, x_{i+1}, \ldots, x_n)},$$

$$\pi(x) = \pi(0) \prod_{i=1}^{n} \frac{\pi_i(x_i|0, 0, \ldots, 0, x_{i+1}, \ldots, x_n)}{\pi_i(0|0, 0, \ldots, 0, x_{i+1}, \ldots, x_n)} \tag{2.53}$$

so that $\pi$ is determined uniquely by its conditional distributions $\pi_i$.

($b$) Let $\{\pi_i, i \in S\}$ be a family of conditional distributions. If the $(\pi_i)$ induce a joint distribution $\pi$, the relationships given by (2.53) must be satisfied for any enumeration of $S$ so that the right hand side of (2.53) is the same for any enumeration of $S$; in order to induce a joint distribution, $\{\pi_i, i \in S\}$ must necessarily satisfy this set of constraints. □

We also have the more general result [71]

**(2.5.2) Theorem.** *Let $\lambda$ be a measure over $(E,\mathcal{E})$ and $\pi$ a specification which for all $i$ has a density $\rho_i > 0$. Then for all $V \in \mathcal{S}$, there exists measurable $\rho_V > 0$ uniquely determined by the $\rho_i$ s.t.:*

(1) $\pi_V = \rho_v \cdot \lambda_V$.

(2) $\mu \in \mathcal{G}(\pi) \Longleftrightarrow \mu_{\{i\}} = \pi_{\{i\}}$.

## Appendix 2: Moebius Inversion Lemma

Consider $S = \{1, 2, \ldots, n\}$, $\Omega = \prod_S E_i$. If $x \in \Omega$, $A \subseteq S$, then $x_A$ is the restriction of $x$ to $A$ and $x^i$ is the configuration equal everywhere to $x$ except at site $i$, where it is 0 (a reference element of $E_i$). Given $f : \Omega \longrightarrow \mathbb{R}$, we define $f_A(x) = f(x_A, 0_{S \setminus A})$.

**(2.5.1) Lemma.** *Let $f{:}\Omega \longrightarrow \mathbb{R}$ satisfy $f(0) = 0$. Then*

($a$) $f(x) = \sum_{A \subseteq S, A \neq \phi} \phi_A(x_A)$ *with*

$$\phi_A(x_A) = \sum_{V \subseteq A} (-1)^{|A \setminus V|} f_V(x_V).$$

($b$) *For all $A \subseteq S$, $i \in A$ we have $\phi_A(x^i_A) = 0$. These conditions assure that the $\{\phi_A, A \subseteq S\}$ are identifiable.*

*Proof:*

(*a*) Given $\phi_A$ defined as in $(a)$, write $g(x) = \sum_{A \subseteq S, A \neq \phi} \phi_A(x_A)$. Then

$$g(x) = f(x) + \sum_{V \subsetneq S} c_V f_V(x_V) \quad \text{and}$$

- if $V \neq \phi$, $c_V = \sum_{P \subseteq S \backslash V} (-1)^{|P|} = \sum_0^m (-1)^k C_m^k = 0$, if $m = |S \backslash V|$.

•• if $V = \phi$, $f_V = 0$ by hypothesis. Thus $f(x) = g(x)$.

(*b*) If $W$ stands for the set of parts of $A \backslash \{i\}$, $i \in A$, $W$ and $W \cup \{i\}$ give the parts of $A$. We also have $f_W(x) = f_{W \cup \{i\}}(x^i)$. Since $W$ and $W \cup \{i\}$ cannot both be even, we find $\phi_A(x^i) = 0$. Assume that $f$ can be written as in $(a)$ in terms of $\psi_A$, the $\psi_A$ satisfying $(b)$. We shall prove by recurrence over $|A|$ that $\phi_A = \psi_A$. If $A = \{i\}$, $f_{\{i\}}(x) = \phi_i(x) = \psi_i(x)$. Consider $i \notin A$ and assume that for all $B \subsetneq A \cup \{i\}$, $\phi_B = \psi_B$. Then

$$f_{A \cup \{i\}}(x) = \sum_{B \subsetneq A \cup \{i\}} \phi_B(x) + \phi_{A \cup \{i\}}(x) = \sum_{B \subsetneq A \cup \{i\}} \psi_B(x) + \psi_{A \cup \{i\}}(x)$$

so that $\phi_{A \cup \{i\}} = \psi_{A \cup \{i\}}$. □

## 2.6. Bibliographical Comments

For the first section on Gibbs fields, we followed Georgii. The results on the existence and uniqueness of Gibbs fields are due to Dobrushin (1968, [50]). Mixing properties are studied by Dobrushin in the stationary case ([51]), and generalized to the case of noninvariant specification models by Künsch ([107]). Simon introduced a unicity condition which can be directly written in terms of the potentials.

The large deviations results were established simultaneously and independently by Comets, Föllmer and Orey, and Olla. For an application of these inequalities in field statistics, we shall follow Comets (cf. Chapter 5). Large deviation inequalities for degenerate fields and a variational principle for the conditional pressure were developed by Comets and Gidas while studying estimation with incomplete data.

The correspondence between a Markov field and a Gibbs field was established by many authors in the early 1970s, the first time by Hammersley–Clifford (cf. Besag 1974). Prum gave the most general form for almost-Markovian fields. The auto models were defined by Besag, and a very nice application for texture modeling was developed by Cross and Jain.

The study of time reversibility and reversibility for a Markov field dynamic is due to Koslov and Vasilyev in the case of a synchronous dynamic.

Künsch ([108]) extended these results to arbitrary dynamics. The idea of considering models that only depend on maximal cliques was taken from Grenander.

Most of the results concerning Gaussian specifications and Gibbs Fields can be found in Künsch's thesis [104]. For the presentation of these results we have followed Georgii.

The Bayesian approach in Pattern Analysis was first considered in the Geman brothers' fundamental article (1984, [67]). They considered *MAP* restoration. The article by Marroquin–Mitter–Poggio that defined *MPM* gave an interesting complementary point of view. We invite the reader interested in stochastic methods for image analysis to consult the book by D. Geman. We also reference several theses with interesting examples of the Markovian methodology. The first colored lattice model is due to Strauss [154]. The range of application of this method is extensive: edge detection and segmentation ([29], [66], [67]), tomographic reconstruction ([47],[96]), texture segmentation ([8], [68]), movement analysis [57] and seismic deconvolution ([6], [7], [110], [111]). Global Bayesian methods and contextual methods are compared experimentally, for the binary segmentation problem, in ([170], [171]).

## 2.7. Exercises

(1) Let $f : E^S \longrightarrow \mathbb{R}$, $E = \{0, 1, \ldots, K-1\}$, $S = \{1, 2, \ldots, n\}$ be such that $f(0) = 0$. By the Moëbius formula, $f$ is associated to certain potentials $\phi_A$, $\emptyset \neq A \subseteq S$. Check that the $(K^n - 1)$ parameters that define $f$ can be rewritten in terms of those of the $\phi_A$ with the identification conditions: $\phi_A(x^i) = 0$ if $i \in A$, with $x^i_i = 0$ and unchanged elsewhere.

(2) Over the two-dimensional triangular lattice consider the six nearest neighbor Markovian stationary model with a 3 point state space, $E = \{a, b, c\}$. Identify the model: cliques, potentials, parametric dimension. What is the conditional distribution at a given site? Repeat for the isotropic model; the model with, at most, pairwise potentials: general, isotropic, reversible (cf. §2.2.4), isotropic, and reversible. Find the distributions if $E = \{0, 1, 2\} \subseteq \mathbb{R}$.

(3) *Conditional distributions do not induce a joint distribution.*

(a) $E = \{1, 2, \ldots, n\}$, $F = \{1, 2, \ldots, m\}$, $n, m \geq 2$. Consider two free families of distributions $(X|Y = y)$ and $(Y|X = x)$, $x \in E$, $y \in F$, over $E$ and $F$. What is the dimension of such a model? What is the dimension of a joint distribution $(X, Y)$ over $E \times F$? Conclude.

(b) Over $\mathbb{Z}$, we consider the Markovian, homogeneous, nearest neighbor kernel: $\pi(y|x, z) = P(X_0 = y|X_{-1} = x, X_1 = z)$. Assume the state space is $E = \{1, 2, \ldots, K\}$. What is the dimension of such a kernel? Under what conditions does this kernel correspond to that of a nearest neighbor Markovian model?

(c) Consider the family of conditional binomial, homogeneous distributions: $\pi(X_0 = x|\cdot) = \mathcal{B}(n, \theta(\cdot))$, $\theta = \frac{\exp A}{1+\exp A}$, $A(\cdot) = \alpha + \beta(x_{-1} + x_1)^2$. Do these distributions induce a joint distribution?

(4) *Causal model and associated bilateral representation.*

(a) Over $\mathbb{Z}$ consider the causal Markovian model $X$ with memory 2 and state space $E = \{0, 1\}$ characterized by the four probabilities:

$$p_{xy} = P(X_n = 1 \mid X_{n-1} = y, X_{n-2} = x),\ x, y = 0 \text{ or } 1, \quad 0 < p_{xy} < 1.$$

Give the Markovian field representation for $X$ and identify the potentials and parameters. Is any binary model over $\mathbb{Z}$ with four nearest neighbors a causal Markovian process with memory 2?

(b) For lexicographical order over $\mathbb{Z}^2$, we consider the causal process $X$ with state space $\{0, 1\}$, whose distribution at site $(i, j)$ depends on the past through $(i-1, j)$ and $(i, j-1)$. Describe the noncausal representation of $X$.

(5) *Markovian modeling for texture segmentation.*
Over $S = \{1, 2, \ldots, n\}^2$ consider two fields defined hierarchically: $\Lambda = \{\lambda_i, i \in S\}$ a label process, $\lambda_i \in \{1, 2, \ldots, K\}$ and $(X|\Lambda)$ a Gaussian "gray level" field conditional on $\Lambda$. We observe $X$ and want to reconstruct $\Lambda$. We adopt the following model:

(a) Over $\Lambda$ we consider model (2.47) for the interactions of the eight nearest neighbors.

(b) (b1) *Gray level texture*: $(X_i|\lambda_i = k) \sim \mathcal{N}(\mu_k, \sigma^2)$.

(b2) *Roughness texture*: $(X_i|\lambda_i = k) \sim \mathcal{N}(0, \sigma_k^2)$. For (b1) and (b2), variables $(X_i|\lambda_i)$ are independent site by site.

(b3) *Covariation textures*: Over the constant label areas $S_k$ for $k = 1, K$, the $X_{S_k}$ are independent and over $S_k$, $X$ is an isotropic Markovian model with four nearest neighbors with parameters $(\alpha_k, \sigma_e^2)$.

Write the different Markovian models with joint distribution $(X, \Lambda)$. We observe $X$. Find distributions $(\Lambda|X)$ and $(\Lambda_i|X, \Lambda_j, j \neq i)$.

(6) *Restoration of a gaussian signal distorted by convolution and additive noise.*
$X = \{X_i, i \in S\}$ is a Gaussian signal. We observe $Y = h * X + \varepsilon$ with $\varepsilon$ a white noise $\varepsilon = \{\varepsilon_i, i \in S\}$, $\varepsilon_i \sim \mathcal{N}(0, \sigma_\varepsilon^2)$. Determine the joint distribution $(X, Y)$ and the conditional distributions $(X|Y)$ and $(X_i|Y, X_j, j \neq i)$ in the following settings:

(a) $S = T = \{1, 2, \ldots, n\}$ is the unidimensional torus with $n$ sites, with the nearest neighbors relationship (1 and $n$ are neighbors),

- $X$ is Gaussian, Markovian with two nearest neighbors.

•• $(h * X)_i = aX_{i-1} + bX_i + cX_{i+1}$.

(b) $X = T^2$ is the two-dimensional torus with $n^2$ points; $X$ is Gaussian Markovian isotropic with four nearest neighbors, $h * X = X$.

(7) *Edge process Markovian model.*
In the same context as § 2.4, Example 3, we are interested in only the edge process $\xi = (\xi_k, k \in S_B)$, $\xi_k \in \{0, 1\}$, $\xi_k = 1$ if there is a vertical (horizontal) edge, and $\xi_k = 0$ if not. A site $k$ is surrounded by six neighbors as indicated in the figure of Example 2.4.3: four neighbors at distance $(\sqrt{2})^{-1}$ and two at unit distance.

(a) Describe the general translation invariant potential.

(b) We shall restrict ourselves to potentials that depend, at most, on two points and to the isotropic case. Show that $\xi$ is auto-logistic with conditional potential at site $k$ given by $U_k(\xi) = \xi_k(\alpha + \beta_1 n_1(k) + \beta_2 n_2(k))$ with $n_1(k)$, $n_2(k)$ to be computed. Interpret parameters $\alpha$, $\beta_1$, and $\beta_2$. How should they be chosen in a real problem?

(c) $\xi$ is transformed into $\delta$ by a transmission noise with error rate $\varepsilon < 1/2$, independent at each site. We observe $\delta = \{\delta_k\}$, $\delta_k \in \{0, 1\}$. Write distributions: $(\xi|\delta)$, $(\xi_k|\delta, \xi_\ell, \ell \neq k)$.

(d) How are these results (and questions) modified if we include two edge directions at 45° and 135°?

(8) *Existence (and uniqueness) of Gibbs measures* [133].
Let $X$ be a Markov chain over $\mathbb{Z}$ with state space $\{-1, +1\}$, and transition matrix

$$M = \begin{pmatrix} p & 1-p \\ 1-q & q \end{pmatrix}, \quad 0 < p, q < 1.$$

(a) Show that the distribution over $[-n, n]$ conditional to $\mathbb{Z}\backslash[-n, n]$ only depends on $X_{-n-1}$, $X_{n+1}$ and is given by

$$\pi_n(x([-n,n])|x_{-n-1} = a,\ x_{n+1} = b) \\ = \frac{M(a, x_{-n})M(x_{-n}, x_{-n+1}) \cdots M(x_n, b)}{M^{2n+2}(a, b)}.$$

Here, $M^m$ stands for the $m$-*th* power of $M$.

(b) Given two sequences $(a_{-n})$, $(b_n)$, define $\mu_n(\cdot) = \pi_n(\cdot|a_{-n-1}, b_{n+1})$. Show that for all finite subset $A$, $\mu_n(x(A)) \longrightarrow \mu(x(A))$ with $\mu$ the chains' stationary law.

(9) *Nonexistence of a Gibbs measure* [133].
Consider the random walk over $S = \mathbb{Z}$ defined by $Y_j - Y_{j-1} = \varepsilon_j$, $\varepsilon_j$ i.i.d. and with uniform distribution over $\{-1, +1\}$. For odd $n$, define $K_n = [-n, n]$.

(a) Calculate probabilities $P_n(Y(K_n), Y_{n+1} = b \mid Y_{-n-1} = a)$.

(b) Determine conditional distribution $\pi_n(Y(K_n) \mid Y_{-n-1} = a, Y_{n+1} = b)$ and check that $\{\pi_n(\cdot|\cdot)\}$ is a well-defined specification.

(c) Let $\mu \in \mathcal{G}(\pi)$ and define $\mu_n = \mu_n(Y(K_n) \mid Y_{-n-1} = Y_{n+1} = 0)$. What are the possible values of $Y_0$? Check that $\mu_n(Y_0 = k) = (C_{n+1}^{\frac{n+1+k}{2}})^2 (C_{2(n+1)}^{n+1})^{-1}$, $|k| \leq n+1$, which tends to 0 if $n \to \infty$ for all $k \leq n+1$ (in this case, we say the mass goes to infinity). Because of Condition (2') of § 2.1.5, this yields the nonexistence of a Gibbs measure: $\mathcal{G}(\pi) = \emptyset$.

(10) *Phase Transition: Cayley's Tree* [133].
We shall consider the graph sequence $\Lambda_n$, $n \geq 1$:

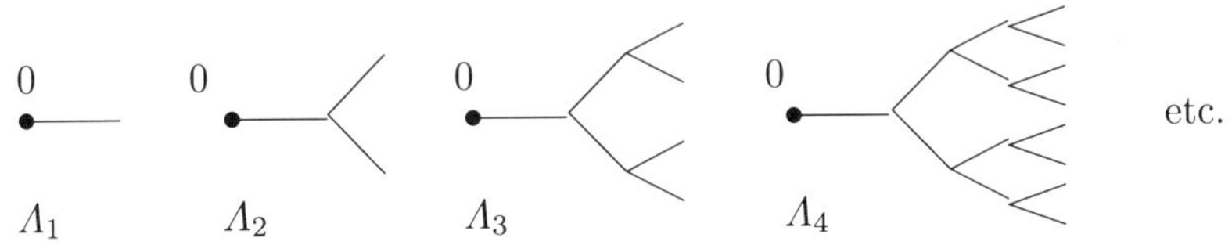


Over $S = \bigcup_{n \geq 1} \Lambda_n$, the boundary of $\Lambda_n$, $\partial \Lambda_n$ is identified with the elements of $\Lambda_{n+1} \backslash \Lambda_n$: they are the $2^n$ end points of $\Lambda_{n+1}$. At each site, the state space is $\{-1, +1\}$ and we define over $\Lambda_n$ the distribution, conditional to $\partial \Lambda_n$,

$$\pi_n(\omega(i),\ i \in \Lambda_n \mid \omega(\partial \Lambda_n)) = Z_n^{-1}(\omega(\partial \Lambda_n)) \Pi^* M(\omega(i), \omega(j)).$$

Here, $\Pi^*$ is the product over $i \in \Lambda_n$, $j \in \Lambda_{n+1}$, such that $i$ and $j$ are neighbors and $M(1,1) = M(-1-1) = a$, $M(1,-1) = M(-1,1) = 1-a$, $0 < a < 1$. $\{\pi_n\}$ is a well-defined specification. Fix $\omega(j) = 1$ if $j \in \partial \Lambda_n$ and define $\mu_n$ the marginal distribution of $\omega(0)$ once this exterior condition is fixed. Also define $R_n(x) = \sum_{\omega} \Pi^* M(\omega(i), \omega(j))$, with $\omega(j) = 1$ if $j \in \partial \Lambda_n$, $\omega(0) = x$. Then $\mu_n(1) = (1 + r_n)^{-1}$ with $r_n = R_n(-1)\ R_n(1)^{-1}$.

(a) Check the following recurrent equations :

$$\begin{aligned} R_{n+1}(1) \quad &= M(1,1) R_n(1)^2 + M(1,-1) R_n(-1)^2, \\ R_{n+1}(-1) \quad &= M(-1,1) R_n(1)^2 + M(-1,-1) R_n(-1)^2. \end{aligned}$$

(b) Deduce that $r_{n+1} = f(r_n)$, with $f(x) = \frac{(1-a)+ax^2}{a+(1-a)x^2}$.

(c) Show that the dynamic system $\{r_n\}$ has only one stable limit given by $r = 1$ if $a \leq \frac{3}{4}$ and a certain value $r$, $0 < r < 1$ if $\frac{3}{4} < a < 1$.

(d) Deduce that if $\frac{3}{4} < a < 1$ (the local dependence is strong), $\mu_n(1) \longrightarrow p > \frac{1}{2}$, and that there is phase transition (consider the other boundary condition, $\omega(j) = -1$ if $j \in \partial \Lambda_n$).

(11) *Sampled Markovian Fields.*

(a) Write the distribution of the Marvov chain $(X_1, \dots, X_n)$ as $\prod_{i=1,n} q_i(x_i, x_{i+1})$. Show that $X$ observed once every two steps is still a Markov chain. Write the distribution (transition, potentials) for the stationary case with binary state space.

(b) Let $X$ be a Markovian field over $S = \{1, 2, \dots, n\}^2$ with four nearest neighbors. Assume we observe $X$ over $S^+ = \{(i,j) \in S, i+j \text{ even}\}$. Show that $X$ over $S^+$ is a Markovian field for the maximal clique system given by $C_{ij} = \{(i,j), (i+2,j), (i+1,j+1), (i+1,j-1)\}$. Determine $X$'s distribution over $S^+$ if the sate space is binary.

(c) Show that the Markov property in (b) is lost if $X$ is Markovian with eight nearest neighbors and is observed over $S^+$, or if $X$ is sampled over $S_2 = \{(i,j),\ i \text{ and } j \text{ even}\}$.

(12) *Bivariate Gaussian Specification.*

(a) Let $S = \{1, 2, \dots, n\}$ be the unidimensional torus with the two nearest neighbor relationship $Z_i = (X_i, Y_i) \in \mathbb{R},\ i \in S$. Consider the quadratic energy

$$-U(X,Y) = \sum_i (X_i^2 + Y_i^2) \; - 2\alpha \sum_{<i,j>} (X_i Y_j + Y_i X_j).$$

Transforming $S$ into a set $S^* = S_X \cup S_Y$ with $2n$ points, and using property (2.7), show that $U(Z)$ defines a Gaussian specification if $0 \leq \alpha < \frac{1}{2}$. Determine the conditional distributions $\mathcal{L}(X_i \mid X_j, j \neq i, Y)$.

(b) Consider the energy $U$ defined by

$$-U(X,Y) = \sum_i (X_i^2 + Y_i^2) \; - 2\alpha \sum_{<i,j>} (X_i Y_j + Y_i X_j)$$

$$-2\beta \sum_{<i,j>} X_i X_j - \; 2\gamma \sum_{<i,j>} Y_i Y_j.$$

Give a sufficient condition that assures that $U$ defines a Gaussian specification. Determine the conditional distributions $\mathcal{L}(X_i \mid X_j, j \neq i, Y)$.

(13) *Marginal distributions of an auto-logistic distribution.*
Let $S$ be finite. We consider the auto-logistic distribution over $S$ ($x_i \in \{0,1\}$) with energy:

$$U(x) = \alpha \sum_S x_i \; + \; \beta \sum_{<i,j>} x_i x_j.$$

Let $k \in S$ be a *fixed site*, $S_k = S \backslash \{k\}$ and $x^k = \{x_i, i \in S_k\}$.

(a) Find distribution $P(x^k \mid x_k)$ (find the potentials, in particular, those of singletons $\{i\}$, with $i$ a neighbor of $k$).

(b) Determine marginal distributions $P(x_k)$ and potentials (it should be remarked that $\partial k$ is a clique for this distribution so the associate potential should be calculated). Is this marginal distribution still auto-logistic?

(14) *Graphical model for a contingency table: Gibbs or log-linear model.* Consider $S = \{1,2,3\}$, $\Omega = E_1 \times E_2 \times E_3$ with $E_1 = \{1,\ldots,I\}, E_2 = \{1\ldots,J\}, E_3 = \{1,\ldots,K\}$.

(a) Show that the general form of

$$p_{ijk} = P(x_1 = i, x_2 = j, x_3 = k)$$

is given by

$$p_{ijk} = Z^{-1}\exp\{\alpha(i)+\beta(j)+\gamma(k)+\alpha\beta(i,j)+\beta\gamma(j,k)+\alpha\gamma(i,k)+\alpha\beta\gamma(i,j,k)\}$$

and that the effect (potentials) $\alpha, \beta, \ldots, \alpha\beta\gamma$ are identifiable under the *analysis of variance constraints*: for all $(i,j,k)$,

$$\begin{aligned}\alpha(\cdot) &= \beta(\cdot) = \gamma(\cdot) = \alpha\beta(i,\cdot)\\ &= \alpha\beta(\cdot,j) = \ldots = \alpha\beta\gamma(i,j,\cdot) = \alpha\beta\gamma(i,\cdot,k) = \alpha\beta\gamma(\cdot,j,k) = 0.\end{aligned}$$

Here, $\cdot$ means adding over the corresponding index. This is the *log-linear* form for the associated $\log p_{ijk}$ model. Taking $(1,1,1)$ as reference state, give the classical *Gibbs constrainsts* and the dimension of the model.

(b) Give the dimension of the following models: model without third-order potentials (interactions); additive in $i,j,k$; additive in $(i,j)$ and $k$. Interpret the associated $p_{i,j,k}$ models.

(c) Define over $S = \{1,2,3,4\}$ the following neighborhood graphs: $\mathcal{G}_1 = \{\{2,3\},\{2,4\},\{3,4\}\}$, $\mathcal{G}_2 = \{\{1,4\}\} \cup \mathcal{G}_1$. Find the associated cliques and the associated log-linear models and interpret them in terms of probabilities $p_{i,j,k,l}$.

(d) Over $S = \{1,2,3,4\}$, determine the graphic models associated with the following $(x,y,z,t)$ models ($\perp$ stands for independent): $x,y,z,t$ pairwise independent; $(x,y) \perp (z,t)$; $(x \mid t) \perp ((y,z) \mid t)$.

(15) *Probability distributions as the marginals of a Boltzmann machine.* A Boltzmann machine $Q$ is a Markovian field with $\{0,1\}$ states over a finite $S$ defined by a quadratic energy $U$:

$$U(x) = \sum_s w_s x_s^2 + \sum_{s,t} w_{st} x_s x_t,\ w_{st} = w_{ts}, s,t \in S.$$

Let $O$ be a finite subset of sites (O for observed), and let $P$ be a positive probability over $\{0,1\}^O$: $P(x^O) > 0$ with $x^O = (x_i, i \in O)$. Sussmann

([157], cf. also [177]) proved that it is always possible to increment the number of sites $S = O \cup H$ ($H$ for hidden) in such a way that $P$ becomes the marginal of a Boltzmann machine over $S$.

(a) If $n = |O|$, give a necessary condition over $h = |H|$ in order for such a solution to exist.

(b) If $n = 3$, show that one hidden site is enough. How many degrees of freedom do we have for choosing $Q$? Give a solution. Ask the same questions for $n = 4$.

(c) If $n = 2$ and $h = 1$, can we find for each $P$ a Boltzmann machine whose energy doesn't have any square terms ($w_s = 0$, for $s \in S$)?

CHAPTER 3

# Limit Theorems and Parametric Estimation for Fields

## 3.1. The Ergodic Theorem for Spatial Processes

In this section we shall recall the ergodic theorem ([124], [160]): let $X$ be a process over $\mathbb{Z}^d$ with values in $\mathbb{R}^m$. We shall assume $X$ is stationary, that is, its distribution $P$ is invariant under translations $\tau_i$, $i \in \mathbb{Z}^d$. The $\sigma$-algebra of invariant sets $\mathcal{I}$ is defined by

$$A \in \mathcal{I} \Longleftrightarrow \tau_i(A) = A \quad \text{for all } i.$$

$X$ will be said to be ergodic if $\mathcal{I}$ is the trivial $\sigma$–algebra: if $A \in \mathcal{I}$, then $P(A) = 0$ or 1. Here are some examples of ergodic processes:

- • $X = \{X_i, i \in \mathbb{Z}^d\}$, $X_i$ *i.i.d.*
- •• $Y = \{Y_i = g(X \circ \tau_i),\ i \in \mathbb{Z}^d\}$ if $X$ is ergodic.
- ••• $X$ such that its distribution $P$ is an extremal stationary point of the set of Gibbs measures.

For dimensions $d \geq 2$, the geometry of the domain over which the process is defined will appear in the formulation of the ergodic theorem. Define $d(C)$ as the interior diameter of a given set $C$:

$$d(C) = \sup\{r : \text{there exists } c \ s.t.\ B(c, r) \subseteq C\}.$$

**(3.1.1) Theorem.** *Ergodic Theorem. Let $X = \{X_i, i \in \mathbb{Z}^d\}$ be a stationary process in $L^p$, $1 \le p < \infty$.*

*(1) If $(D_n)$ is a sequence of bounded convex sets s.t. $d(D_n) \to \infty$, then if $\overline{X_n} = |D_n|^{-1} \sum_{D_n} X_i$,*

$$\lim_n \overline{X}_n = E(X_0|\mathcal{I}) \quad \text{in } L^p.$$

*(2) If the sequence $(D_n)$ is increasing, the limit occurs $P$ – a.e.*

**(3.1.1) Corollary.** *Let $X$ be an ergodic process, with sample paths in $\Omega$ and $g{:}\Omega \to \mathbb{R}$ integrable. Then*

*(a) If $(D_n)$ is a sequence of bounded convex sets s.t. $d(D_n) \to \infty$,*

$$|D_n|^{-1} \sum_{D_n} g(X \circ \tau_i) \xrightarrow{L^1} E(g(X)).$$

*(b) If $(D_n)$ is increasing, the limit is true a.s.*

# 3.2. Strong Law of Large Numbers and Quadratic Mean Convergence

## *3.2.1. Strong Law of Large Numbers Under $L^2$ Conditions*

Let $X = \{X_i, i \in S\}$ be a real, centered process over $S$, a numerable, infinite set. Consider $(D_n)$ an increasing sequence of finite subsets of $S$ and define $\overline{X}_n$ as the mean of $X$ over $D_n$. Assume also that the sequence of domains satisfy the following conditions:

$(D)$: There exists $\alpha > 0$ and $(m_n)$ a strictly increasing sequence of integers such that

$$\sum_{n \ge 1} n^\alpha |D_{m_n}|^{-1} < \infty, \tag{3.1}$$

$$\sum_{n \ge 1} n^\alpha \left(\frac{|D_{m_{n+1}} \backslash D_{m_n}|}{|D_{m_n}|}\right)^2 < \infty. \tag{3.2}$$

**(3.2.1) Example.** $S = \mathbb{Z}^d$, $D_n = \mathbb{Z}^d \cap nA$ where $A$ is an open convex set with finite measure $a$ and which contains the origin. Then $|D_n| \sim an^d$, $|D_m \backslash D_n| \sim a(m^d - n^d)$ if $m \ge n$. Choose $m_n = n$ and $\alpha < 1$ if $d \ge 2$; $m_n = n^\beta$ with $\beta > 1$ and $0 < \alpha < (\beta - 1) \wedge 2$ if $d = 1$.

**(3.2.1) Theorem.** *Under condition* $(D)$ *and if*

$$(i) \ \sup_n E(X_i^2) = \|X\|_2^2 < \infty, \ (ii) \ \sup_n E(|D_n|\overline{X}_n^2) = M < \infty,$$

$$then \quad \overline{X}_n \xrightarrow{a.e.} 0.$$

*Proof:* For all $\varepsilon > 0$,

$$P(|\overline{X}_{m(n)}| \geq \varepsilon n^{-\alpha/2}) \leq n^\alpha \varepsilon^{-2} |D_{m(n)}|^{-1} M.$$

(3.2) Borel–Cantelli's Lemma yields

$$\lim_n \overline{X}_{m(n)} = 0 \quad a.e.$$

Define $Z_n = \sup\{\overline{X}_k, m(n)+1 \leq k < m(n+1)\}$. Then

$$Z_n \leq \overline{X}_{m(n)} + |D_{m(n)}|^{-1} \sum_{i \in D_{m(n+1)} \backslash D_{m(n)}} |X_i|,$$

so that because of $(i)$ and $(ii)$,

$$E(Z_n^2) \leq u_n = 2M|D_{m(n)}|^{-1} + 2\|X\|_2^2 (|D_{m(n+1)} \backslash D_{m(n)}|/|D_{m(n)}|)^2,$$

$$P(|Z_n| > \varepsilon n^{-\alpha/2}) \leq n^\alpha \varepsilon^{-2} u_n,$$

and condition $(D)$ gives the $a.e.$ convergence of $Z_n$ toward 0. This ends the proof. □

APPLICATION: *Almost everywhere convergence of the empirical mean and covariance of a second-order stationary field.*

$$X = \{X_i, i \in \mathbb{Z}^d\}, \quad \mu = E(X_i), \quad \gamma_\ell = Cov(X_i, X_{i+\ell}).$$

$(a)$ Consider $D_n = [1,n]^d$; then, if

$$\sum |\gamma_\ell| < \infty, \tag{3.3}$$

we have

$$\overline{X}_n \xrightarrow{a.e.} \mu.$$

If $X$ has a spectral density $f$ such that its $d$-$th$ derivative $f^{(d)}_{\lambda_1,\ldots,\lambda_d}$ is square integrable, then (3.3) holds. Indeed, let $I = \{i_1, i_2, \ldots, i_\ell\}$ be a finite subset of $\{1, 2, \ldots, d\}$ : $f_I^{(\ell)}$ is in $L^2$ and thus, so is $R_I$:

$$R_I(k) = \int_{T^d} e^{i<\lambda,k>} f_I^{(\ell)} d\lambda = i^\ell k_{i_1} \cdots k_{i_\ell} \gamma(k).$$

This gives the summability of the $(\gamma(k))$.

(*b*) Consider $Y = \{Y_i = (X_i - \mu)(X_{i+\ell} - \mu), i \in \mathbb{Z}^d\}$. If $X$ is fourth-order stationary and if

$$\sum_k |\gamma_Y(k)| < \infty, \tag{3.4}$$

then $C_n(\ell) = n^{-d} \sum_{i \in D_n : i+\ell \in D_n} (X_i - \overline{X})(X_{i+\ell} - \overline{X}) \xrightarrow{a.e.} \gamma_\ell$.

If $X$ is Gaussian, $\gamma_Y(k) = \gamma(k)^2 + \gamma_{k+\ell}\gamma_{k-\ell}$, and if $X$ has a square integrable spectral density, then (3.4) holds.

REMARK: Assume that for some $\delta > 0$

$$\|X\|_{2+\delta} = \sup_i \|X_i\|_{2+\delta} < \infty, \quad \|X\|_{4+2\delta} = \sup_i \|X_i\|_{4+2\delta} < \infty;$$

then we have the following bounds for the covariances in terms of the mixing coefficients $\alpha_{a,b}(k)$ of $X$ defined in (1.40) ([52], [86])

$$|Cov(X_i, X_j)| \leq 8 \ \|X\|^2_{2+\delta}\alpha_{1,1}(|i-j|)^{\delta/2+\delta},$$

$$|Cov(X_i X_{i+\ell}, X_j X_{j+\ell})| \leq 8 \ \|X\|^4_{4+2\delta}\alpha_{2,2}(|i-j| - \ell)^{\delta/2+\delta}$$

if $(i-j)$ is sufficiently big. From this we deduce that conditions

$$\sum_{m \geq 1} m^{d-1}\alpha_{1,1}(m)^{\delta/2+\delta} < \infty, \quad \sum_{m \geq 1} m^{d-1}\alpha_{2,2}(m)^{\delta/2+\delta} < \infty$$

yield, respectively, (3.3) and (3.4). In particular, in order for $\overline{X}_n$ to converge toward $\mu$, it is not necessary to assume second-order stationarity, nor is it necessary to assume fourth-order stationarity for the convergence of the empirical covariance.

### *3.2.2. Quadratic Mean Convergence for Rectangular $D_n$*

Assume $X$ is a second-order stationary field over $\mathbb{Z}^d$ with spectral measure $F$. Call the absolutely continuous part of $F$, $f(\lambda)d\lambda$. Assume also that $D_n = \prod_{1=1,\alpha} [1, n_i]$ and $n \longrightarrow \infty$ if each $n_i \longrightarrow \infty$.

**(3.2.2) Theorem.**

(1) $\overline{X}_n \longrightarrow \mu$ *in* $L^2$ *if and only if* $F$ *does not charge* $\{0\}$.

(2) *If* $F$ *is absolutely continuous and if* $f$ *is bounded and continuous in* 0,

$$\lim_{n \to \infty} n_1 \cdots n_d Var(\overline{X}_n) = (2\pi)^d f(0).$$

(3) *If additionally $X$ is Gaussian,*

$$\sqrt{n}(\overline{X}_n - \mu) \xrightarrow{\mathcal{D}} \mathcal{N}(0, (2\pi)^d f(0)).$$

*Proof*:

(1) Because of the spectral isometry,

$$Var\overline{X}_n = \frac{(2\pi)^d}{n_1 \cdots n_d} \int_{T^d} \phi_n(\lambda) F(d\lambda)$$

with $\phi_n$ the product of unidimensional Fejer kernels

$$\phi_n(\lambda) = \prod_{i=1}^{d} \varphi_{n_i}(\lambda_i), \quad \varphi_n(\lambda) = \frac{1}{2\pi n}\left(\frac{\sin\frac{n\lambda}{2}}{\sin\frac{\lambda}{2}}\right)^2.$$

From this, we can deduce that for a $\delta > 0$ and sufficiently big $n$,

$$F(\{0\}) \le Var\overline{X}_n \le F([-\delta,\delta]^d) + \frac{(2\pi)^d}{(n_1 \cdots n_d(\sin\frac{\delta}{2})^d)^2}.$$

(2) Assume $f \le M$, with modulus of continuity $\delta_\varepsilon$ in 0,

$$\int_{T^d} \phi_n(\lambda) f(\lambda) d\lambda = \int_{T^d} \phi_n(\lambda)[f(\lambda) - f(0)] d\lambda + (2\pi)^d f(0).$$

The first term tends to 0 with $\varepsilon$ because it is bounded by

$$\varepsilon \int_{[-\delta(\varepsilon),\delta(\varepsilon)]^d} \phi_n(\lambda) d\lambda + M \int_{T^d \setminus [-\delta(\varepsilon),\delta(\varepsilon)]^d} \phi_n(\lambda) d\lambda.$$

(3) Is direct, as $\overline{X}_n$ is Gaussian. □

If $X$ is not Gaussian, but suitably mixing, the asymptotic normality remains true (cf. 3.3).

## 3.3. Central Limit Theorem for Fields

Let $X = \{X_i, i \in \mathbb{Z}^d\}$ be a centered, real field (we shall not assume stationarity as yet) and $(D_n)$ a sequence of strictly increasing finite domains of $\mathbb{Z}^d$. Let $S_n = \sum_{D_n} X_i$ and let $\sigma_n^2$ be the variance of $S_n$. For $k, \ell \in \mathbb{N} \cup \{\infty\}$, we consider the mixing coefficient $\alpha_{k,\ell}(n)$ of $X$ defined in (1.40):

$$\alpha_{k,\ell}(n) = \sup\{ \ |P(A \cap B) - P(A)P(B)|, \ A \in \mathcal{F}(X, \Lambda_1),$$
$$B \in \mathcal{F}(X, \Lambda_2), \ |\Lambda_1| \le k, \ |\Lambda_2| \le \ell, \ \text{dist}(\Lambda_1, \Lambda_2) \ge n\}.$$

The stationary version of the following theorem is due to Bolthausen [21].

**(3.3.1) Theorem.** *Assume that the following conditions are satisfied:*

(*i*) $\sum_{m\geq 1} m^{d-1}\alpha_{k,\ell}(m) < \infty$ *if* $k+\ell \leq 4$ *and* $\alpha_{1,\infty}(m) = o(m^{-d})$.

(*ii*) *There exists* $\delta > 0$ *s.t.*

$$\|X\|_{2+\delta} = \sup_i \|X_i\|_{2+\delta} < \infty \quad \text{and} \sum_{m\geq 1} m^{d-1}\alpha_{1,1}(m)^{\delta/2+\delta} < \infty.$$

*Then* $\limsup_n |D_n|^{-1} \sum_{i,j\in D_n} |cov(X_i,X_j)| < \infty$. *Assume additionally that*

(*iii*) $\liminf_n |D_n|^{-1}\sigma_n^2 > 0$. *Then*

$$\sigma_n^{-1} S_n \xrightarrow{\mathcal{D}} \mathcal{N}(0,1).$$

REMARKS:

(1) If we want to consider only one mixing coefficient, we choose $\alpha_{2,\infty}$ and condition $\sum_{m\geq 1} m^{d-1}\alpha_{2,\infty}(m)^{\delta/2+\delta} < \infty$. Notice that these conditions are satisfied by a Gibbs field under Dobrushin's unicity condition (Theorem 2.1.3) and by a stationary Gaussian field with a sufficiently regular density (Theorem 1.7.2).

(2) If $S \subseteq \mathbb{R}^d$ is a nonfinite numerable, not necessarily regular set such that for each $i$ the ball centered in $i$ with radius $m$ satisfies $|B(i,m) \cap S| = 0(m^d)$ uniformly in $i$, then the Theorem remains true.

(3) If $X$ is multidimensional, $\Sigma_n = Var\ S_n$ and if (*iii*) is replaced by

$$\liminf_n |D_n|^{-1}\Sigma_n \geq I_0 > 0$$

with $I_0$ a positive definite matrix, then

$$\Sigma_n^{-1/2} S_n \xrightarrow{\mathcal{D}} \mathcal{N}(0,I).$$

*Proof:* The first result follows from

$$|cov(X_i,X_j)| \leq 8\alpha_{1,1}(|i-j|)^{\frac{\delta}{2+\delta}} \|X\|_{2+\delta}^2. \tag{3.5}$$

Consider for $N > 0$ the truncated functions (cf. [93])

$$f_N(x) = (x \wedge N) \vee (-N),\ \widetilde{f}_N(x) = x - f_N(x).$$

If $N \longrightarrow \infty$,

$$E[\sigma_n^{-1}\sum_{D_n}(\widetilde{f}(X_i) - E\widetilde{f}(X_i))]^2 = \sigma_n^{-2} \sum_{i,j\in D_n} cov(\widetilde{f}_N(X_i),\widetilde{f}_N(X_j)) \longrightarrow 0.$$

Thus, it is enough to consider only the case of bounded variables (additionally for bounded variables, condition $(i)$ is sufficient). We have the following Lemma:

**(3.3.1) Lemma.** (Stein, [152]) *Let $(\nu_n)$ be a sequence of probabilities over $\mathbb{R}$ which satisfies*

$(i)$ $\sup_n \int x^2 \nu_n(dx) < \infty$;

$(ii)$ *for all* $\lambda \in \mathbb{R}$, $\lim_n \int (i\lambda - x)e^{i\lambda x}\nu_n(dx) = 0$; *then*

$$\nu_n \xrightarrow{\mathcal{D}} \mathcal{N}(0,1).$$

*Proof of Lemma:* Condition $(i)$ assures that $(\nu_n)$ is tight. Let $\nu$ be an accumulation point; then

$$\int x^2 \nu(dx) \leq \sup_n \int x^2 \nu_n(dx) < \infty$$

and if $\nu_{n_k} \longrightarrow \nu$, $\sup_n \int x\nu_{n_k}(dx) = \int x\nu(dx)$. So that

$$\int (i\lambda - x)e^{i\lambda x}\nu(dx) = 0,$$

which ends the proof. □

*Continuing with the Proof of the Theorem:* $\alpha_{k,\ell}(m)$, decreasing in $m$, is $o(m^{-d})$ if $k + \ell \leq 4$; we can choose a sequence $(m_n)$ such that for $n \to \infty$,

$$\alpha_{k,\ell}(m_n)|D_n|^{1/2} \to 0 \quad \text{and} \quad m_n^{-d}|D_n|^{1/2} \to \infty.$$

For $i \in \mathbb{Z}^d$, define

$$S_{i,n} = \sum_{j \in D_n, d(i,j) \leq m_n} X_j \ , \quad S_{i,n}^* = S_n - S_{i,n},$$

$$a_n = \sum_{i \in D_n} E(X_i S_{i,n}) \ , \ \overline{S}_n = a_n^{-1/2} S_n \ , \ \overline{S}_{i,n} = a_n^{-1/2} S_{i,n}.$$

We have $\sigma_n^2 = Var\ S_n = a_n + \sum_{i \in D_n} E(X_i \cdot S_{i,n}^*)$, where

$$\left|\sum_{i \in D_n} E(X_i \cdot S_{i,n}^*)\right| \leq \sum_{\substack{i,j \in D_n \\ d(i,j) > m_n}} |Cov(X_i, X_j)|$$

$$\leq 8\|X\|_\infty^2 |D_n| \sum_{m > m_n} m^{d-1} \ \alpha_{1,1}(m)^{\delta/2+\delta},$$

is a $o(|D_n|)$ so that because of $(iii)$, $\sigma_n^2 = a_n(1+o(1))$. Thus, it is enough to show the asymptotic normality of $\overline{S}_n$; since $\sup_n E\overline{S}_n^2 < \infty$, this will follow from condition

$$\lim_{n\to\infty} E((i\lambda - \overline{S}_n)e^{i\lambda\overline{S}_n}) = 0.$$

However, we observe that $(i\lambda - \overline{S}_n)e^{i\lambda\overline{S}_n} = A_1 - A_2 - A_3$, with

$$\begin{cases} A_1 = i\lambda e^{i\lambda\overline{S}_n}\Big(1 - a_n^{-1}\sum_{j\in D_n} X_j S_{j,n}\Big), \\ \\ A_2 = a_n^{-1/2} e^{i\lambda\overline{S}_n}\sum_{j\in D_n} X_j(1 - i\lambda\overline{S}_{j,n} - e^{-i\lambda\overline{S}_{j,n}}), \\ \\ A_3 = a_n^{-1/2}\sum_{j\in D_n} X_j \cdot e^{i\lambda(\overline{S}_n - \overline{S}_{j,n})}. \end{cases}$$

*Asymptotic behavior of* $A_1$:

$$\begin{aligned} E|A_1|^2 &= \lambda^2 a_n^{-2} Var\Big(\sum_{j\in D_n} X_j S_{j,n}\Big) \\ &= \lambda^2 a_n^{-2} \sum_{\substack{j,j',\ell,\ell'\in D_n \\ d(j,\ell) \text{ and } d(j',\ell')\le m_n}} cov(X_j X_\ell, X_{j'}X_{\ell'}). \end{aligned}$$

If $d(j,j') = k \geq 3m_n$, $|cov(X_jX_\ell, X_{j'}X_{\ell'})| \leq 8\ \alpha_{2,2}(k - 2m_n)$.
If $\inf\{d(j,j'), d(j,\ell), d(j,\ell')\} = i$,

$$\begin{aligned} |cov(X_jX_\ell, X_{j'}X_{\ell'})| &\leq |E(X_jX_\ell X_{j'}X_{\ell'})| + |E(X_jX_\ell)| + |E(X_{j'}X_{\ell'})| \\ &\leq 8\ \alpha_{1,3}(i) \end{aligned}$$

Thus, we find

$$\begin{aligned} E|A_1|^2 &\leq c'\lambda^2 a_n^{-2}|D_n|\Big(m_n^{2d}\sum_{k=3m_n}^{\infty} k^{d-1}\alpha_{2,2}(k - 2m_n) + m_n^{2d}\sum_{i=0}^{3m_n} i^{d-1}\alpha_{1,3}(i)\Big) \\ &\leq \lambda^2 O(a_n^{-2}|D_n|m_n^{2d}) = \lambda^2 O(|D_n|^{-1}m_n^{2d}) = \lambda^2 o(1) \end{aligned}$$

since because of $(iii)$, $a_n^{-1} \leq c|D_n|^{-1}$ for a certain $c > 0$ and $n$ sufficiently big.

*Asymptotic behavior of* $A_2$: $|\overline{S}_{j,n}| \leq c|D_n|^{-1/2}m_n^d \longrightarrow 0$ if $n \longrightarrow \infty$ so that,

$$|1 - i\lambda\overline{S}_{j,n} - e^{-i\lambda\overline{S}_{j,n}}| \leq c\lambda^2\overline{S}_{j,n}^2.$$

This in turn yields

$$E|A_2| \leq c\lambda^2 a_n^{1/2} \sup_{j\in D_n} E(\overline{S}_{j,n}^2) \leq c' a_n^{-1/2} \sum_{\substack{i,i'\in D_n \\ d(i,j) \text{ and } d(i',j)\leq m_n}} |E(X_i X_{i'})|$$

$$\leq c'' a_n^{-1/2} m_n^d = o(1).$$

*Asymptotic behavior of* $A_3$: $E|A_3| \leq c a_n^{-1/2} |D_n| \alpha_{1,\infty}(m_n) = o(1)$. This ends the proof. □

APPLICATION: *Asymptotic Normality of the Moran Indicator* [33].

Consider $S \subset \mathbb{R}^d$ an infinite nonnecessarily regular lattice without accumulation points. Let $W$ be a matrix of known bounded weights over $S^2$, such that

$$W = (w_{ij}, i, j \in S), \quad w_{i,i} = 0, \quad w_{i,j} = 0 \quad \text{if } \|i - j\| > R. \tag{3.6}$$

We would like to test, on the basis of $W$, the spatial noncorrelation of a field $X$ over $S$. In order to do this, consider the following index defined over an increasing sequence of finite subsets $(D_n)$:

$$I_n = \sum_{i\in \overset{\circ}{D}_n} \sum_j w_{ij} X_i X_j. \tag{3.7}$$

Here, $\overset{\circ}{D}_n = \{k \in D_n \ t.q. \text{ for all } \ell \ t.q. \ w_{k,\ell} \neq 0, \ \ell \in D_n\}$. Consider:

$(H_0)$ : The variables $X_i$ are centered, independent and such that there exists $\delta > 0$ with $\|X\|_{2+\delta} = \sup_i \|X_i\|_{2+\delta} < \infty$.

Finally, for $c_n = \sum_{i\in \overset{\circ}{D}_n} \sum_j (w_{ij}^2 + w_{ij} w_{ji}) \sigma_i^2 \cdot \sigma_j^2$ with $\sigma_i^2 = Var\ X_i$, assume

$$\liminf_n c_n | \overset{\circ}{D}_n |^{-1} > 0. \tag{3.8}$$

**(3.3.2) Theorem.** *Under conditions* (3.6), (3.8) *and* $(H_0)$,

(*a*) $c_n^{-1/2} I_n \xrightarrow{\mathcal{D}} \mathcal{N}(0,1)$.

(*b*) *Define* $c_n^* = \sum_{i\in \overset{\circ}{D}_n} \sum_j (w_{ij}^2 + w_{ij} w_{ji})$; *if* $\sigma_i^2 = \sigma^2$ *for all* $i$, *then*

$$(c_n^*)^{-1/2} \left(\sum_{D_n} X_i^2\right)^{-1} I_n \xrightarrow{\mathcal{D}} \mathcal{N}(0,1).$$

*Proof:* $I_n = \sum_{\overset{\circ}{D}_n} X_i V_i$ with $V_i = \sum_j w_{ij} X_j$. The variables

$$\{Z_i = X_i V_i,\ i \in S\}$$

are independent if $\|i-j\| > 2R$, centered, belong to $L_{2+\delta}$ and $Var\ I_n = c_n$. The result is a direct consequence of Theorem 3.3.1. □

REMARKS: If the mean $\mu$ of variables $X_i$ is unknown, consider

$$I_n^M = \frac{|D_n| \sum\limits_{i,j \in D_n} w_{ij} Z_i Z_j}{\sum\limits_{D_n} Z_i^2}, \qquad Z_i = (X_i - \overline{X}).$$

An analogous result can be obtained for Geary's index (cf. [33]):

$$J_n = \sum_{i,j \in D_n} w_{ij} (X_i - X_j)^2.$$

If $X$ is mixing, the asymptotic normality of $I_n$ or $J_n$ can be easily obtained. This is not so with their asymptotic variance.

*Convergence Rate for the Central Limit Theorem (CLT).* Assume that for $k \le \ell$, $k < \infty$ we have, for finite $M$, $N$,

$$\alpha_{k,\ell}(m) \le M k^N \exp(-am), \quad N \ge 0, \quad a > 0$$

and that there exists $\delta > 0$ such that

(3.9) $$\|X\|_{8+\delta} = \sup_i \|X_i\|_{8+\delta} < \infty.$$

If $\Phi(x)$ is the standard normal distribution function,

(3.10) $$\Delta_n = \sup_{x \in \mathbb{R}} |P(\sigma_n^{-1} S_n \le x) - \Phi(x)| = 0(\sigma_n^{-1} (\log \sigma_n)^d).$$

([159], $N \ge 1$; [84] $N \ge 0$, and $\|X\|_{4+\delta} < \infty$). We can also calculate the convergence rate when the mixing is polynomially decreasing ([84]). If the field is $m$-dependent ($X_i$ and $X_j$ are independent if $\|i-j\| > m$), we have ([84])

$$\Delta_n = 0(\sigma_n^{-1} (\log \sigma_n)^{\frac{d-1}{2}}).$$

If for $d = 1$ this is the optimal rate, we would like to know if for $d \ge 2$, $\sigma_n^{-1}$ is also the optimal rate.

*CLT for Functionals of Ergodic Markovian Fields.* Consider $X$ an ergodic Markovian field over $\mathbb{Z}^d$ with range $m$, $X_t \in E \equiv \mathbb{R}$. The next property provides a $CLT$ for certain functionals of $X$ based only on the ergodicity condition. Define

$$V_t = (X_s, s \in \partial t) \in E^M, \ \partial t = \{s \in \mathbb{Z}^d, \ \|s - t\| \leq m\},$$

with cardinal M . Let $a$ be a bounded measurable function $a : E^{M+1} \longrightarrow \mathbb{R}$, such that if

(3.11) $$Y_t = a(X_t, V_t),$$

we have

(3.12) $$E(Y_t \mid X_s, \ s \neq t) = 0.$$

Let $(D_n)$ be an increasing sequence of bounded convex sets of $\mathbb{Z}^d$, and

$$S_n = \sum_{D_n} Y_s, \quad S_{t,n} = \sum_{\substack{s \in D_n \\ \|s-t\| \leq m}} Y_s.$$

**(3.3.3) Theorem.** ([83]) *Let $X$ be a real ergodic Markovian field and $Y$ the field defined by* (3.11) *that satisfies* (3.12)*. Then, if*

$$(i) \quad \sigma^2 = \sum_{\|t\| \leq m} E(Y_0 Y_t) > 0, \quad (ii) \quad \frac{|\partial D_n|}{|D_n|} \longrightarrow 0,$$

*we have*

$$|D_n|^{-1/2} S_n \xrightarrow{\mathcal{D}} \mathcal{N}(0, \sigma^2).$$

*Proof:* We define $a_n$, $\overline{S}_n$, and $\overline{S}_{n,t}$ as in Theorem 3.3.1. Under $(ii)$, $a_n = Var\ S_n \sim |D_n| \sum_{\|t\| \leq m} E(Y_0 Y_t)$. On the other hand, $Y_u$ is a function of $X_s$, $s \neq t$ if $\|t - u\| > m$. The proof consists of modifying the proof of Theorem 3.3.1 as follows: whenever we use the mixing property, we change this for the "martingale type" condition (3.12). We have to show that each $E(A_i)$, $i = 1, 3$ tends to 0.

- $$E(A_3) = a_n^{-1/2} \sum_{D_n} E[Y_t \exp i\lambda(\overline{S}_n - \overline{S}_{t,n})] = 0$$

is obtained by first conditioning, for each $t$, by $(X_s, s \neq t)$.

•• $$E(|A_2|) \leq \ C \ a_n^{-3/2} \sum_{D_n} E[S_{t,n}^2] \longrightarrow 0$$

since $S_{t,n}$ is bounded and $a_n \sim \sigma^2 |D_n|$.

••• $$E(|A_1|) \leq \ C \ E[|a_n^{-1} \sum_{D_n} (X_t S_{t,n} - E(X_t S_{t,n}))|]$$

$$\leq \ C \ E\Big[|D_n|^{-1} \Big| \sum_{D_n} (X_t R_t - E(X_t R_t)) \Big| \Big] + o(1)$$

with $R_t = \sum_{s: \|s-t\| \leq m} Y_s$ because $R_t = S_{t,n}$, except over the boundary of $D_n$.

Using the $L^1$ version of the Ergodic Theorem, we get $E(|A_1|) \longrightarrow 0$. □

REMARK: Condition $(i)$ is generally difficult to prove; it is satisfied by the isotropic Ising model over $\mathbb{Z}^2$ with four nearest neighbors (cf. [83]).

*CLT for Gibbs Fields.*
Asymptotic independence and a $CLT$ can be obtained for real valued positively dependent fields (Newman [121], [122], [123]). For attractive Gibbs fields, the Fortuin–Kastelyn–Ginibre (FKG) inequality conveys exactly this positive dependence [133].

*CLT for a Positively Dependent Real Field.*
Consider $x, y \in \mathbb{R}^m$ with the order relationship $x \leq y$ if and only if $x_k \leq y_k$, $k = 1, m$. Let $X = \{X_i, i \in \mathbb{Z}^d\}$ be a strictly stationary real field of $L^2$; we will say it is *positively dependent* if for every finite subset $S$ of $\mathbb{Z}^d$, and all increasing functions $f$, $g$ from $\mathbb{R}^S$ to $\mathbb{R}$ , $cov(f(X(S)),\ g(X(S))) \geq 0$. Let $B_0^n = \{1, 2, \ldots, n\}^d$, $B_k^n = nk + B_0^n$ for $k \in \mathbb{Z}^d$:

$$X_k^{(n)} = \frac{S_k^n - ES_k^n}{n^{d/2}}, \text{ where } S_k^n = \sum_{B_k^n} X_j.$$

**(3.3.4) Theorem.** *Let $X$ be a real, strictly stationary, positively dependent field such that if $A = \sum_{\mathbb{Z}^d} cov(X_0, X_k)$, $0 < A < \infty$. Then*

$$\{X_k^{(n)}, k \in \mathbb{Z}^d\} \xrightarrow{\mathcal{D}} \{Z_k, k \in \mathbb{Z}^d\}$$

*in weak finite dimensional convergence. Here, $\{Z_k\}$ stands for a Gaussian white noise $N(0,A)$.*

*Example of a Real Positively Dependent Field: The FKG Inequality* [133].
Let $\mu$ be a probability measure over $E^S$ with $E \subseteq \mathbb{R}$. Let $f$ be its density with respect to a given reference measure. Then $\mu$ is said to be *attractive* if its density $f$ satisfies

$$f(x \vee y) f(x \wedge y) \geq f(x) f(y),$$

where $x \vee y$ (resp. $x \wedge y$) is the configuration of the supremum (the infimum). If $f = Z^{-1} \exp(H)$, the *potential is attractive* if

$$H(x \vee y) + H(x \wedge y) \geq H(x) + H(y).$$

*Fundamental Example: Quadratic Energy.* If

$$H(x) = \sum_{<i,j>} J_{ij} x_i x_j + \sum_i h_i x_i,\ x_i \in \mathbb{R},$$

the potential is attractive if for every $i, j$, $J_{ij} \geq 0$.

**(3.3.5) Theorem.** FKG Inequalities. *If $\mu$ is attractive over $\Omega(S)$, then $\mu$ is positively dependent, that is, for any two real increasing functions $f$, $g$ over $\Omega(S)$, we have*

$$cov_{\mu}(g(X(S)), h(X(S))) \geq 0.$$

Combining the last two results, we can obtain a functional $CLT$ for real, attractive, square integrable and strictly stationary Gibbs fields. Newman [122] generalized this result to nonmonotonal functions that satisfy the FKG inequality.

## 3.4. Quasi-Likelihood or Minimum Contrast Estimation

Let $X$ be a field over $S$, observed over a sequence $(D_n)$ of increasing finite domains. Call $X(n) = \{X_i, i \in D_n\}$. Consider a parametric model of distributions $P_\theta$, $\theta \in \Theta$, a compact subset of $\mathbb{R}^p$. Assume the true value $\theta_0 \in \overset{\circ}{\Theta}$, the interior of $\Theta$. Define $P_0 = P_{\theta_0}$.

• A *contrast function* for $\theta_0$ is a deterministic application $K(\theta_0, \theta)$ $K_{\theta_0} : \Theta \longrightarrow \mathbb{R}^+$, which has a unique minimum in $\theta = \theta_0$.

•• A *contrast process* for $K_{\theta_0}$ is a sequence of random variables $(U_n(\theta), n \in \mathbb{N})$ adapted to $(X(n))$ defined for all $\theta \in \Theta$ and such that

$$\liminf_{n\to\infty} [U_n(\theta) - U_n(\theta_0)] \geq K(\theta_0, \theta) \quad \text{in } P_0 - \text{probability}. \tag{3.13}$$

The *minimum contrast estimator* is a value which realizes that minimum value of $U_n$,

$$\widehat{\theta}_n = \operatorname*{Arg\,min}_{\theta\in\Theta} U_n(\theta) \tag{3.14}$$

### 3.4.1. Consistency

**(3.4.1) Theorem.** *Assume $\theta \longmapsto K(\theta_0,\theta)$, $\theta \longmapsto U_n(\theta)$ are $P_0 -$ a.e. continuous. If $W_n(\eta)$ is the modulus of continuity of $U_n(\theta)$,*

$$W_n(\eta) = \sup\{|U_n(\alpha) - U_n(\beta)|, \alpha, \beta \in \Theta,\ \|\alpha - \beta\| \leq \eta\},$$

*assume there exists $\varepsilon_k \longrightarrow 0$ s.t. for each $k$,*

$$\lim_{n\to\infty} P_0\Big(W_n\Big(\frac{1}{k}\Big) \geq \varepsilon_k\Big) = 0. \tag{3.15}$$

*Then $(\widehat{\theta}_n)$ is consistent: $\lim_n \widehat{\theta}_n = \theta_0$ in $P_0$-probability.*

REMARK: When (3.13) is replaced by the ergodic condition,

$$\lim_n [U_n(\theta) - U_n(\theta_0)] = K(\theta_0, \theta),$$

this corresponds to Theorem 3.2.8 of Dacunha–Castelle and Duflo [41]. For Gibbs fields, ergodicity in general cannot be assured.

*Proof:* We proceed as in [41], replacing a *lim* by a *lim inf*. As $\Theta$ has a dense numerable subset $D$, $\inf_{\theta\in\Theta} U_n(\theta) = \inf_{\theta\in D} U_n(\theta)$ and $W_n(\eta)$ are measurable. Let $B$ be a nonempty open ball centered in $\theta_0$, $\varepsilon > 0$ *s.t.* $K(\theta_0,\theta) \geq 2\varepsilon$ over $\Theta\backslash B$ and $k$ *s.t.* $\varepsilon_k < \varepsilon$. Cover $\Theta\backslash B$ with $N$ balls centered in $\theta_i \in \Theta\backslash B$ with radius $1/k$, then

$$\begin{aligned}\{\widehat{\theta}_n \notin B\} &\subset \{\inf_{\Theta\backslash B} U_n(\theta) < U_n(\theta_0)\}\\ &\subset \{\inf_{i=1,N} (U_n(\theta_i) - U_n(\theta_0)) < W_n\Big(\frac{1}{k}\Big)\}\\ &\subset \{\inf_{i=1,N} (U_n(\theta_i) - U_n(\theta_0)) < \varepsilon\} \cup \Big\{W_n\Big(\frac{1}{k}\Big) > \varepsilon\Big\}.\end{aligned}$$

From (3.13) we can deduce the following lower bound:

$$\begin{aligned}\liminf_n \big(\inf_{i=1,N} (U_n(\theta_i) - U_n(\theta_0))\big) &= \inf_{i=1,N} \big(\liminf_{n\to\infty} (U_n(\theta_i) - U_n(\theta_0))\big)\\ &\geq \inf_{i=1,N} K(\theta_0,\theta_i) \geq 2\varepsilon \quad \text{in } P_0\text{-probability.}\end{aligned}$$

Thus, with probability tending to 1, $\inf_{i=1,N} (U_n(\theta_i) - U_n(\theta_0)) \geq \varepsilon$ and condition (3.15) over $W_n$ ends the proof. □

**(3.4.1) Corollary.** *If* $U_n = \sum_{k=1}^{p} a_{n,k} U_{n,k}$, $a_{n,k} \geq 0$ *is such that*

(1) $U_{n,k}$, $k = 1, p$ *satisfy* (3.15).

(2) $U_{n,1}$ *is a contrast process with respect to a continuous contrast function* $\theta \longmapsto K(\theta_0,\theta)$.

(3) $\liminf a_{n,1} = a > 0$.

*Then the minimum contrast estimator for* $U_n$ *is consistent.*

*Proof:* It is enough to check that $U_n$ is a contrast process with respect to $aK(\theta_0,\theta)$, which satisfies (3.15). □

This result allows us to understand why the consistency of the conditional pseudo-likelihood estimator for a Markovian field is a consequence of the consistency of coding estimation (cf. § 5.2.3).

**(3.4.2) Theorem.** Two-Stage Estimation. *Assume* $\Theta = \mathcal{A} \times \mathcal{B}$, $\theta = (a,b)$, $\theta_0 = (a_0,b_0)$. *Let* $(\widehat{a}_n)$ *be a sequence of consistent estimators of* $a$ *and*

$$\widehat{b}_n = \operatorname*{ArgMin}_{b \in \mathcal{B}} U_n(\widehat{a}_n, b).$$

*Then, under the hypothesis of Theorem* 3.4.1,

$$\lim \widehat{b}_n = b_0 \quad \text{in } P_0\text{-probability.}$$

*Proof:* $V_n(b) = U_n(\widehat{a}_n, b)$ is a continuous contrast process with respect to the continuous contrast function

$$b \longmapsto K_{a_0}(b_0, b) = K((a_0, b_0), (a_0, b)).$$

On the other hand, with $P_0$-probability,

$$\begin{aligned}\liminf_n \, (V_n(b) - V_n(b_0)) &= \liminf_n \, (U_n(\widehat{a}_n, b) - U_n(\widehat{a}_n, b_0)) \\ &= \liminf_n \, (U_n((a_0, b)) - U_n((a_0, b_0))) \geq K((a_0, b_0), (a_0, b)) = K_{a_0}(b_0, b)\end{aligned}$$

because of the continuity of the $U_n$, the consistency of $\widehat{a}_n$, and condition (3.13) satisfied by $U_n$. Finally, if $W_n^V$ is the modulus of continuity of $V_n$, then $0 \leq W_n^V \leq W_n$. This ends the proof. □

The standard context for the application of this kind of two-stage estimation procedure is that of regression (linear or not, parameters $a$), with conditional residuals (parameters $b$).

*Strong Consistency.* Define conditions (3.13) and (3.15) in their *a.e.* version:

(3.13)′ $$\liminf_{n \to \infty} \, (U_n(\theta) - U_n(\theta_0)) \geq K(\theta_0, \theta) \; P_0 - a.e.$$

(3.15)′ $$P_0(\limsup_{n \to \infty} \, (W_n\left(\frac{1}{k}\right) \geq \varepsilon_k)) = 0$$

**(3.4.3) Theorem.** Strong Consistency. *If in Theorem* 3.4.1 *conditions* (3.13) *and* (3.15) *are replaced by their* $P_0 - a.e.$ *versions* (3.13)′ *and* (3.15)′, *then* $(\widehat{\theta}_n)$ *is strongly consistent*

$$\lim \widehat{\theta}_n = \theta_0 \quad P_0 - a.e.$$

*Proof:* Reconsider the proof of Theorem 3.4.1.

$$(3.13)' \Longrightarrow P_0(C) = 1 \text{ where } C = \{\liminf_n \left( \inf_{i=1,N} (U_n(\theta_i) - U_n(\theta_0)) \right) \geq 2\varepsilon\},$$

$$(3.15)' \Longrightarrow P_0(\overline{D}_k) = 0 \text{ with } \overline{D}_k = \{W_n\left(\frac{1}{k}\right) \geq \varepsilon_k, \text{ infinitely often } (i.o.)\}.$$

Let $\omega \in C \cap D_k$ ($D_k$ is the complement of $\overline{D}_k$); then there exists $N(\omega, \varepsilon)$ *s.t.* for all $n \geq N(\omega, \varepsilon)$,

$$\inf_{i=1,N} (U_n(\theta_i) - U_n(\theta_0)) \geq \varepsilon > \varepsilon_k > W_n\left(\frac{1}{k}\right).$$

Also, we have

$$\begin{aligned}\{\widehat{\theta}_n \notin B \; i.o.\} &\subseteq \{\liminf_{i=1,N} (U_n(\theta_i) - U_n(\theta_0)) \leq W_n\left(\frac{1}{k}\right) \; i.o.\} \\ &\subseteq \overline{C \cap D_k}.\end{aligned}$$

That is, $P(\widehat{\theta}_n \notin B \; i.o.) = 0$. Since $B$ is arbitrary, this yields the strong consistency. □

**(3.4.4) Theorem.** Case of a Convex Contrast ([148]). *Assume that* $\Theta$ *is an open convex of* $\mathbb{R}^p$, *and*

(*a*) $\lim_n (U_n(\theta) - U_n(\theta_0)) = K(\theta_0, \theta) \; P_0 - a.e.$

(*b*) $\theta \longmapsto U_n(\theta)$ *is convex for every* $\omega$.

*Then* $\widehat{\theta}_n \longrightarrow \theta_0 \quad P_0 - a.e.$

This result is a consequence of the following Proposition:

PROPOSITION. — *Let* $\Theta$ *be an open convex of* $\mathbb{R}^p$ *and* $f_n : \Theta \to \mathbb{R}$ *a sequence of convex functions which converge simply toward* $f$. *Assume that the minimum of* $f$ *is reached only at* $\theta_0$.

*Then if the minimum of* $f_n$ *is reached at* $\theta_n$, $\theta_n \to \theta_0$.

*Proof:* We know that the convergence of $f_n$ toward $f$ is uniform over compacts and that $f$ is convex [138]. Let $r > 0$ be *s.t.* $B(\theta_0, r) \subset \Theta$. Assume that $\theta_n \not\longrightarrow \theta_0$; then there exists a subsequence with $\|\theta_n - \theta_0\| \to a$, $0 < a \leq \infty$. Define $\beta_n = u_n\theta_n + (1 - u_n)\theta_0$ with $u_n = \rho / \|\theta_n - \theta_0\|$; then $\|\beta_n - \theta_0\| = \rho$ where we have chosen $\rho = r$ if $a = \infty$, $0 < \rho < \inf(r, a)$ if $a < \infty$. For $n$ sufficiently big, $0 < u_n < 1$. Let $\beta$ be an accumulation point of $(\beta_n) : \beta \neq \theta_0$, $0 < u_n < 1$

$$\begin{aligned}f_n(\beta_n) &\leq u_n f_n(\theta_n) + (1 - u_n) f_n(\theta_0) \\ &\leq u_n f_n(\theta_0) + (1 - u_n) f_n(\theta_0) = f_n(\theta_0).\end{aligned}$$

Since the convergence of $f_n$ toward $f$ is uniform in $B(\theta_0, r)$, we have

$$\lim f_n(\beta_n) = f(\beta) \quad \text{and} \quad f(\beta) \leq \lim f_n(\theta_0) = f(\theta_0)$$

for the subsequence which converges toward $\beta$. In other words, $f(\beta) \leq f(\theta_0)$, $\|\beta - \theta_0\| = \rho$, which contradicts the unicity of the point where $f$ reaches its minimum. □

## *3.4.2. Normality and Asymptotic Tests*

If $h$ belongs to $\mathcal{C}^2$ at $\theta$, we will define $h^{(1)}(\theta)$ as the vector of first derivatives in $\theta$, and $h^{(2)}(\theta)$ as the matrix of second derivatives in $\theta$. If $A$ and $B$ are two $p \times p$ symmetric matrices, define

- $\|A - B\| = \sum_{1 \leq i,j \leq p} |A_{ij} - B_{ij}|$.

•• $A \geq B$ $(A > B)$ if $(A - B)$ is semipositive definite (*s.p.d.*) (positive definite (*p.d.*).

••• Assume $A \geq 0$ is a symmetric *p.d.* matrix which can be written as $A = PDP'$ with $P$ orthogonal and $D$ diagonal. Define $A^{1/2}$ to be the square root $PD^{1/2}P'$.

Consider the following conditions:

(H1) There exists a neighborhood $V$ of $\theta_0$ over which $U_n$ is twice continuously differentiable and a $P_0$–integrable random variable $h$ *s.t.* for all $i, j = 1, p$ and all $\alpha \in V$, $|U^{(2)}_{n;\alpha_i,\alpha_j}(\alpha, x)| \leq h(x)$.

(H2) There exists a sequence $(a_n) \to \infty$ such that $J_n = Var(\sqrt{a_n} U_n^{(1)}(\theta_0))$ exists and satisfies

(1) There exists $J > 0$ with $J_n \geq J$ from a certain point on.

(2) $\sqrt{a_n} J_n^{-1/2} U_n^{(1)}(\theta_0) \xrightarrow{\mathcal{D}} \mathcal{N}(O, I_p)$.

(H3) There exists a sequence of nonstochastic $p \times p$ matrices $(I_n)$ such that

(1) There exists $I > 0$, with $I_n \geq I$ from a certain point on.

(2) $\lim_n (U_n^{(2)}(\theta_0) - I_n) = 0$ in $P_0$-probability.

**(3.4.5) Theorem.** Asymptotic Normality of $\widehat{\theta}_n$. *Assume that the estimator $\widehat{\theta}_n$ of the minimum contrast $U_n$ is consistent and that (H1) to (H3) are satisfied. Then*

$$\sqrt{a}_n J_n^{-1/2} I_n(\widehat{\theta}_n - \theta_0) \xrightarrow{\mathcal{D}} \mathcal{N}(0, I_p).$$

*Proof:* Since $\widehat{\theta}_n$ is consistent, $\widehat{\theta}_n \in V$ with $P_0$-probability tending to 1. From Taylor's development with integral residual, we get

$$U_n^{(1)}(\widehat{\theta}_n) = 0 = U_n^{(1)}(\theta_0) + \Delta_n(\theta_0, \widehat{\theta}_n)(\widehat{\theta}_n - \theta_0)$$

$$\text{with} \quad \Delta_n(\theta_0, \widehat{\theta}_n) = \int_0^1 U_n^{(2)}(\theta_0 + t(\widehat{\theta}_n - \theta_0))dt.$$

First we will show that: $\lim_n (\Delta_n(\theta_0, \widehat{\theta}_n) - U_n^{(2)}(\theta_0)) = 0$ in $P_0$-probability. Let $B \subset V$ be a closed ball centered in $\theta_0$. Then

$$\sigma(B, x) = \sup_{\alpha \in B} \|U_n^{(2)}(\theta_0, x) - U_n^{(2)}(\alpha, x)\|$$

is a $r.v.$ bounded by $2p^2 h(x)$. Also, $\sigma(B, x) \to 0$ if $B$'s radius goes to 0. By Lebesgue's Theorem, for all $\varepsilon > 0$, we can choose $B$ $s.t.$:

$$E(\sigma(B, \cdot)) \le \varepsilon.$$

Thus, for big enough $n$,

$$\limsup_n \|U_n^{(2)}(\theta_0) - U_n^{(2)}(\alpha)\| \le \varepsilon \quad \text{in } P_0 \text{ -probability.}$$

Since $\varepsilon$ is arbitrary, this ends the first part of the proof.

Now $\Delta_n(\theta_0, \widehat{\theta}_n)$ is invertible over a set whose probability tends to 1. Over this set we can define

$$A_n = J_n^{-1/2} I_n \Delta_n(\theta_0, \widehat{\theta}_n)^{-1} J_n^{-1/2}.$$

Define $X_n = \sqrt{a_n} J_n^{-1/2} U_n^{(1)}(\theta_0)$, $Y_n = \sqrt{a_n} J_n^{-1/2} I_n(\widehat{\theta}_n - \theta_0) = -A_n X_n$. The Theorem is a consequence of conditions $(H2)$ and $(H3)$. □

In practice, we shall use approximation

$$\sqrt{n}(\widehat{\theta}_n - \theta_0) \sim \mathcal{N}(0, V_n), \quad \text{with} \quad V_n = \widehat{I}_n^{-1} \widehat{J}_n \widehat{I}_n^{-1}$$

$$\widehat{I}_n = U_n^{(2)}(\widehat{\theta}_n), \quad \widehat{J}_n = a_n U_n^{(1)}(\widehat{\theta}_n)\ {}^t U_n^{(1)}(\widehat{\theta}_n).$$

*Asymptotic Difference of Contrasts Tests*

Let $(H_p)$ stand for the hypothesis $\theta \in \Theta \subseteq \mathbb{R}^p$ of dimension $p$ and $(H_q)$, $q < p$, the $q$ dimensional subhypothesis of $(H_p)$ : $(H_q)$ $\theta = r(\alpha)$, $\alpha \in \Lambda$ an open set of $\mathbb{R}^q$, where $r$ belongs to class $\mathcal{C}^2$ over $\Lambda$ and such that

$$\theta_0 = r(\alpha_0), \quad R_0 = \frac{\partial r}{\partial \alpha}(\alpha_0) \text{ is of full rank.}$$

We will define $\overline{U}_n(\alpha) = U_n(r(\alpha))$ the contrast under $(H_q)$, and consider $\widehat{\alpha}_n$ a minimum contrast estimator, $\overline{\theta}_n = r(\widehat{\alpha}_n)$, $I_{n,0} = I_n(\theta_0)$, $J_{n,0} = J_n(\theta_0)$, $\overline{I}_{n,0} = I_n(\alpha_0)$, $\overline{J}_{n,0} = \overline{J}_n(\alpha_0)$ and $\overline{I}_0, \overline{J}_0$ are the matrices in $(H2)$ and $(H3)$ relative to $\overline{U}_n(\alpha)$.

If $(G_n)$ and $(F_n)$ are two sequences of distribution functions, we will say $(G_n)$ and $(F_n)$ are asymptotically equivalent ($G_n \overset{\mathcal{L}}{\sim} F_n$) if for all $x$, $\lim_n (G_n(x) - F_n(x)) = 0$.

**(3.4.6) Theorem.** Asymptotic Difference of Contrasts Test. *Assume $\widehat{\alpha}_n$ is consistent and that $(U_n)$ and $(\overline{U}_n)$ satisfy hypotheses $(H1)$ to $(H3)$. Then*

$$2a_n[U_n(\overline{\theta}_n) - U_n(\widehat{\theta}_n)] \overset{\mathcal{D}}{\sim} F_{n,0} \text{ under } (H_q).$$

*Here, $F_{n,0}$ is in distribution a weighted sum of $(p-q)$ independent $\chi_1^2$,*

$$F_{n,0} \overset{\mathcal{D}}{\sim} \sum_{i=1,p-q} \lambda_{i,n}\chi_{i,1}^2$$

*where $\lambda_{i,n}$, $i = 1,p-q$ are the $(p-q)$ nonzero eigenvalues of*

$$A_{n,0} = J_{n,0}^{1/2} Q_{n,0} J_{n,0}^{1/2}$$

*or equivalently of $J_{n,0}Q_{n,0}$ where $Q_{n,0} = [I_{n,0}^{-1} - R_0\overline{I}_{n,0}^{-1}R_0']$.*

*Proof:* Since $U_n^{(1)}(\widehat{\theta}_n) = \overline{U}_n^{(1)}(\widehat{\alpha}_n) = 0$, there exist $\theta_n^* \in [\theta_0, \widehat{\theta}_n]$, $\alpha_n^* \in [\alpha, \widehat{\alpha}_n]$ such that:

$$U_n(\theta_0) - U_n(\widehat{\theta}_n) = \tfrac{1}{2}(\widehat{\theta}_n - \theta_0)'U_n^{(2)}(\theta_n^*)(\widehat{\theta}_n - \theta_0),$$
$$\overline{U}_n(\alpha_0) - \overline{U}_n(\widehat{\alpha}_n) = \tfrac{1}{2}(\widehat{\alpha}_n - \alpha_0)'U_n^{(2)}(\alpha_n^*)(\widehat{\alpha}_n - \alpha_0).$$

We also have

$$\widehat{\theta}_n - \theta_0 = -(\varDelta_n(\theta_0, \widehat{\theta}_n))^{-1}U_n^{(1)}(\theta_0), \quad \widehat{\alpha}_n - \alpha_0 = -(\varDelta_n(\alpha_0, \widehat{\alpha}_n))^{-1}U_n^{(1)}(\alpha_0).$$

But if $(H_q)$, $U_n(\theta_0) = \overline{U}_n(\alpha_0)$, $\overline{U}_n^{(1)}(\alpha_0) = R_0'U_n^{(1)}(\theta_0)$. So we have:

$$2\, a_n(\overline{U}_n(\widehat{\alpha}_n) - U_n(\widehat{\theta}_n)) = \varepsilon_n' C_n \varepsilon_n \quad \text{with}$$
$$\varepsilon_n = \sqrt{a_n} J_n^{-1/2} \quad U_n^{(1)}(\theta_0), \quad C_n = J_n^{1/2}\, B_n\, J_n^{1/2} \quad \text{and}$$
$$B_n = \varDelta_n^{-1} U_n^{(2)}(\theta_n^*)\varDelta_n^{-1} - R_0\overline{\varDelta}_n^{-1}\overline{U}_n^{(2)}(\alpha_n^*)\overline{\varDelta}_n^{-1}R_0'$$

where $\varDelta_n = \varDelta_n(\theta_0, \widehat{\theta}_n)$, $\overline{\varDelta}_n = \overline{\varDelta}_n(\alpha_0, \widehat{\alpha}_n)$. The Proposition is then a consequence of the following properties:

(*i*) $\lim_n (C_n - A_{n,0}) = 0$ in $P_0$-probability,

(*ii*) $\varepsilon_n \xrightarrow{\mathcal{D}} \mathcal{N}(0, I_p)$.

On the other hand, under $(H_q)$, $\overline{U}_n^{(2)}(\alpha_0) = R_0' U_n^{(2)}(\theta_0) R_0$, and thus $\overline{I}_{n,0} = R_0' I_{n,0} R_0$. So that matrix $(I - I_{n,0}^{1/2} R_0 \overline{I}_{n,0}^{-1} R_0' I_{n,0}^{1/2})$ is idempotent of rank $(p-q)$, which ends the proof. □

REMARKS:

(1) If for all $n$, $I_{n,0} = J_{n,0}$, then $A_{n,0}$ is idempotent of rank $(p-q)$ and we have the usual $\chi^2_{p-q}$ test, even in a *nonnecessarily stationary* and/or *nonergodic* situation. In particular, this is what happens if we consider independent, nonnecessarily equidistributed observations or nonnecessarily homogeneous Markov chains over $\mathbb{Z}$ and likelihood contrasts. This equality is also true for coding contrasts of a Markovian field (cf. § 5.3.3).

(2) If the model is ergodic, defining $I_0$, $J_0$ and $\overline{I}_0$ as the limits of associated sequences $I_n$, $J_n$, $\overline{I}_n$, and

$$A_0 = J_0^{1/2}[I_0^{-1} - R_0 \overline{I}_0^{-1} R_0'] J_0^{1/2},$$

we have, under $(H_q)$,

$$2\, a_n [U_n(\overline{\theta}_n) - U_n(\widehat{\theta}_n)] \xrightarrow{\mathcal{D}} \sum_1^{p-q} \lambda_i \chi^2_{i,1}.$$

Here, $\lambda_i$, $i = 1, p-q$ are the nonzero eigenvalues of $A_0$.

*Additive Contrasts for Mixing Processes* [88]. We shall assume the lattice $S \subseteq \mathbb{R}^d$ satisfies

• (*A*0) Uniformly for $s \in S$, $|B(s,m) \cap S| = 0(m^d)$ where $B(s,m)$ is the ball of center $s$ and radius $m$.

$X$ is assumed to be $\alpha$-mixing ($\alpha = \alpha_{\infty,\infty}$, cf. (3.3) and (1.40)) and the contrast $U_n$ is defined additively:

$$U_n(\alpha) = \frac{1}{d_n} \sum_{\overset{\circ}{D}_n} g_t(Y_t, \alpha). \tag{3.16}$$

with $Y_t = X(V_t)$, and $V_t$ a neighborhood of bounded range in $D$, $\overset{\circ}{D}_n$ is the $D$-interior of $D_n$, $d_n = |\overset{\circ}{D}_n|$. Let $\delta > 0$

•• (*A*1) *On the mixing coefficients:*

$$\sum_{i,j \in \overset{\circ}{D}_n} \alpha(d(i,j))^{\frac{\delta}{2+\delta}} = O(d_n) \quad \text{and} \quad \sum_{m=1}^{\infty} m^{d-1} \alpha(m) < \infty.$$

••• *(A2) On functions* $(g_t, t \in S)$:

(1) Condition $(H1)$ is satisfied uniformly for each $(g_t)$, $t \in S$ (in a neighborhood $V$ of $\theta_0$).

(2) For all $t \in S$, $E_{\theta_0}(g_t^{(1)}(\theta_0)) = 0$ and for $\alpha \in V$, $k = 1, 2$:

$$\|g_t^{(k)}(\alpha)\|_{2+\delta} \leq M < \infty.$$

•••• $(A3)$ There exist two $p \times p$ symmetrical, positive definite matrices $I$ and $J$ such that for big enough $n$,

$$J_n = Var_{\theta_0}(\sqrt{d_n} U_n^{(1)}(\theta_0)) \geq J > 0, \quad I_n = E_{\theta_0}(U_n^{(2)}(\theta_0)) \geq I > 0 \ .$$

**(3.4.7) Theorem.** *Under conditions* $(A0)$ *to* $(A3)$, *conditions* $(H1)$ *to* $(H3)$, *with respect to the additive contrast* $U_n(\alpha)$ *defined in* (3.16), *are satisfied.*

*Proof:* $(H1)$ is clearly true. $(H2)$ $J_n$ is well defined. Indeed,

$$Var_{\theta_0}(\sqrt{d_n} U_n^{(1)}(\theta_0)) \leq \frac{8M^2}{d_n} \sum_{i,j \in \overset{\circ}{D}_n} \alpha((d(i,j) - 2D)^+)^{\frac{\delta}{2+\delta}}$$

with $a^+ = \sup\{0, a\}$ because of the bounds over the covariance given in 3.2.1. The normal convergence is a consequence of Theorem 3.3.1. $(H3)$ is immediately obtained, checking that the variance of $U_n^{(2)}(\theta_0)$ tends to zero. □

Classically, the $g_t$ that define the contrast (3.16) are local, conditional or marginal, functionals. In Chapters 4 and 5 we shall be particularly interested in the conditional pseudo-likelihood of a Markov field with conditional density $\pi_t$ at $t$ for which $g_t$ is given by

$$g_t(Y_t, \alpha) = -\log \pi_t(x_t \mid x_{\partial t}, \alpha).$$

### *3.4.3. Identification by Penalized Contrasts*

Assume the space of parameters $\Theta \subseteq \mathbb{R}^M$ where $\mathbb{R}^M$ corresponds to the dominating model. Let $\mathcal{E}$ be a finite family of subspaces of $\mathbb{R}^M$, $\delta \in \mathcal{E}$ a generic element of $\mathcal{E}$, $|\delta|$ the dimension of $\delta$, and $\Theta_\delta = \Theta \cap \delta$ the associated parametric subspace (submodel). Assume the true value $\theta_0 \in \Theta_{\delta_0}$, $\delta_0 \in \mathcal{E}$ where $\delta_0$ is the minimal subspace associated to $\theta_0$: if $\delta \not\subseteq \delta_0$ and $\delta \in \mathcal{E}$, then $\theta_0 \notin \Theta_\delta$. A standard choice for $\mathcal{E}$ is the family of all nonempty subsets of $M = \{1, 2, \ldots, m\}$

$$\delta = \{\theta = (\theta_i)_{i \in M}, \ \theta_i = 0 \text{ if } i \notin \delta\}$$

or a growing sequence of subspaces $\delta$. Other choices indirectly associated to the canonical basis of $\mathbb{R}^M$ are sometimes useful, for example, choices related to the subhypothesis of isotropy for a given field.

Having observed $X(n)$, to identify the model is to determine its support $\widehat{\delta}_n \in \mathcal{E}$. If $a_n$ is the rate associated to $U_n$ (cf. (H2)), we can use a *penalized contrast with rate* $c(n)$ as a decision function for the model's selection:

$$W_n(\alpha) = U_n(\alpha) + \frac{c(n)}{a_n}|\delta(\alpha)|.$$

Define

$$\text{(3.17)} \quad \overline{W}_{n,\delta} = \overline{U}_{n,\delta} + \frac{c(n)}{a_n}|\delta|, \text{ with } \overline{U}_{n,\delta} = U_n(\widehat{\alpha}_{n,\delta}) = \operatorname*{Arg\ Min}_{\alpha \in \Theta_\delta} U_n(\alpha).$$

We shall choose $\widehat{\delta}_n$ which minimizes $\overline{W}_{n,\delta}$ over $\mathcal{E}$, that is, that responds to *Akaike's principle* with penalization rate $c(n)$:

$$\text{(3.18)} \qquad \widehat{\delta}_n = \operatorname*{Arg\ Min}_{\delta \in \mathcal{E}} \overline{W}_{n,\delta}.$$

We shall reformulate and complete the hypothesis over $U_n$ which shall be required to prove the a.e. consistency of criterion (3.18). Call $\theta_0$ the true value of the parameter, $\delta_0$ its support, and $P_0$ the distribution under $\theta_0$.

(I1) (1) $\liminf_n \, [U_n(\alpha) - U_n(\theta_0)] \geq K(\theta_0, \alpha)$ $P_0$. *a.e.* (i.e., (3.13)′).

(2) If $\delta \in \mathcal{E}$, then $\alpha \longmapsto K(\theta_0, \alpha)$ from $\Theta_\delta$ to $\mathbb{R}^+$ has a unique minimum $\alpha_\delta$ over $\Theta_\delta$.

(3) $U_n(\alpha)$ satisfies (H1) (cf. 3.4.2); $\alpha \longmapsto K(\theta, \alpha)$ is continuous.

(4) $P_0(\limsup_n \, (W_n(\frac{1}{k}) \geq \varepsilon_k)) = 0$ for a sequence $\varepsilon_k \to 0$ (i.e., (3.15)′).

(I2) There exists a sequence $(\ell_n)$ that tends toward $+\infty$, with $\dfrac{a_n}{\ell_n} \to \infty$ and finite $c \geq 0$ such that

$$\limsup_n \sqrt{\frac{a_n}{\ell_n}} \|U_n^{(1)}(\theta_0)\| \leq c < \infty \quad P_0 - a.e.$$

(I3) There exists a symmetric, positive definite $I$ such that

$$\liminf_n \, (U_n^{(2)}(\theta_0)) \geq I \quad P_0 - a.e.$$

Also assume the following conditions over the penalization rate $c(n)$:

$$(C1) \qquad \lim_n \frac{c(n)}{a_n} = 0; \qquad (C2) \quad \liminf_n \frac{c(n)}{\ell(n)} > \frac{c^2}{2\lambda},$$

where $\lambda$ is the smallest eigenvalue of $I$.

(I2) is an iterated logarithm-type condition that reinforces (H2). The following result is an adaptation to the nonergodic case of a result of Senoussi [148].

**(3.4.8) Theorem.** Almost Everywhere Model Identification. *Under hypothesis* (I1) *to* (I3) *for* $U_n$ *and conditions* (C1) *and* (C2) *over the penalization rate,*

$$(\widehat{\delta}_n, \widehat{\alpha}_{n,\widehat{\delta}_n}) \longrightarrow (\delta_0, \theta_0) \quad P_0 - a.e.$$

*Proof:* Recalling the proof of the strong consistency (Theorem 3.4.3), we can prove that for all $\delta \in \mathcal{E}$, $\widehat{\alpha}_{n,\delta} \longrightarrow \alpha_\delta \quad P_0 - a.e.$ so that because of (I1–4),

$$\lim_n |U_n(\alpha_\delta) - U_n(\widehat{\alpha}_{n,\delta})| = 0 \quad P_0 - a.e.$$

It can be deduced that $N_\delta = \{\omega : \liminf_n \, [U_n(\widehat{\alpha}_{n,\delta}) - U_n(\theta)] \not\geq K(\theta_0, \alpha_\delta)\}$ has probability zero. If $F$ is the event over which inequality (I2) is true, then $\overline{N} \cap F$ has probability 1 if $\overline{N}$ is the complement of $N = \bigcup_{\delta \in \mathcal{E}} N_\delta$. From now on, we shall assume $\omega \in \overline{N} \cap F$. Since $(\widehat{\delta}_n)$ belongs to (finite) $\mathcal{E}$, it has at least one accumulation point which we shall call $\overline{\delta}$. So there exists a subsequence $(m(n))_n$ such that, after a certain point, $\widehat{\delta}_{m(n)} = \overline{\delta}$. Consider this subsequence and $n$ big enough so that, in fact, $\widehat{\delta}_n = \overline{\delta}$. We shall show that $\overline{\delta} = \delta_0$ which ends the proof for if all the accumulation points are equal to $\delta_0$, $\widehat{\delta}_n \to \delta_0$. From the definition of $\widehat{\delta}_n$, we have

$$\overline{W}_{n,\widehat{\delta}_n} - \overline{W}_{n,\delta_0} \leq 0 \tag{3.19}$$

*(i) Under (C1),* $\overline{\delta} \supseteq \delta_0$*: a.e. there is not underparametrization.*

$$\begin{aligned} 0 \geq (\overline{W}_{n,\widehat{\delta}_n} - \overline{W}_{n,\delta_0}) &= (\overline{W}_{n,\overline{\delta}} - \overline{W}_{n,\delta_0}) \\ &= [U_n(\widehat{\alpha}_{n,\overline{\delta}}) - U_n(\alpha_{\overline{\delta}})] + [U_n(\alpha_{\overline{\delta}}) - U_n(\theta_0)] \\ &+ [U_n(\theta_0) - U_n(\widehat{\theta}_n)] + \frac{c(n)}{a_n}(|\delta_0| - |\delta|). \end{aligned}$$

The first, third and last terms of this sum go to 0, so that

$$0 \geq \liminf(\overline{W}_{n,\widehat{\delta}_n} - \overline{W}_{n,\delta_0}) = \liminf[U_n(\alpha_{\overline{\delta}}) - U_n(\theta_0)] \geq K(\theta_0, \alpha_\delta),$$

which gives $\alpha_\delta = \theta_0$, that is, $\delta \supseteq \delta_0$.

*(ii) Under (C2),* $\delta_0 \subsetneq \overline{\delta}$ *is impossible: a.e. there is not overparametrization.*

Assume $\delta_0 \subsetneq \overline{\delta}$. Over the subsequence we have

$$0 \geq \frac{a_n}{\ell_n}[\overline{W}_{n,\widehat{\delta}_n} - \overline{W}_{n,\delta_0}] = \frac{a_n}{\ell_n}[U_n(\widehat{\alpha}_{n,\overline{\delta}}) - U_n(\theta_0)]$$

$$+ \frac{a_n}{\ell_n}[U_n(\theta_0) - U_n(\widehat{\theta}_n)] + \frac{c(n)}{\ell_n}(|\delta| - |\delta_0|).$$

Since the second term is $\geq 0$, and $|\delta| - |\delta_0| \geq 1$, condition (C2) yields

$$0 \geq \liminf_n \frac{a_n}{\ell_n}[\overline{W}_{n,\widehat{\delta}_n} - \overline{W}_{n,\delta_0}] \geq \liminf_n \frac{a_n}{\ell_n}[U_n(\widehat{\alpha}_{n,\overline{\delta}}) - U_n(\theta_0)] + b.$$

We complete the proof using the following Lemma:

**(3.4.1) Lemma.** *Under* (I2), *assuming model* $P_0$, *and if* $\lambda$ *is the smallest eigenvalue of* $I$, *then*

$$\liminf_n \frac{a_n}{\ell_n}(U_n(\widehat{\theta}_n) - U_n(\theta_0)) \geq -\frac{1}{2}\frac{c^2}{\lambda} \quad P_0 - a.e.$$

Observing that if $\overline{\delta} \supseteq \delta_0$, then $\theta_0 = \alpha_{\delta}$, we can apply the Lemma so that

$$0 \geq \liminf \frac{a_n}{\ell_n}[\overline{W}_{n,\widehat{\delta}_n} - \overline{W}_{n,\delta_0}] \geq -\frac{1}{2}\frac{\Lambda_0}{\lambda_0}c^2 + b,$$

which is impossible because of the choice of $b$. □

*Proof of the Lemma:* Recall $\Delta_n$, the integral matrix defined in the proof of Theorem 3.4.5. Then,

$$U_n^{(1)}(\theta) = \Delta_n(\theta_0, \widehat{\theta}_n)(\widehat{\theta}_n - \theta_0),$$

$$U_n(\widehat{\theta}_n) - U_n(\theta_0) = -\tfrac{1}{2}{}^t(\widehat{\theta}_n - \theta_0)U_n^{(2)}(\theta_n^*)(\widehat{\theta}_n - \theta_0).$$

These two identities, the regularity condition (H1) over $U_n$ and conditions (I2) and (I3), thus yield

$$\liminf_n \frac{a_n}{\ell_n}[U_n(\widehat{\theta}_n) - U_n(\theta)] \geq -\frac{1}{2}\frac{c^2}{\lambda} \quad P_0 - a.e.$$

□

REMARK: In the ergodic situation, if, for a symmetric, positive definite matrix $I_0$, we have

(I2)′ $\left\{\sqrt{\frac{a_n}{\ell_n}} I_0^{-1/2} U_n^{(1)}(\theta_0),\ n \geq 1\right\}$ is a.e. relatively compact with all its accumulation points in the ball $B(0,1)$ of $\mathbb{R}^M$, and

(I3)′ $$U_n^{(2)}(\theta_0) \longrightarrow I_0 \quad P_0 - a.e.,$$

then the result is still true under (C1) and (C2)′ [148]:

$$(\text{C2})' : \liminf \frac{c(n)}{\ell_n} > 1.$$

### *3.4.4. Identification and Law of the Iterated Logarithm (LIL) for Fields*

*First case: Independent Fields.* Recall Cantelli's result (cf. Feller [55]).

THEOREM. *Law of the Iterated Logarithm for an Independent Field. — Let $Y = \{Y_i, i \in S\}$ be a real, centered, independent field such that there exists $a > 0$ with*

$$\sup_i E|Y_i|^{2+\delta} < \infty, \quad \text{and} \quad \inf_i E(Y_i^2) \geq a > 0. \tag{3.20}$$

*Let $S_n = \sum_{D_n} Y_i$, $s_n^2 = Var(S_n)$. If $s_n \to \infty$, then, defining $\log_3$ as the third iteration of the logarithm, we have for any constant $c > 3$:*

$$\limsup_n \frac{S_n}{\sqrt{2 \log\log s_n + c \log_3 s_n} \cdot s_n} = 0 \quad a.e.$$

*In particular, for all $b > 2$,*

$$\limsup \frac{S_n}{\sqrt{b s_n^2 \log\log s_n}} = 0 \quad a.e.$$

Assume $X = \{X_i, i \in S\}$ is an independent field, $X_i$ with density $\pi_i(x_i, \alpha)$, $\theta_0$ is the true value of the parameter, $U_n(\alpha)$ is the likelihood contrast over $D_n$, and $d_n = |D_n|$ :

$$U_n(\alpha) = -d_n^{-1} \sum_{D_n} \log \pi_i(x_i, \alpha).$$

Define $Z_i$ as the gradient of the opposite of the log density in $i$ and in $\theta_0$, and $s_{n,k}^2 = \sum_{D_n} Var(Z_{ik})$, $k = 1, p$, $s_n^2 = \sum_{k=1,p} s_{n,k}^2$.

**(3.4.9) Corollary.** *Let $X = \{X_i, i \in S\}$ be an independent field with density in $\mathcal{C}^2$ in a certain neighborhood of $\theta_0$. Assume that the estimation by maximum likelihood $\widehat{\theta}_n$ is strongly consistent and that the centered field $\{Z_i\}$ satisfies (3.20) for each $i \in S$. Assume also that*

$$A = \sup_i \sum_{k=1,p} Var(Z_{i,k}) < \infty.$$

*Then, if each $s_{n,k} \longrightarrow \infty$ as $n \to \infty$, the identification is a.e. if for $a_n = d_n = |D_n|$, we have*

$$\lim_n \frac{c(n)}{d_n} = 0, \quad \liminf_n \frac{c(n)}{\log\log d_n} > A.$$

*Proof:* $\{Z_i\}$ satisfies the Law of the Iterated Logarithm; then, if $b > 2$ and for a given norm over $\mathbb{R}^p$,

$$\limsup_n \frac{\|\sum_{D_n} Z_i\|}{\sqrt{bs_n^2 \log\log s_n}} = 0 \quad a.e.$$

However,

$$\frac{\|\sum_{D_n} Z_i\|}{\sqrt{bs_n^2 \log\log s_n}} = \frac{\sqrt{d_n}\|U_n^{(1)}(\theta_0)\|}{\sqrt{b \cdot \frac{s_n^2}{d_n} \log\log s_n}}.$$

Since $s_n^2 \leq d_n A$ and because (3.20), $(\log\log s_n)/(\log\log d_n) \longrightarrow 1$. Choosing $\ell_n = bA \log\log d_n$, $a_n = d_n$ in the proof of Theorem (3.4.8), we get

$$\limsup_n \|\sqrt{\frac{a_n}{\ell_n}} U_n^{(1)}(\theta_0)\| = 0 \quad a.e.$$

Condition (I3) is a consequence of (3.20). □

This result shows that it is possible to identify a Markovian field using the Penalized Coding Contrast (cf. § 5.3.3, Theorem 5.3.5).

*Second case: Model Identification for Stationary Mixing Fields.*

We shall start by recalling the strong Invariance Principle for mixing fields (Berkes and Morrow [12]). Let $Y = \{Y_j, j \in \mathbb{Z}^d\}$ be a centered, second-order stationary mixing (cf. (1.40)) field over $\mathbb{R}^p$. Assume that it belongs to $L^{2+\delta}$ for some $\delta > 0$,

$$\sup_j E\|Y_j\|^{2+\delta} < \infty, \tag{3.21}$$

and that the mixing coefficients satisfy for a certain $0 < \varepsilon < \frac{1}{2}$,

$$\alpha_Y(A, B) \leq C[dist(A, B)]^{-d(1+\varepsilon)(1+\frac{2}{\delta})}. \tag{3.22}$$

(3.21) and (3.22) yield the absolute convergence of the covariance series

$$J = \sum_{\ell \in \mathbb{Z}^d} Cov(Y_0, Y_\ell) < \infty. \tag{3.23}$$

Let $\eta$, $0 < \eta < 1$ and $G_\eta$ be the subsets of indexes of $\mathbb{N}^d$ defined by

$$G_\eta = \{n = (n_1, \ldots, n_d) : \text{ for } k = 1, d, \; n_k \geq \prod_{\substack{i=1,d \\ i \neq k}} n_i^\eta\}. \tag{3.24}$$

Let $S_n = \sum_{1 \leq i \leq n} Y_i$ where $1 = {}^t(1, \ldots, 1)$ and $i \leq j$ is the strong order relationship $i_k \leq j_k$ for $k = 1, d$. We have the following strong approximation result:

PROPERTY: Strong Invariance Principle for Mixing Fields. — *Assume $Y$ is a second-order stationary field that satisfies (3.21) and (3.22). Then, without changing the distribution of $Y$, we can define a new probability space and a $p$-dimensional Brownian $W$ indexed over $(\mathbb{R}^+)^d$, with covariance $J$, defined as in (3.23), that satisfies for a certain $\lambda = \lambda(Y, \eta) > 0$,*

$$\sup_{n \in G_\eta} |n|^{\lambda - \frac{1}{2}} \sup_{1 \le m \le n} \|S_m - W([0, m])\| < \infty \quad a.e.$$

*Law of the Iterated Logarithm for $S_n$.* Based on the Law of the Iterated Logarithm for the $p$-dimensional Brownian $W$ [148], it can be deduced that under conditions (3.21) and (3.22),

$$\left\{ \frac{S_n}{\sqrt{2|n| \log \log |n|}}, \quad n \in G_\eta \right\}$$

is a.e. relatively compact. Also its closure is the ellipsoid $\mathcal{C}_p(0, J^-) \cap \operatorname{Im}(J)$ (for any of the choice of the inverse $J^-$ of $J$), where

$$\mathcal{C}_p(0, J^-) = \{u \in \mathbb{R}^p : {}^t u J^- u \le 1\},$$
$$\operatorname{Im}(J) = \{v \in \mathbb{R}^p : \exists u \in \mathbb{R}^p \text{ such that } v = Ju\}.$$

*Application:* Let $X = \{X_i, i \in \mathbb{Z}^d\}$ be a strictly stationary ergodic field with distribution $P_\theta$, and let $U_n(\alpha)$ be an additive contrast over rectangular domain $D_n = [1, n]^d$ of $\mathbb{Z}^d$:

$$(3.25) \quad U_n(\alpha) = d_n^{-1} \sum_{D_n} g(X(V_i), \alpha), \quad V_i = V + i, \quad V \text{ finite}, \quad d_n = |D_n|.$$

Assume $g$ belongs to $\mathcal{C}^2$ in $\alpha$ over a neighborhood of $\theta_0$ and

$$(3.26) \qquad Y_j = g_\theta^{(1)}(X(V_j), \theta_0), \quad j \in \mathbb{Z}^d.$$

If there exists $\delta > 0$ such that $Y \in L^{2+\delta}$ and if for that same $\delta$, $X$ is mixing and satisfies (3.21), then $Y$ is also mixing and satisfies (3.21). Write

$$(3.27) \qquad J = \sum_{\mathbb{Z}^d} Cov_{\theta_0}(Y_0, Y_\ell), \quad I = E_{\theta_0}[g^{(2)}(X(V_i), \theta_0)]$$

**(3.4.10) Corollary.** *Let $X$ be a strictly stationary and ergodic field, $U_n$, the contrast defined in* (3.25). *Assume $g$ satisfies condition* (H1) (*cf.* 3.4.2) *and that the minimum contrast estimator $\widehat{\theta}_n$ is strongly consistent.*

*Assume there exists $\delta > 0$ such that $X$ satisfies* (3.21) *and* (3.22), *$Y$ satisfies* (3.21), *and $I$ is positive definite. Then identification of the model by penalized contrast* (3.18) *is a.e. over $G_\eta$, $\eta > 0$ for $n \to \infty$ if*

$$\lim_n \frac{c(n)}{d_n} = 0, \quad \liminf \frac{c(n)}{\log\log d_n} > \lambda,$$

*where* $\lambda$ *is the greatest eigenvalue of* $JI^{-1}$.

*Proof:* We shall follow the proof of the Lemma used in Theorem 3.4.7. Because of the ergodicity of $X$ and the consistency of $\widehat{\theta}_n$, we have

$$\lim_n \Delta_n(\theta_0, \widehat{\theta}_n) = I \quad a.e.$$

Consider $Z_j = I^{-1/2} Y_j$ and $S_n^* = \sum_{D_n} Z_j$. Under the Corollary's hypothesis, we have the $LIL$ for $S_n^*$:

$$\limsup_n \sqrt{\frac{d_n}{2\log\log d_n}} \| I^{-1/2} U_n^{(1)}(\theta) \| = \sqrt{\lambda}$$

in the supremum norm. This yields (I2). □

## 3.5. Bibliographical Comments

The Ergodic Theorem is considered in a very general setting (including not only statistical mechanics' models but also processes with values over the subsets of $R^d$, percolation models) by Nguyen and Zessin, and based on the results of Tempel'man. For the various regularity concepts for sequences of spatial domains and the type of convergence associated, the reader may consult Prum. Properties 3.1 and 3.2 are generalizations to the spatial case of Lemma 1.4.34 and Theorem 1.4.5 of [41].

Theorem 3.2 is the nonstationary version of Bolthausen's result, using Lemma 3.1 due to Stein. The spatial correlation indexes are described in the book of Cliff and Ord (cf. also Moran). Sen gives a Martingale approach to the asymptotic normality of these indexes, whereas Cliff and Ord explicitly use the normality of the observations.

The $CLT$ for a functional of an ergodic Markovian field is due to Künsch [83]. The results of convergence rates for the $CLT$ are due to Takahata, and to Guyon and Richardson. The $CLT$ for positively dependent real fields is due to Newman. As a consequence of the F.K.G. inequality (cf. Prum), these results can be applied to attractive Gibbs fields.

The consistency Theorem 3.3 generalizes Theorem 3.2.8 of Dacunha–Castelle and Duflo to a nonergodic setting. For convex contrasts, the result is due to Senoussi. The asymptotic normality is standard except that it doesn't require the ergodicity hypothesis. The asymptotic test of difference of contrasts based on a mixture of 1 degree of freedom (d.f.) $\chi^2$ is less classical and appears in the context of time series statistics in Yao [171]. For the results, extensions and applications of § 3.4, the reader can consult

Hardouin. The *a.e.* identification by penalized likelihood result is a nonergodic adaptation of the result obtained by Senoussi.

## 3.6. Exercises

(1) *Conditional least squares (CLS) for the Ising model.*
Consider $X$ the Ising model over $\mathbb{Z}^2$. Assume it is isotropic, with four nearest neighbors (parameters $\beta$) and an exterior field ($\alpha$). Assume $X$ is ergodic. For $D_n = \{1, 2, \ldots, n\}^2$, consider the contrast

$$U_n(\alpha, \beta) = \sum_{D_n} (X_i - E_{\alpha,\beta}(X_i|\cdot))^2.$$

Using Theorem 3.4.1 and Theorem 3.3.3, show that the estimation of $(\alpha, \beta)$ by $CLS$ is consistent and asymptotically normal. How can we generalize this result for general Markovian ergodic models?

(2) Consider the regression: $y_i = f(x_i, \theta) + \varepsilon_i$, $i = 1, n$, $\varepsilon_i$ independent $(0, \sigma^2)$. The $\{x_i\}$ are i.i.d. and their common distribution $g$ over $\mathbb{R}^k$ has compact support. We have $\theta \in \Theta$ a compact of $\mathbb{R}^p$, and the true value $\theta_0$ belongs to the interior of $\Theta$. Assume $f$ is continuous in $(x, \theta)$. What is the condition for consistency of least squares estimation? If $var(\varepsilon_i) = \sigma^2 h(x_i)$, $h$ known, under what conditions are the weighted least squares consistent?

(3) *Minimum $\chi^2$ estimation.*
Assume $y$ stands for the subject's transportation choice ($y \in \{0, 1, 2\}$: individual, metro, bus), $x$ stands for the income level (we consider three possible levels, $x_1$, $x_2$, $x_3$) and we consider the following conditional model:

$$P_{yt} = P(y_{it} = j|x_t) = a_j(x_t)\Big[\sum_{\ell=0,2} a_\ell(x_t)\Big]^{-1}$$

with $i = 1, n_t$ (the repetition index under condition $x_t$), $t = 1, 3$ and $a_0(x) \equiv 1$, $a_j(x) = \exp(\alpha_j + \beta_j x)$ for $j = 1, 2$. For $n = n_1 + n_2 + n_3$ independently observed subjects, call $n_{jt}$ the number of realizations of $i$ $\{i : y_{it} = j\}$.

(a) Evaluate $\log(P_{1t} \cdot P_{0t}^{-1})$ and $\log(P_{2t} \cdot P_{0t}^{-1})$. Under which conditions is the estimator that minimizes $U_n = \sum_{t=1}^{3} \sum_{j=1}^{2} \Big(\frac{n_{jt}}{n_{ot}} - a_j(x_t)\Big)^2$ consistent? What is the best weighted least-squares contrast? Study the asymptotic normality of the test $\beta_1 = \beta_2$.

(b) Validate the model proposed in the complete conditional model.

(4) *Estimation of an inhomogeneous Markov chain.*
Let $Y = \{Y_i, i \in \mathbb{N}\}$ be a Markov chain with finite state space $E$ and transitions $P_i(y_i \mid y_{i-1}, \alpha) = p(y_i \mid y_{i-1}; \alpha, x_i) > 0$, with $\alpha \in \Theta$ a compact

subset of $\mathbb{R}^p$ and $x_i$ an explicative variable which belongs to a compact measurable space $(X, \mathcal{X})$. Assume $p$ is continuous and of class $\mathcal{C}^2$ in $\alpha$. Assume, $\theta$ the true parameter value, belongs to the interior of $\Theta$. Also assume we have a measure $\mu > 0$ over $X$ such that

$$\alpha \longmapsto \sum_{y \in E} \int_X p(\cdot|y;\alpha,x)\mu(dx)$$

is injective,

$$\liminf_n \left(\frac{1}{n}\sum_1^n \mathbb{1}(x_i \in A)\right) \geq \mu(A).$$

(a) Show that the $ML$ estimator conditional to $y_0$ is consistent. Study the asymptotic normality and asymptotic test for the likelihood ratio.

(b) Example: $X \subseteq \mathbb{R}$; $F$ is $\mathcal{C}^2$; $f = F' > 0$, $Y_i \in \{0,1\}$; and $P_i(y_i \mid y_{i-1}, \alpha) = F(\alpha x_i y_{i-1} + \beta y_{i-1})$. If $\{x_i\}$ are i.i.d. realizations with common distribution $G$ over $X$, give the consistency conditions for the $ML$ estimator. Determine the distribution of the $ML$ estimator and test $\alpha = 0$.

(5) *Estimation by marginal pseudo-likelihood.*
$X$ is a real ergodic and exponentially mixing field over $\mathbb{Z}^d$, whose distribution depends on a parameter $\alpha \in \Theta$, a compact subset of $\mathbb{R}^p$. Let $A$ be a finite subset of $\mathbb{Z}^d$, $A_i = A + i$ and $g : \mathbb{R}^{|A|} \times \Theta \to \mathbb{R}$ continuous. For an increasing sequence of domains $(D_n)$ of $\mathbb{Z}^d$, $d_n = |D_n|$, we define the marginal contrast

$$U_n(\alpha) = d_n^{-1} \sum_{D_n} g(X(A_i), \alpha).$$

(a) Specify what conditions over $(X, g)$ entail that the minimum contrast estimator is consistent, as well for normality and for asymptotic test.

(b) Example: Let $Y = \{Y_i, i \in \mathbb{N}\}$ be a homogeneous chain with states $\{-1,+1\}$ and transition $p = P(Y_i \neq Y_{i-1} \mid Y_{i-1})$. Assume that for each $i$ we distort the field with the i.i.d. noise, $P(X_i = Y_i) = 1 - \varepsilon = 1 - P(X_i \neq Y_i)$. Can $\theta = (p, \varepsilon)$ be identified from distribution $(X_0, X_1)$? Estimate $\theta$ from the marginal contrast for $A = \{0,1,2\}$. Test the independence of chain $Y$.

(6) *Asymptotic normality for the Geary index.*

(a) Consider the same context as in the application of Theorem 3.2 to the Moran index. For $X = \{X_i, i \in S\}$, a centered independent process, $\sigma_i^2 = Var X_i$, we define the Geary index: $J_n = \Sigma_{i,j \in D_n} w_{ij}(X_i - X_j)^2$. Specify what conditions assure asymptotic normality of $J_n$ and identify the limiting distribution.

(b) For $S = \mathbb{Z}^2$, $w_{ij} = 1$ if $|i - j| = 1$, 0 if not, and $X$ *i.i.d.* $(0, \sigma^2)$, give the asymptotic distributions of Moran's (3.7) and Geary's indexes.

(c) (following [135]) For the Gaussian AR model, $X = \rho W X + \varepsilon$, show that the Neymann–Pearson test of $\rho = 0$ against $\rho = \rho_0 > 0$ is based on ${}^t x(I - \rho_0 W)^2 x \, ({}^t x x)^{-1}$ (here, $\varepsilon$ is i.i.d. and $W$ is symmetric). Show that the development of this statistic at $\rho_0 = 0$ gives the Moran index.

CHAPTER 4

# Estimation for Second-Order Processes

In this chapter $X = \{X_j, j \in \mathbb{Z}^d\}$ will be a real, second-order stationary field with mean $\mu$, spectral density $f$ and covariance $R$ (only § 4.3.2. and 4.4 will consider the nonstationary case). $X$ shall be observed over an increasing sequence of convex domains $(D_n)$, whose interior diameter increases to infinity: typically, $D_n$ will be the rectangle $\prod_{i=1}^d [1, n_i]$, with $n_i \to \infty$ for each $i = 1, d$. We will call $N = |D_n|$.

## 4.1. Empirical Estimators, Periodogram, and Tapered Data

One of the important differences between $\mathbb{Z}$ and $\mathbb{Z}^d$ is that the boundary $\partial D_n$ of a domain $D_n$ in $\mathbb{Z}^d$, $d \geq 2$ is such that $|\partial D_n|$ tends to infinity if $D_n$ is increasing. If $D_n$ is *spherical*, $|\partial D_n|$ has order $|D_n|^{1-\frac{1}{d}}$. This entails that the bias of classical estimators will also have order $|D_n|^{1-\frac{1}{d}}$, which contributes significantly when studying convergence in distribution whenever $d \geq 2$ [81]. This will lead us to consider *unbiased estimators,* but also to estimate the second-order characteristics by *tapering data* over the boundary of domain $D_n$.

When $X$ is ergodic, the *a.e.* convergence results for the empirical estimators remain unchanged with respect to $d = 1$ [5].

### *4.1.1. Almost Everywhere Convergence of Empirical Estimators*

The empirical estimators of $\mu$, $R(k)$, $k \in \mathbb{Z}^d$ are given by

$$\overline{X_n} = N^{-1}\sum_{D_n} X_i, \quad \widehat{R}_n(k) = N^{-1}\sum_{D_{n,k}} X_j X_{j+k} - \overline{X}_n^2, \tag{4.1}$$

with $D_{n,k} = \{i \in D_n,\ i+k \in D_n\}$. The spectogram is defined as

$$\begin{aligned} I_n(\lambda) &= (2\pi)^{-d}\sum_{\Delta_n} \widehat{R}_n(k) \exp i < \lambda, k > \\ &= (2\pi)^{-d} N^{-1}\Big|\sum_{D_n} X_j \exp i < \lambda, j >\Big|^2 \end{aligned} \tag{4.2}$$

with $\Delta_n = D_n - D_n = \{k : k = i - j \text{ for } i, j \in D_n\}$. If $\mu$ is known (assume $\mu = 0$), $\hat{R}_n(k) = N^{-1}\sum_{D_{n,k}} X_j X_{j+k}$. The unbiased estimators of the covariance and of the spectral density are, respectively,

$$R_n^*(k) = |D_{n,k}|^{-1}\sum_{D_{n,k}} X_j X_{j+k} \tag{4.3}$$

$$I_n^*(\lambda) = (2\pi)^{-d}\sum_{\Delta_n} R_n^*(k) \exp i < \lambda, k > . \tag{4.4}$$

Consider $\varphi \in \mathcal{C}(T^d, \mathbb{R})$, $J(\varphi) = \int_{T^d} f(\lambda)\varphi(\lambda)d\lambda$ and the associated *spectographic estimators*:

$$J_n(\varphi) = \int_{T^d} I_n\, \varphi d\lambda, \quad J_n^*(\varphi) = \int_{T^d} I_n^* \varphi d\lambda. \tag{4.5}$$

**(4.1.1) Theorem.** *If $X$ is strictly stationary and ergodic, the following expressions converge a.e.:*

$$\overline{X}_n \longrightarrow \mu, \quad \widehat{R}_n(k) \text{ and } R_n^*(k) \longrightarrow R(k), \quad k \in \mathbb{Z}^d,$$
$$J_n(\varphi) \text{ and } J_n^*(\varphi) \longrightarrow J(\varphi) \quad \text{for} \quad \varphi \in \mathcal{C}(T^d).$$

*Proof:* The first three results are a consequence of $X$'s ergodicity. If we define $F_n(d\lambda) = I_n(\lambda)d\lambda$, the weak convergence of $F_n$ to $F$ ($F(d\lambda) = f(\lambda)d\lambda$) is assured since for all $k$

$$\widehat{R}_n(k) = \int \exp i < \lambda, k > F_n(d\lambda) \longrightarrow \int \exp i < \lambda, k > f(\lambda)d\lambda = R(k).$$

□

*Bias:* We shall assume $\mu = 0$ and that the sequence $(D_n)$ satisfies

$$\lim_n \frac{|D_{n,k}|}{|D_n|} = 1, \quad k \in \mathbb{Z}^d. \tag{4.6}$$

Then the Dominated Converge Theorem yields

$$E(I_n(\lambda)) = \sum_{\Delta_n} \frac{|D_{n,k}|}{|D_n|} R(k) \exp i < \lambda, k > \longrightarrow f(\lambda)$$

whenever $\Sigma|R(k)| < \infty$. We have the same result for $I_n^*$. $R_n^*(k)$ estimates $R(k)$ without a bias. In particular, for $D_n$ the cube $[1, n]^d$, the bias has the following form:

**(4.1.2) Theorem.** Bias Evaluation. *If* $D_n = [1,n]^d$, *then*

(*a*) $E[J_n(\varphi) - J(\varphi)] = \dfrac{H(f,\varphi)}{n}(1 + o(1))$, *with*

$$H(f, \varphi) = -\sum_s |s|\widehat{\varphi}_s \cdot R(s), \quad |s| = \sum_1^d |s_i|.$$

*The bias behaves like* $n^{-1} = |D_n|^{-\frac{1}{d}}$ *(*$\widehat{\varphi}_s$ *is the Fourier coefficient of* $\varphi$*).*

(*b*) *If for* $h = f$ *or* $h = \varphi$, $h^{(m)}_{\lambda_i^m}$ *is continuous,* $i = 1,d$ *with* $m = \left[\frac{d}{2}\right] + 1$, *then*

$$E[J_n^*(\varphi) - J(\varphi)] = o(n^{-\frac{d}{2}}).$$

*Proof:*

(*a*) Results of $|D_{n,k}| = |D_n|\left(1 - \dfrac{|k|}{n}\right)(1 + o(1))$.

(*b*) The bias is exactly equal to $\sum_{s:\inf_i|s_i|\geq n} \widehat{\varphi}_s R(s)$. Under the Theorem's assumptions, for example, for $\varphi$, if $s \neq 0$, then for $i$, $s_i \neq 0$, $|\widehat{\varphi}_s| = \frac{1}{|s_i|^m}|\widehat{\varphi}_{m,i}|$ with $\widehat{\varphi}_{m,i}$ the (summable) Fourier transform of $\varphi^{(m)}_{\lambda_i}$. This gives the desired result. □

The more regular functions $f$ or $\varphi$ are, the faster the bias $b$ will tend to zero. A consequence of part (*a*) is that

$$\lim_n \sqrt{N} E(\widehat{R}_n(k) - R(k)) = \begin{cases} 0 & \text{if } d = 1, \\ -|k| & \text{if } d = 2, \\ +\infty & \text{if } d \geq 3, \end{cases}$$

and analogous results for $J_n(\varphi)$. Over $\mathbb{Z}$, the edge effect of order $0(1)$ is of no consequence; over $\mathbb{Z}^2$, the edge effect, of order $0(\sqrt{N})$, creates a bias of order $\frac{1}{\sqrt{N}}$, precisely that of the rate of the standard results of convergence in distribution. If $d \geq 3$, $\sqrt{N}(\widehat{R}_n(k) - R(k))$ doesn't even converge in mean. A natural idea would be to consider the unbiased estimators $R_n^*$, $J_n^*$. However, $(R_n^*(k),\ k \in \mathbb{Z}^d)$ is no longer positive definite as is $(\widehat{R}_n(k), k \in \mathbb{Z}^d)$. This positivity is important since it guarantees the existence of a solution in certain estimation problems (cf. § 4.3, Remark 1).

## *4.1.2. Empirical Estimators with Tapered Data*

In order to find an alternative solution to the inconveniences described above (the bias of $\widehat{R}$ and the nonpositive definiteness of $R^*$), consider the same estimators but now *tapering the data* at the edges of the observation domain (Tuckey [166]). For dimensions $d = 1, 2, 3$, these estimators do not have the defects of $\widehat{R}$ or $R^*$ and if the tapering is reasonably done there is no efficiency loss (Dahlhaus and Künsch [44]). From a practical point of view, tapering can improve statistical analysis as it lessens the importance of observations close to the edges, which can be influenced by nonstationary exterior factors. Tapering is also a way to deal with missing values (if at site $i$, $X_i$ is missing, then $i$ belongs to the boundary of $D_n$). For time series, yet another advantage is that it permits a better estimation of the peaks of the spectral density (leakage effect, cf. Dahlhaus [42], [43]).

*Tapering over a Rectangular Domain and Associated Estimators.* Consider $D_n = \prod_{i=1}^d [1, n_i]$, $N = |D_n| = n_1 \times \cdots \times n_d$. Instead of working with the observed data $\{X_j,\ j \in D_n\}$, we shall consider the tapered data:

$$\{a_n(j) X_j,\ j \in D_n\}, \quad n \geq 0, \tag{4.7}$$

where the *tapering weights* are defined in the following way. Consider $w : [0,1] \longrightarrow [0,1]$, $w(0) = 0$, $w(1) = 1$, increasing and belonging to $\mathcal{C}^2$. The *w-taper*, which tapers $100(1-\rho)\%$ of the edge values, is defined for $0 \leq \rho \leq 1$ by

$$h(u) = \begin{cases} w\left(\dfrac{2u}{\rho}\right) & \text{if } \ 0 \leq u \leq \dfrac{1}{2}\rho, \\ 1 & \text{if } \ \dfrac{1}{2}\rho \leq u \leq \dfrac{1}{2}, \\ h(1-u) & \text{if } \ \dfrac{1}{2} \leq u \leq 1. \end{cases}$$

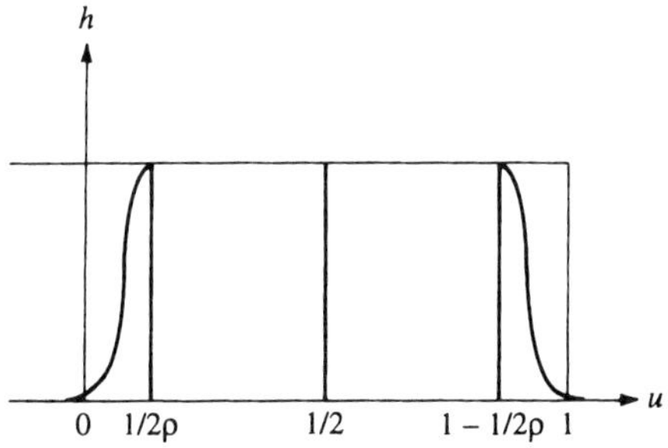


*The Tuckey–Hanning Taper:* $w(u) = \frac{1}{2}(1 - \cos \pi u)$

Over $D_n$, $a_n$ is given by

$$a_n(t) = a\Big(\frac{t-\frac{1}{2}}{n}\Big) = \prod_{i=1}^{d} h\Big(\frac{t_1 - \frac{1}{2}}{n_i}\Big), \quad t \in D_n$$

The discrete Fourier transform associated to the tapered data $\{X_j^h, j \in D_n\}$ and the tapered spectogram, are, respectively,

$$d_n(\lambda) = \sum_{D_n} a\Big(\frac{t-\frac{1}{2}}{n}\Big) \cdot X_t \ \exp -i < \lambda, t >,$$

(4.8) $$I_n^h(\lambda) = ((2\pi)^d H_{2,n}(0))^{-1} |d_n(\lambda)|^2, \quad \text{with}$$

(4.9) $$H_{k,n}(\lambda) = \sum_{D_n} a\Big(\frac{t-\frac{1}{2}}{n}\Big)^k \exp -i < \lambda, t > .$$

The tapered empirical covariance and estimator of $J(\varphi)$ are given by

(4.10) $$\widehat{R}_n^h(k) = H_{2,n}(0)^{-1} \sum_{D_{n,k}} X_t^h X_{t+k}^h, \quad J_n^h(\varphi) = \int_{T^d} I_n^h \ \varphi \ d\lambda.$$

**(4.1.3) Theorem.** Bias Control for Tapered Estimators. *Assume that the taper $h$ and $f$ belong to $\mathcal{C}^2$ and that $\frac{n_i}{n} \longrightarrow a_i \in \, ]0,\infty[$, $i = 1,d$. Then*

$$E[J_n^h(\varphi) - J(\varphi)] = C \ N^{-\frac{2}{d}}(1 + o(1))$$

*with*

$$2C = \Big(\prod_{i=1}^{d} a_i\Big)^{\frac{d}{2}} \cdot \frac{\int_0^1 h'(x)^2 dx}{\int_0^1 h(x)^2 dx} \cdot \sum_{1}^{d} a_i^{-2} \Big(\int_{T^d} \varphi \cdot f_{\lambda_i^2}^{(2)} d\lambda\Big).$$

*Here, $\rho$ can be a function of $n$ as long as $\rho^{-1} = o(N^{\frac{1}{4d}})$. The term $o(1)$ is uniform in $\rho$.*

The proof is given in Appendix 1.

Let's analyze this result. If $\rho$ is fixed, the bias is of order $N^{-\frac{2}{d}}$ thus smaller than $N^{-\frac{1}{2}}$ for $d = 1, 2$ and 3. If $\rho = \rho_n \to 0$, $C$ behaves like $\rho_n^{-1}$ so that the bias is like $o(N^{-\frac{2}{d}+\frac{1}{4d}})$, which is still smaller than $N^{-\frac{1}{2}}$. That is, for $d = 1, 2, 3$,

$$\lim_n E(\sqrt{N}(J_n^h(\varphi) - J(\varphi)) = 0.$$

The asymptotic covariance between $J_n^h(\varphi_1)$ and $J_n^h(\varphi_2)$ is considered in the following property:

**(4.1.4) Theorem.** *Assume $X$ is a fourth-order stationary field, with $f_4$, the fourth-order cumulant spectral density, continuous. Then, if $\varphi_1, \varphi_2 \in \mathcal{C}(T^d)$ are even,*

$$\lim_n N \operatorname{cov}(J_n^h(\varphi_1), J_n^h(\varphi_2)) = (2\pi)^d e(h)\Big[2\int_{T^d} \varphi_1\varphi_2 f^2 d\lambda + \int_{T^{2d}} \varphi_1(\alpha)\varphi_2(\beta) f_4(\alpha, -\alpha, \beta) d\alpha d\beta\Big] \equiv \Delta.$$

*The taper factor is given by*

$$e(h) = \Big[\Big(\int_0^1 h^2(x)dx\Big)^{-2}\Big(\int_0^1 h^4(x)dx\Big)\Big]^d \geq 1.$$

We have $e(h) = 1$ if there isn't a taper. It is possible to choose $\rho_n$ *s.t.* $e(h_n) \to 1$. The proof is given in Appendix 2.

**(4.1.1) Corollary.** *Assume $\varphi_1$,$\varphi_2$, and $f$ are continuous. Theorem* 4.1.4 *can be applied in the following cases:*

(*a*) *$X$ is fourth-order stationary, there exists $\delta > 0$ such that*

$$\|X\|_{4+2\delta} = \sup_i \|X_i\|_{4+2\delta} < \infty,$$

*and $X$ is $\alpha$-mixing, with mixing coefficient such that for $\alpha = \alpha_{1,3}$ and $\alpha = \alpha_{2,2}$*

$$\sum_{m\geq 1} m^{3d-1}\alpha(m)^{\frac{\delta}{2+\delta}} < \infty.$$

(*b*) *$X$ is a linear process: $X_t = \sum_{\mathbb{Z}^d} a_s\varepsilon_{t-s}$, $a = (a_s) \in \ell^1$, and $\varepsilon$ is an i.i.d. sequence of centered r.v. with fourth-order cumulant, $\kappa_4 < \infty$.*

*Proof*: It is enough to check the continuity of $f_4$. If $X$ is $k$-th order stationary, the $k$-th order cumulant spectral density is given by

$$f_k(\lambda_1, \ldots, \lambda_{k-1}) = (2\pi)^{-d(k-1)} \sum_{u_1,\ldots,u_{k-1}\in\mathbb{Z}^d} c_k(u_1, \ldots, u_{k-1}) \exp\Big(-i\sum_{j=1}^{k-1} < u_j, \lambda_j >\Big),$$

with $c_k(u_1, \ldots, u_{k-1}) = cum(X_0, X_{u_1}, \ldots, X_{u_{k-1}})$. The cumulants' summability assure the existence and continuity of $f_k$. This summability is proved in Appendix 3.

In the case $(b)$ of a linear process, $\sigma^2 = var(\varepsilon_t)$, $f_4$ evaluated at $(\alpha, -\alpha, \beta)$ is

$$f_4(\alpha, -\alpha, \beta) = (2\pi)^{-d} \frac{\kappa_4}{\sigma^4} f(\alpha) f(\beta).$$

□

**(4.1.5) Theorem.** Asymptotic Normality of $J_n^h(\varphi)$. *Let $X$ be fourth-order stationary, with spectral density $f$ and the taper function in $\mathcal{C}^2$. Assume $f_4$ is continuous and $\frac{n_i}{n} \to a_i \in \,]0,\infty[$, $i = 1,d$. Let $\varphi_1, \ldots, \varphi_p$ be $p$ continuous even functions over $T^d$ such that the matrix*

$$\Delta = \lim_{n\to\infty} (|N| \operatorname{Cov}(J_n^h(\varphi_i),\ J_n^h(\varphi_j)),\ i,j = 1,p)$$

*is invertible. Also assume that there exists $\delta > 0$ s.t. $\|X\|_{4+2\delta} < \infty$, $X$ $\alpha$-mixing with*

$$\sum_{m\geq 1} m^{d-1} \alpha_{4,\infty}(m)^{\frac{\delta}{2+\delta}} < \infty.$$

*Then, for dimensions $d = 1,2,3$ of the lattice $\mathbb{Z}^d$, we have*

$$(\sqrt{N}(J_n^h(\varphi_i) - J(\varphi_i)),\ i = 1,p) \xrightarrow{\mathcal{L}} \mathcal{N}_p(0, \Delta).$$

*Proof:* It is enough to check the Theorem for a linear combination $\varphi$ of $\varphi_i$ for which

$$\sigma^2 = \lim(N \cdot Var\ J_n^h(\varphi)) > 0.$$

As $\varphi$ is continuous, it can be uniformly approached by Cesaro–Fejer sums: for all $\varepsilon > 0$, there exists $K = (K_1, \ldots, K_d)$, $K_i > 0$ for $i = 1$, $d$ such that

$$\varphi_K(\lambda) = \sum_{[-K,K]} \widehat{\varphi}_k \exp i < \lambda, k > \prod_1^d \left(1 - \frac{|k_i|}{K_i}\right) \text{ and } \|\varphi - \varphi_K\|_\infty \leq \varepsilon.$$

Define

$$Z_n^\varepsilon = N^{1/2}(J_n^h(\varphi_K - \varphi) - J(\varphi_K - \varphi)).$$

Uniformly in $n$, $(Z_n^\varepsilon)$ is small in probability (Theorem 4.1.3 yields $|EZ_n^\varepsilon| \leq C\varepsilon$ and Property 4.4, $Var\ Z_n^\varepsilon < C'\varepsilon^2$). On the other hand, if $\sigma_K^2 = \lim_n(N\ Var\ J_n^h(\varphi_K))$, then $\lim_{\varepsilon\to 0} \sigma_K^2 = \sigma^2$. So that it is enough to check for fixed $K$ the asymptotic normality of

$$J_n^h(\varphi_K) = \sum_{[-K,K]} \widehat{\varphi}_k \widehat{R}_n^h(k) \prod_1^k \left(1 - \frac{|k_i|}{K_i}\right).$$

Theorem 3.3.1 ends the proof. □

*Empirical Estimators $\widehat{R}_n(k)$, Unbiased Estimators $R_n^*(k)$, or Tapered Estimators $\widehat{R}_n^h(k)$: Which One?*

We shall summarize the advantages and disadvantages of each one of these estimators.

(1) *Empirical Estimation* $\widehat{R}_n(k)$: If the dimension of the lattice is $d = 1$ (time series), the bias does not contribute to the asymptotics of $J_n(\varphi)$. However, it does contribute for $d = 2$, and for $d \geq 3$ it becomes the dominating term. Matrix $(\widehat{R}_n)$ is always positive definite.

(2) *Unbiased Estimation* $R_n^*(k)$: By construction, whatever the dimension of the lattice, there is not a bias problem for $J_n^*(\varphi)$. These estimators have two drawbacks: $R_n^*$ is not *a.e.* positive definite, and for big $k$, $R_n^*(k)$ varies more than $\widehat{R}_n(k)$. Both problems are lessened in the following circumstances:

- $n$ big: indeed, $P(R_n^*$ is positive definite $) \longrightarrow 1$ if $n \longrightarrow \infty$.
- $k \in V$ a bounded set of neighbors of 0.

(3) *Tapered Estimators* $\widehat{R}_n^h(k)$: Up to $d \leq 3$, there are no bias problems for $J_n^h(\varphi)$. There is, however, a possible efficiency loss introduced by coefficient $e(h)$ (cf. Theorem 4.1.4). These results are proved only for $d \leq 3$.

An additional advantage of data tapering is to provide better estimators for spectra with peaks and to lessen any nonstationarity effect which could be present on the boundary of the observation domain.

## 4.2. Parametric Models and Estimation by Gaussian Contrasts

Let $X$ be a second-order stationary, centered field with spectral density $f_\theta$, with $\theta \in \Theta$ a compact of $\mathbb{R}^p$. If $X$ is Gaussian, $f_\theta$ specifies the distribution of $X$ completely. If not, $f_\theta$ is a *second-order specification* of $X$. We define $P_{\theta_0} = P_0$ as the distribution of $X$ under $\theta_0$. Consider the following hypothesis:

(H1) *Over the taper and the sequence* $(D_n)$:

(1) $D_n = \prod_{i=1,d}[1, n_i]$, $\frac{n_i}{n} \rightarrow a_i \in ]0, \infty[$, $i = 1, d$ and $N = n_1 \cdots n_d$.

(2) The taper function $w$ belongs to $\mathcal{C}^2$ and if $\rho = \rho_n$, $\rho_n = o(N^{1/4d})$.

(H2) *Over the parametric model*:

(1) $\Theta$ is relatively compact and $\theta_0 \in \overset{\circ}{\Theta}$ interior of $\Theta \subseteq \mathbb{R}^p$.

(2) There exists $0 < m \leq M < \infty$ such that over $\Theta$, $m \leq f_\theta \leq M$ and $f_{\theta^2}^{(2)}(\lambda)$ exists and is continuous in $(\theta, \lambda)$.

(3) If $\theta \neq \alpha$, $f_\theta \neq f_\alpha$ in $L^1(T^d, d\lambda)$.

(H3) *Over the field* $X$:

(1) $X$ is fourth-order stationary and $f_{4,\theta_0}$ is continuous.

(2) There exists $\delta > 0$ such that $\|X\|_{4+2\delta} < \infty$ and $X$ is strongly mixing with

$$\sum_{m \geq 0} m^{d-1} \alpha_{4,\infty}(m)^{\delta/2+\delta} < \infty$$

The continuity of $f_{4,\theta_0}$ holds if, for example (cf. Appendix 3(a)),

$$\sum_{m \geq 1} m^{3d-1} \alpha_{2,\infty}(m)^{\delta/2+\delta} < \infty.$$

The *Gaussian Contrast* (Whittle [168]) $U_n$ with tapered data is defined by, writing $I_n = I_n^h$,

$$2 \cdot (2\pi)^d \cdot U_n(\theta) = \int_{T^d} [\log f_\theta(\lambda) + f_\theta^{-1}(\lambda) I_n(\lambda)] d\lambda, \tag{4.11}$$

and an associated estimator is

$$\widehat{\theta}_n = \operatorname{Arg} \operatorname*{Min}_{\Theta} U_n(\theta).$$

Define the following matrices

$$\Gamma(\theta)_{ij} = \frac{1}{2}(2\pi)^{-d} \int_{T^d} \frac{\partial}{\partial \theta_i} \log f_\theta \cdot \frac{\partial}{\partial \theta_j} \log f_\theta d\lambda,$$

$$B(\theta)_{ij} = \frac{1}{4}(2\pi)^{-d} \int_{T^{2d}} \frac{f_{4,\theta}(\lambda, -\lambda, \mu)}{f_\theta(\lambda) f_\theta(\mu)} \frac{\partial}{\partial \theta_i} \log f_\theta(\lambda) \frac{\partial}{\partial \theta_j} \log f_\theta(\mu) d\lambda d\mu.$$

**(4.2.1) Theorem.** *Assume that hypotheses* (H1), (H2) and (H3) *hold. Then*

(1) $\widehat{\theta}_n$ *is consistent.*

(2) *If* $\Delta_0 = \Gamma(\theta_0)^{-1}[\Gamma(\theta_0) + B(\theta_0)]\Gamma(\theta_0)^{-1}$ *is invertible, then for* $d =$ 1,2,3 *and for* $e(h)$ *as in Property* 4.4, *we have*

$$N^{1/2}(\widehat{\theta}_n - \theta_0) \xrightarrow{\mathcal{D}} \mathcal{N}_p(0, e(h)\Delta_0).$$

(3) *If* $X$ *is Gaussian and if* $\rho_n \to 0$, *then* $\widehat{\theta}_n$ *is asymptotically efficient for* $d =$ 1,2,3.

REMARKS:

(1) The Theorem remains true for all $d \geq 1$ if we consider the unbiased periodogram $I_n^*$ instead.

(2) If $X$ has a causal linear representation, with i.i.d. innovation $(\varepsilon_j)$ with variance $\sigma^2$, and fourth-order cumulant $\kappa_4$, then

$$f_{4,\theta_0}(\lambda, -\lambda, \mu) = (2\pi)^{-d}\frac{\kappa_4}{\sigma^4} f(\lambda) f(\mu).$$

Considering parametrization $\theta = (\sigma^2, \varphi)$ with $f_\theta(\lambda) = \sigma^2 g_\varphi(\lambda)$, asymptotically there is no correlation between $\widehat{\sigma_n^2}$ and $\widehat{\varphi_n}$ since

$$\int \log g_\varphi(\lambda) d\lambda = 0 \quad \text{and} \quad \Delta_0 = \begin{pmatrix} 2\sigma^4 + \kappa_4 & 0 \\ 0 & \Gamma_g^{-1}(\varphi_0) \end{pmatrix}.$$

*Proof:*

(1) *Consistency:* It is enough to check conditions (3.13) and (3.15) of Theorem 3.3. Let $\eta(\varepsilon)$ be the modulus of continuity of $(\lambda, \theta) \longrightarrow f_\theta^{-1}(\lambda)$. Defining

$$V_n(\alpha) = \int_{T^d} f_\alpha^{-1}(\lambda) I_n(\lambda) d\lambda,$$

we have $\sup\{|V_n(\alpha) - V_n(\theta)|,\ \alpha, \theta \in \Theta,\ |\alpha - \theta| \leq \eta(\varepsilon)\} \leq \varepsilon \cdot (2\pi)^d \widehat{R}_n(0)$. This gives (3.15). On the other hand, $V_n(\theta) \xrightarrow{P_0-\text{prob.}} J(f_\theta^{-1})$, so that

$$U_n(\theta) - U_n(\theta_0) \xrightarrow{P_0-\text{prob.}} K(\theta_0, \theta), \quad \text{with}$$

$$K(\theta_0, \theta) = \frac{1}{2(2\pi)^d} \int_{T^d} \Big[\Big(\frac{f_\theta}{f_{\theta_0}} - 1\Big) - \log\Big(\frac{f_\theta}{f_{\theta_0}}\Big)\Big] d\lambda \geq 0.$$

Hypothesis (H2(3)) assures $K(\theta_0, \theta) > 0$ if $\theta \neq \theta_0$.

(2) *Asymptotic Normality:* With tending to 1 probability, $\widehat{\theta}_n \in \mathring{\Theta}$ so that, following the proof of Theorem 3.4.5:

$$U_n^{(1)}(\widehat{\theta}_n) = 0 = U_n^{(1)}(\theta_0) + \Delta_n(\theta_0, \widehat{\theta}_n)(\widehat{\theta}_n - \theta_0).$$

• Since $2(2\pi)^d U_n^{(1)}(\theta_0) = (J_n(\varphi_i) - J(\varphi_i), i = 1, p)$, $\varphi_i = -f_{\theta_0}^2 \cdot \dfrac{\partial}{\partial \theta_i} f_{\theta_0}$, Theorem 4.1.5 yields

$$2(2\pi)^d N^{1/2} U_n^{(1)}(\theta_0) \xrightarrow{\mathcal{D}} \mathcal{N}_p(0, e(h)(\Gamma(\theta_0) + B(\theta_0))).$$

•• On the other hand,

$$2 \cdot (2\pi)^d U_n^{(2)}(\theta) = \int_{T^d} \Big\{\Big(2\frac{f_\theta^{(1)} f_\theta^{(1)'}}{f_\theta^3} - \frac{f_\theta^{(2)}}{f_\theta^2}\Big) I_n(\lambda) + \Big(\frac{f_\theta^{(2)}}{f_\theta} - \frac{f_\theta^{(1)} f_\theta^{(1)'}}{f_\theta^2}\Big)\Big\} d\lambda.$$

Since $\widehat{\theta}_n \xrightarrow{\text{prob.}} \theta_0$, we can deduce that $U_n^{(2)}(\theta_n^*) \xrightarrow{\text{prob.}} \Gamma(\theta_0)$, which gives (2).

(3) *Efficiency in the Gaussian Case:* If $\rho_n \to 1$ under (H1(2)), $e(h_n) \to 1$. In the case of a Gaussian field, $\Delta_0 = \Gamma(\theta_0)^{-1}$. Still under the

hypothesis of Gaussianity, if we define $(-NU_n^*(\theta))$ as the log likelihood if $\theta$, then the efficiency holds as a consequence of

$$\sup_{V(\theta_0)} \|U_n^{(k)}(\theta) - U_n^{*(k)}(\theta)\|_1 = o_{P_0}(1), \quad \text{for } k = 0, 1, 2$$

with $V(\theta_0)$ a bounded neighborhood of $\theta_0$. □

The heuristics that allow us to approximate $NU_n^*(\theta) = -\frac{1}{2}\log|\Sigma_n(\theta)| -\frac{1}{2}{}^tX(n)\Sigma_n(\theta)^{-1}X(n)$, the log likelihood of observation $X(n)$ by $U_n(\theta)$ in the Gaussian case are equivalent to those considered for $d = 1$ [38]. If $d > 1$, it is necessary to consider the lexicographic order and the causal $AR(\infty)$ representation of $X$.

*Hypothesis Testing by Difference of Contrasts.*

Let $\Lambda$ be a compact of $\mathbb{R}^q$, $q < p$, $\alpha_0$ in the interior of $\Lambda$, $r : \Lambda \longrightarrow \Theta$ a function of class $\mathcal{C}^2$ over the interior of $\Lambda$, such that

$$R_0 = \frac{\partial r}{\partial \alpha}(\alpha_0)$$

has full rank: $\theta_0 = r(\alpha_0)$ if $(H_q)$ is true. With the notation of Property 3.8, in particular that of conditions (H2(2)) and (H3(2)), we have, following the proof of the asymptotic normality of $\widehat{\theta}_n$:

$$\text{Under } (H_p): \lim_n I_{n_0} = I_0 = \Gamma(\theta_0), \quad \lim_n J_{n_0} = J_0 = e(h)(\Gamma(\theta_0) + B(\theta_0)),$$

$$\text{Under } (H_q): \lim_n \overline{I}_{n_0} = \overline{I}_0 = \Gamma(\alpha_0) = \overline{\Gamma}_0,$$

Since $I_0$ and $J_0$ are invertible, we have that both matrices

$$A_0 = e(h)(\Gamma_0 + B_0)^{\frac{1}{2}}[\Gamma_0^{-1} - R_0\overline{\Gamma}_0^{-1}R_0'](\Gamma_0 + B_0)^{\frac{1}{2}}$$

and $P_0Q_0$ with

$$P_0 = (I_p - \Gamma_0^{\frac{1}{2}}R_0\overline{\Gamma}_0^{-1}R_0'\Gamma_0^{\frac{1}{2}}), \; Q_0 = e(h)\Gamma_0^{-\frac{1}{2}}(\Gamma_0 + B_0)\Gamma_0^{\frac{1}{2}}$$

have the same eigenvalues. Let $\widehat{\alpha}_n$ be the minimum contrast estimator under $(H_q)$, then we have the following result:

**(4.2.2) Theorem.** *Under the conditions of Theorem* 4.2.1, *the difference of contrast statistic* $\Delta_n$,

$$\Delta_n = 2N(U_n(\widehat{\alpha}_n) - U_n(\widehat{\theta}_n))$$

*converges in distribution under* $(H_q)$ *to a mixture of independent* $\chi_1^2$,

$$Z = \sum_{i=1,p-q} \lambda_i \chi_{i,1}^2$$

*with* $\lambda_i$ *the* $(p-q)$ *nonzero eigenvalues of* $A_0$ *(or* $P_0Q_0$*).*

REMARKS:

(1) If $X$ is Gaussian, $B(\theta_0) = 0$, $Q_0 = e(h)I$ and since $P_0$ is idempotent, the statistic behaves as $e(h)\chi^2_{p-q}$.

(2) More generally, we have the same result in the following context. Assume $X$ is parametrized with respect to its causal $AR$ representation, and its innovation $(\varepsilon_i)$ an independent process with variance $\sigma^2$. Consider parametrization $\theta = (\sigma^2, \varphi)$, with

$$f_\theta(\lambda) = \sigma^2 g_\varphi(\lambda), \qquad \int \log g_\varphi(\lambda) d\lambda = 0.$$

If $(H_q)$ is a subhypothesis involving $\varphi$ and not $\sigma^2$, we have

$$Q_0 = \begin{pmatrix} 1 + \frac{1}{4}\frac{\kappa_4}{\sigma^4} & 0 \\ 0 & I_{p-1} \end{pmatrix},$$

$$P_0 = \begin{pmatrix} 0 & 0 \\ 0 & I_{p-1} - \Gamma_g(\varphi_0)^{\frac{1}{2}} R_0 \Gamma_g(\alpha_0)^{-1} R_0' \Gamma_g(\varphi_0)^{\frac{1}{2}\prime} \end{pmatrix} = P_0 Q_0.$$

The statistic also behaves as $e(h)\chi^2_{p-q}$.

(3) A multidimensional version of Theorems 4.7 and 4.8 is considered by Yao [171].

(4) Specific results for separable processes are given in [118'].

## 4.3. Parametric Estimation: Gaussian Markovian Fields

Let $X$ be the stationary, $L$-Markovian field (cf.1.3.4):

$$\begin{cases} X_t = \sum_L c_s X_{t+s} + e_t, & e_t \perp X_s \quad \text{if } s \neq t, \quad t \in \mathbb{Z}^d, \\ 0 \notin L, \quad c_s = c_{-s}, \quad L = -L. \end{cases}$$

Calling $L^+$ the positive part of $L$ and $M = \{0\} \cup L^+$, $p = |M|$, the spectral density is given by

(4.12)
$$f_\theta(\lambda) = (\sum_M \theta_s \cos <\lambda, s>)^{-1}, \ \theta_0 = \sigma_e^{-2}, \text{ and } \theta_s = -2\sigma_e^{-2} c_s \ \textit{for } s \neq 0.$$

Model (4.12) is defined over the set

$$(4.13)\quad \Theta = \{\theta \in \mathbb{R}^p, \ s.t. \ f_\theta^{-1}(\lambda) = \sum_M \theta_s \cos <\lambda, s> \geq 0, \ f_\theta \text{ integrable}\}.$$

**(4.3.1) Theorem.**
(1) *If* $d \leq 2$, $\Theta = \{\theta \in \mathbb{R}^p : f_\theta^{-1}(\lambda) = \sum_M \theta_s \cos(\lambda, s) > 0$, for each $\lambda\}$.
(2) *If* $d \geq 3$, *there exist* $\theta$ *in* $\Theta$ *such that* $f_\theta^{-1}$ *is zero for some* $\lambda$.

*Proof*: Assume $f_\theta^{-1}(\lambda_0) = 0$. Since $\lambda_0$ is a minimum of $f_\theta$, its Taylor development in a neighborhood of $\lambda_0$ yields $f_\theta^{-1}(\lambda) \leq a\|\lambda - \lambda_0\|_2^2, \quad a < \infty$. Now $\|\lambda - \lambda_0\|^{-2}$ is only integrable for $d \geq 3$. This gives (1). Over $T^d$ and if $d \geq 3$, $f(\lambda) = \left(1 - d^{-1}\Sigma_{1=1,d} \cos \lambda_i\right)^{-1}$ is positive and integrable, and is infinite in $\lambda = 0$. This gives (2). □

Let $W$ be the trace over $M$ of all the covariances defined over $\mathbb{Z}^d$:

$$\text{(4.14)} \qquad \begin{cases} W = \{R_k,\ k \in M,\ R_k = (2\pi)^{-d} \displaystyle\int g(\lambda) \cos < \lambda, k > d\lambda, \\ \qquad\qquad\qquad \text{for every spectral density } g \not\equiv 0\}. \end{cases}$$

Consider $\phi : \Theta \longrightarrow W$ defined by

$$\text{(4.15)} \qquad \phi_k(\theta) = R_k(\theta) = (2\pi)^{-d} \int f_\theta(\lambda) \cos < \lambda, k > d\lambda,\ k \in M,$$

and consider the following equation for $(R_k)_{k \in M}$ of $W$, at $\theta \in \Theta$:

$$\text{(4.16)} \qquad \phi_k(\theta) = R_k, \quad k \in M.$$

The phenomenon described in Theorem 4.3.1 yields the following result:

**(4.3.2) Theorem.** (Künsch [105])

(1) *If* $d \leq 2$, *Equation* (4.16) *has one and only one solution: given* $(R_k)_{k \in M}$ *the trace of a covariance over* $M$, *there exists one and only one* $\theta$ *such that the covariance defined over* $M$ *coincides with that defined by* $f_\theta$.

(2) *If* $d \geq 3$, (4.16) *does not always have a solution.*

*Proof:*

(1) If $d \leq 2$, $\Theta$ is open and $\phi$, which is differentiable over $\Theta$ satisfies

$$\sum_{j,k \in M} c_j \frac{\partial \phi_k}{\partial \theta_j} c_k = -(2\pi)^d \int \Big[\sum_{k \in M} c_k \cos < k, \lambda >\Big]^2 f_\theta(\lambda)^2 d\lambda < 0.$$

The Jacobian of $\phi$ is strictly negative, which assures the uniqueness (if it exists) of the solution of (4.16). We have to show that $\phi(\Theta) = W$. In order to do this, we shall prove

$$\text{(4.17)} \qquad \partial\phi(\Theta) \cap W = \emptyset.$$

This condition entails the existence of a solution; indeed, $W$ is connected since it is convex and if $\phi(\Theta) \subsetneq W$, $W$ could be written as the union of two nonempty open sets $W = \phi(\Theta) \cup ((\overline{\phi(\Theta)})^c \cap W)$ which is impossible. Assume that (4.17) is not true and let $(R_k)_{k\in M} \in \partial\phi(\Theta) \cap W$. Thus, there exists a sequence $(\theta^n)$ of $\Theta$ such that

$$\lim_n R_k(\theta^n) = R_k, \quad \lim_n \frac{\theta_k^n}{\theta_0^n} = \rho_k, \quad k \in M.$$

Let $P(\lambda) = \sum_M \rho_k \cos < \lambda, k >$ and $g \not\equiv 0$ be the spectral density associated to $(R_k)_{k\in M}$, $\varepsilon > 0$:

$$\frac{\int P(\lambda) g(\lambda) d\lambda}{\int g(\lambda) d\lambda} = \frac{\sum_M \rho_k R_k}{R_0} = \lim_n \frac{\sum_M \rho_k R_k(\theta^n)}{R_0(\theta^n)}$$

$$= \lim_n \frac{\int P(\lambda) \left[\sum_M \frac{\theta_k^n}{\theta_0^n} \cos < \lambda, k >\right]^{-1} d\lambda}{\int \left[\sum_M \frac{\theta_k^n}{\theta_0^n} \cos < \lambda, k >\right]^{-1} d\lambda}$$

$$\leq \varepsilon + \left(\liminf \int \left[\sum_M \frac{\theta_k^n}{\theta_0^n} \cos < \lambda, k >\right]^{-1} d\lambda\right)^{-1}$$

since if $n$ is sufficiently big, $P(\lambda) \leq \left(\sum_M \frac{\theta_k^n}{\theta_0^n} \cos < \lambda, k >\right) + \varepsilon$. As $(R_k)_{k\in M} \notin \phi(\Theta)$, and $R_0 \neq 0$, $P(\lambda)$ has at least one zero. By Fatou's Lemma and since $\varepsilon$ is arbitrary, we have $\int_{T^d} P(\lambda) g(\lambda) d\lambda = 0$. But this is impossible since $P$ has only a finite number of zeros. Hence, (4.17) is satisfied, which ends part (1) of the Theorem.

(2) $d \geq 3$: Consider the Markovian model with $d$ nearest neighbors,

$$M = \{k \in \mathbb{Z}^d, \ k > 0, \ \sum_1^d |k_i| \leq 1\}.$$

Choose $(R_k)_{k\in M}$ isotropic

$$R_k = \rho \ R_0 \text{ if } k \neq 0, \ k \in M, \ R_0 > 0, \ \rho \neq 0. \tag{4.18}$$

If $\theta$ is a solution of (4.6), by symmetry we have that $\theta_k \equiv \theta_1$ if $k \in M$, $k \neq 0$ so that we can restrict $\Theta$ to $\Theta = \{\theta = (\theta_0, \theta_1), \ \theta_0 > 0, \ |\theta_1| \leq d^{-1}\theta_0\}$. As the model's correlation $\rho_\theta$ is a continuous function of $\theta_1 \cdot \theta_0^{-1}$, the set of all possible correlations for such an isotropic model is a closed subset $F$ of $]-1, 1[$. In order to establish (2), it is sufficient to check that in (4.18), we can choose $\rho \neq 1$ as close to 1 as we wish to. Indeed, consider the $MA$ process $Z_t = \sum_{s\in[-\ell,\ell]^d} \varepsilon_{t+s}$; since its correlation with its neighbors is given by $\rho = \dfrac{2\ell}{2\ell+1}$, now choose $\ell$ big enough so that $\rho \notin F$. □

### *4.3.1. Maximum Likelihood Estimation*

The contrast is given by

$$2(2\pi)^d U_n(\theta) = \int_{T^d} \Big[ -\log\Big(\sum_M \theta_s \cos < \lambda, s >\Big) + \Big(\sum_M \theta_s \cos < \lambda, s >\Big) I_n^h(\lambda)\Big] d\lambda.$$

If $d \leq 2$, and if the taper is regular, hypotheses (H1–H3) of Theorem 4.7 hold (for (H3), since $f_{\theta_0}$ is analytical in a neighborhood of the torus, the mixing is exponentially decreasing because of Theorem 1.7.2). Hence, with a probability converging to 1, $\widehat{\theta}_n \in \Theta$ is an open set and satisfies $U_n^{(1)}(\widehat{\theta}_n) = 0$. Also consider

$$R_{\widehat{\theta}_n}(k) = \widehat{R}_n^h(k), \quad k \in M. \tag{4.19}$$

Since $(\widehat{R}_n^h(k), k \in M)$ is the trace over $M$ of a covariance over all $\mathbb{Z}^d$, $(\widehat{R}_n^h, k \in \mathbb{Z}^d)$, we have the following Corollary:

**(4.3.1) Corollary.** *If $X$ is a Markovian field with density* (4.12) *and if $d \leq 2$, there exists one and only one solution for the Gaussian minimum contrast estimation problem.*

REMARKS:

(1) For any given $d$, if $\theta_0$ is in the interior of $\Theta$ and $X$ is ergodic, $\widehat{\theta}_n$ is consistent and satisfies (4.19) with tending to 1 probability.

(2) Consider $\Theta_1 = \{\theta \in \mathbb{R}^p : \sum_{k \in M, k \neq 0} |\theta_k| < \theta_0\}$. $\theta \in \Theta_1$ is a sufficient condition in order for $f_\theta$ to be a spectral density; however, it is not necessary: $\Theta_1 \not\subseteq \Theta$.

**(4.3.1) Example.** $d = 1$, $f^{-1}(\lambda) = 1 - \cos\lambda + a\cos 2\lambda > 0$ if $a > 0$.

**(4.3.2) Example.** $d \geq 2$, and let $f^{-1}(\lambda) = 1 - \frac{1}{d}\sum_1^d \cos\lambda_i + \gamma\cos\Big(\sum_i^d \lambda_i\Big)$. If $\gamma > 0$ is small enough, then $f^{-1}(\lambda) > 0$. Indeed, pick $0 < \varepsilon < \frac{1}{2d^2}$ so that if $\theta_0 \in ]0, \frac{\pi}{2}]$, $\cos\theta_0 = 1 - \varepsilon$. Thus, $\cos(d\theta_0) > 1 - 2d^2\varepsilon$. Now pick $0 < \gamma < \varepsilon$.

- If $\lambda_i \in [-\theta_0, \theta_0]$, $i = 1, d$ : $f^{-1}(\lambda) \geq \gamma\cos(d\theta_0) > 0$.
- If for $k$ values, $\lambda_i \in [-\theta_0, \theta_0]$, and the other $(d-k)$ $\lambda_i \notin [-\theta_0, \theta_0]$,

then

$$f^{-1}(\lambda) \geq 1 - \frac{k}{d}(1-\varepsilon) - \frac{d-k}{k} - \gamma = \frac{d\varepsilon}{d} - \gamma \geq \frac{\varepsilon}{d} - \gamma > 0.$$

**(4.3.3) Example.** $d \geq 3$, $f^{-1}(\lambda) = 1 - \dfrac{1}{d}\sum_1^d \cos\lambda_i$

*$X$ is Nonstationary: Choosing the Parametrization, Maximum Likelihood and Least Squares.*

If $X \sim \mathcal{N}_N(\mu, \Sigma)$, the log likelihood is given by

$$(4.20) \qquad \ell_N(\mu, \Sigma) = -\frac{N}{2}\log(2\pi) - \frac{1}{2}\log|\Sigma| - \frac{1}{2}(X-\mu)'\Sigma^{-1}(X-\mu).$$

It is natural to consider the parametrization of $\Sigma$ by $\Sigma^{-1}$. Let: $\Sigma = \kappa\, Q^{-1}$; $\kappa$ will have different interpretations according to the underlying model.

- *CAR Modeling* (cf. 1.2.5): $\kappa$ is the variance of the conditional residuals in the spatial bilateral representation

$$X_i = \sum_{j \neq i} c_{i,j} X_j + e_i, \quad \kappa = Var\ e_i = \sigma_e^2, \quad e_i \perp X_j \text{ if } j \neq i\ , \quad c_{ij} = c_{ji},$$

$$Q = (I - C).$$

- *Causal AR modeling* (cf. (1.24)): For any given order of $S = \{1, 2, \ldots, n\}$, there exists a causal $AR$ representation, $Q = {}^t AA$ with

$$\sum_{j=1,i} a_{i,j} X_j = \varepsilon_i, \quad a_{i,i} = 1, \quad Var\ \varepsilon_i = \sigma_\varepsilon^2 = \kappa \quad i = 1, n.$$

Rewriting (4.20) in terms of $\kappa$ and $Q$,

$$(4.21)\ \ell_N(\mu, \kappa, Q) = -\frac{N}{2}\log(2\pi\kappa) + \frac{1}{2}\log|Q| - \frac{1}{2}\kappa^{-1}(X-\mu)'Q(X-\mu).$$

We observe the following:

(1) *For any causal model,* $|Q| = 1$, only the least squares part (i.e., the quadratic form) will appear. This is the standard situation for time series with a causal representation: in terms of $Q$, the $ML$ (Maximum Likelihood) and the $LS$ (Least Squares) methods are equivalent.

(2) *For a spatial bilateral CAR representation,* $\log|Q|$ appears explicitely in the estimation: $|Q|$ and its derivatives must be calculated at each step of the optimization algorithm. In this case, the $ML$ and the $LS$ methods differ from each other.

(3) For a *spatial CAR modeling,* we can consider another least squares estimation procedure, *conditional least squares (CLS),* which minimize :

$$U_n = \sum_1^n e_i^2 = \|QX\|^2.$$

Like we shall see below, the *estimation by CLS is consistent.* If $p_i(x_i|\cdot)$ is the conditional Gaussian distribution of $X_i$, we observe that

$$p\ell_N = \prod_{i=1,n} p_i(x_i \mid \cdot) = \prod_{i=1,n} \left(\frac{1}{2\pi\kappa}\right)^n e^{-\frac{1}{2}\kappa^{-1}U_n}.$$

$CLS$ is also called estimation by *conditional pseudo-likelihood.* This method shall be considered in length in Chapter 5.

(4) For a *simultaneous, noncausal AR modeling,* $|Q|$ is no longer a constant and appears in the likelihood. In particular, the $ML$ and $LS$ methods do not behave alike in the estimation of the residuals: more so, *$LS$ estimation is not consistent.* This is a well-known result in econometric simultaneous equations. Indeed, consider the simplest example in $\mathbb{Z}$:

$$X_i = a(X_{i-1}X_{i+1}) + \varepsilon_i, \quad i \in \mathbb{Z},\ \{\varepsilon_i\} \text{ a white noise, } |a| < \tfrac{1}{2}.$$

Estimation by $LS$, which minimizes $\sum_2^{n-1} \varepsilon_i^2$, yields, with $r_k = E(X_0 X_k)$,

$$\widehat{a}_n = \Big[\sum_2^{n-1} X_i(X_{i-1} + X_{i+1})\Big]\Big(\sum_2^{n-1}(X_{i-1} + X_{i+1})^2\Big)^{-1},$$

$$P\ \lim(\widehat{a}_n - a) = \frac{r_1 - a(r_0 + r_2)}{r_0 + r_2} \neq 0 \text{ if } a \neq 0.$$

Consider again the estimation by $ML$ of a $CAR$ model. If

$$\sup_i \sum_{j \in \partial i} |c_{i,j}| \leq \rho < 1, \tag{4.22}$$

the field will be exponentially $\alpha$-mixing. This provides a good setting for the asymptotics. If $\mu = D\theta$, the $ML$ estimators are given by

$$\widehat{\theta} = (D'\widehat{Q}D)^{-1}D'\widehat{Q}X, \quad \widehat{\kappa} = N^{-1}(X - D\widehat{\theta})'\widehat{Q}(X - D\widehat{\theta}),$$

where $\widehat{Q}$ minimizes

$$\mathcal{M}(Q) = -N^{-1}\log|Q| + \log(N^{-1}X'Q\{I - D(D'QD)^{-1}D'Q\}X).$$

Classically considered in geography and spatial economy, the one-parameter $CAR$ models (cf. [16], [33], [80]) are defined by

$$X_i = \alpha \sum_{\partial i} w_{ij}X_j + e_i,\ i = 1, n, \quad e_i \perp X_j \text{ if } i \neq j,\ Var\ e_i = \sigma^2, \tag{4.23}$$

with given weights $w_{ij}$ $(= w_{ji})$ and $w_{ii} = 0$. Here, $Q = I - \beta W$ with $W$ the symmetrical weight matrix. Notice that

- If $\xi_1, \ldots, \xi_n$ are the eigenvalues of $W$, $\log|Q| = \sum_1^n \log(1 - \beta\xi_i)$.
- $D'QD = (I - \beta E)D'D$, with $E = D'HD(D'D)^{-1}$; $(D'QD)^{-1}$

and its derivatives can be evaluated by means of the spectral representation of $E$.

## *4.3.2. Estimation of Markovian Fields by Conditional Least Squares (CLS)*

Let $X$ be the centered Markovian field

$$X_i = \sum_{\partial i} c_{i,j} X_j + e_i, \quad e_i \perp X_j, \quad i \neq j, \quad Var\ e_i = \sigma^2.$$

Estimation by $CLS$ minimizes $U_n(\alpha) = \sum_{D_n} e_i^2$. It also minimizes the product of the conditional distributions $(X_i|\cdot)$, called the *Conditional Pseudo-Likelihood (CPL)*,

$$p\ell_n = \prod_{D_n} p_i(X_i|\cdot) = (\sqrt{2\pi}\sigma)^{-N} \exp\Big(-\frac{1}{2}\frac{U_n(\alpha)}{\sigma^2}\Big).$$

The advantage of this contrast is that it provides an explicit linear and consistent estimator.

### *4.3.2.1. Estimation by CLS: The Stationary Case*

Consider a centered, Gaussian Markovian field over $\mathbb{Z}^d$, with $p$ neighbor types defined by the partition of $L = L_1 \cup L_2 \cup \cdots \cup L_p$, $L_k = -L_k$ finite, $k = 1, p$

(4.24) $$X_i = \sum_{j \in L} c_j\ X_{i+j} + e_i, \ \ c_j = a_k \text{ if } j \in L_k, \quad \sigma^2 = Var\ e_i.$$

The parameter $\theta = (\sigma^2, a_k, k = 1, p)$ has dimension $(p+1)$. We shall assume that

(4.21′) $$\sum_L |c_j| \leq \rho < 1.$$

Under this condition, there exists a unique Gaussian, centered field specified by (4.24): this field is ergodic and $\alpha$-exponentially mixing.

*Some Definitions:*

$$\mathcal{X}_i = {}^t\Big(\sum\nolimits_{L_k} X_{i+j}, k = 1, p\Big) \in \mathbb{R}^p,$$

$${}^t\mathcal{X} = (\mathcal{X}_i, i \in \overset{\circ}{D}_n), \ \overset{\circ}{D}_n = L\text{−interior of } D_n, \ N = |\ \overset{\circ}{D}_n\ |,$$

$$X(n) = (X_i, i \in \overset{\circ}{D}_n), \ e(n) = (e_i, i \in \overset{\circ}{D}_n), \quad a = {}^t(a_k, k = 1, p).$$

(4.24) observed over $\overset{\circ}{D}_n$ can then be written

(4.25) $$X(n) = \mathcal{X}\ a + e(n).$$

The $CLS$ estimator for $a$ and the resulting one for $\sigma^2$ are given by

$$\widehat{a}_n = ({}^t\mathcal{X}\mathcal{X})^{-1}\,{}^t\mathcal{X}X(n), \quad \widehat{\sigma}^2_n = \frac{1}{N}\sum_{\overset{\circ}{D}_n} e_i^2(\widehat{a}_n). \tag{4.26}$$

We shall assume that $(D_n)$ is an increasing sequence of convex sets whose interior diameter tends to infinity. Also define

$$\begin{cases} \Delta = V^{-1}GV^{-1} \quad V = Cov(\mathcal{X}_0) \quad (V \text{ is invertible under } (4.21')), \\ G = \sigma^2[V + \sigma^2 IL_p - \sum_L C_\ell T_\ell], \\ IL_p = \text{Diag}(|L_k|,\ k = 1,p),\ T_\ell = Cov(\mathcal{X}_0, \mathcal{X}_\ell). \end{cases}$$

**(4.3.3) Theorem.** *Let $X$ be the Gaussian, centered Markovian field defined by* (4.24). *Then under* $(4.21)'$,

(a) $\widehat{\theta}_n = (\widehat{a}_n, \widehat{\sigma}^2_n)$ *is consistent, and*

(b) $\sqrt{N}(\widehat{a}_n - a) \xrightarrow{\mathcal{D}} \mathcal{N}_p(0,\Delta)$.

*Proof:*

(a) Because of $X$'s ergodicity, $\frac{1}{N}{}^t\mathcal{X}\mathcal{X} \xrightarrow{P} V$. Since (4.24) is a conditional specification, we have $E({}^t\mathcal{X}e(n)) = 0$ and as we shall see in the second part, $N^{-1}Var({}^t\mathcal{X}e(n)) \longrightarrow 0$. Then $\widehat{a}_n$ is consistent since

$$\widehat{a}_n - a = \left(\frac{1}{N}{}^t\mathcal{X}\mathcal{X}\right)^{-1}\left(\frac{1}{N}{}^t\mathcal{X}e(n)\right).$$

That $\widehat{\sigma}^2_n$ is consistent is a consequence of the convergence of $N^{-1}\sum_{D_n} X_i X_{i+j}$ toward $\gamma_j = E(X_0X_j)$, $\widehat{a}_n \xrightarrow{P} a$, and of the first Yule–Walker equation: $\sigma^2 = \gamma_0 - \sum_L c_s\gamma_s$.

(b) The asymptotic variance of $\widehat{a}_n$ is given by $\Delta = V^{-1}GV^{-1}$ with $G = \lim \frac{1}{N}Var({}^t\mathcal{X}e(n))$. Since $X$ is Gaussian, we have

$$Var({}^t\mathcal{X}e(n)) = A_n + B_n + C_n \text{ with } A_n = \sum_{i,j\in\overset{\circ}{D}_n} E(e_i\mathcal{X}_i)E(e_j\mathcal{X}_j) = 0,$$

$$\frac{1}{N}B_n = \frac{1}{N}\sum_{i,j\in\overset{\circ}{D}_n} E(e_i\mathcal{X}_j)E(e_j{}^t\mathcal{X}_i) \longrightarrow \sigma^4 IL_p,$$

$$\frac{1}{N}C_n = \frac{1}{N}\sum_{i,j\in\overset{\circ}{D}_n} E(e_ie_j)E(\mathcal{X}_i{}^t\mathcal{X}_j) \longrightarrow \sigma^2\Big(V - \sum_L c_\ell T_\ell\Big).$$

As $X$ is mixing, $\widehat{a}_n$ is asymptotically normal. □

We also have joint normality of $\widehat{a}_n$ and $\widehat{\sigma}_n^2$, but it must be noted that $\widehat{a}_n$ and $\widehat{\sigma}_n^2$ are asymptotically correlated.

**(4.3.4) Example.** Isotropic Processes with $\nu$–Nearest Neighbors over a Regular Lattice:

$$X_i = \alpha \sum_{\partial i} X_j + e_i, \quad |\partial i| = \nu.$$

$Var\ \widehat{\alpha}_n \simeq \frac{1}{N}\frac{2\alpha^2\sigma^2}{\nu\gamma_1}$ ($\gamma_1$ is the covariance between two nearest neighbors).

(1) Consider a causal $AR(1)$ over $\mathbb{Z}$:

$$X_i = \rho X_{i-1} + \varepsilon_i, \quad |\rho| < 1, \quad i = 1, n.$$

This model admits the equivalent $CAR$ representation:

$$X_i = \alpha(X_{i-1} + X_{i+1}) + e_i, \quad e_i \perp X_j \text{ si } i \neq j, \quad \alpha = \frac{\rho}{1+\rho^2}, \quad \sigma_e^2 = \frac{\sigma_\varepsilon^2}{1+\rho^2}.$$

Estimation by coding is obtained by maximizing the product of the conditional distributions $(X_i|\cdot)$ in one out of each two sites (cf. 5.1.3). Asymptotically, we have:

| Method | $\lim n\ Var(\widehat{\alpha}_n)$ |
|---|---|
| Coding | $(1-\rho^2)(1+\rho^2)^{-2}$ |
| $PML$ | $(1-\rho^2)^2(1+\rho^2)^{-4}$ |
| $ML$ | $(1-\rho^2)^3(1+\rho^2)^{-4}$ |

(2) For the four nearest neighbors Markovian model over $\mathbb{Z}^2$, and if $\rho$ is the correlation between two nearest neighbors, then (Besag [15], [16]; [83])

| Method | $\lim n\ Var(\widehat{\alpha}_n)$ |
|---|---|
| Coding | $\alpha(1-4\alpha\rho)/2\rho$ |
| $PML$ | $\alpha^2(1-4\alpha\rho)^2/4\rho^2$ |
| $ML$ | $2J^{-1}$ |

where

$$J = (\pi)^{-2} \int_{T^2} \left[\frac{\cos\lambda + \cos\mu}{1 - 2\alpha(\cos\lambda + \cos\mu)}\right]^2 d\lambda d\mu.$$

If $\alpha = 0$, all three methods have the same asymptotic efficiency. If $\alpha \longrightarrow \left(\frac{1}{4}\right)_-$, the $ML$ method is slightly more efficient than the $PML$ method is; as the latter with respect to the coding method.

*4.3.2.2. Estimation by CLS: The Nonstationary Case*

We shall assume that $S$ is an infinite numerable subset of $\mathbb{R}^d$ without accumulation points and that $X$ is Gaussian, Markovian and specified by

$$(4.27) \qquad \begin{cases} X_i - m_i(\alpha) = \sum_{\partial i} c_{i,j}(\beta)(X_j - m_j(\alpha)) + e_i, \\ e_i \perp X_j \text{ if } i \neq j, \; Var\ e_i = a_i\sigma^2. \end{cases}$$

We shall assume that the model satisfies the following hypothesis:

(C1-1) $\{a_i\}$ is positive and doubly bounded: $0 < a \leq a_i \leq A < \infty$ if $i \in S$.

(C1-2) $\sup_i \sum_{\partial i} |c_{i,j}(\beta)| \leq \rho < 1$ for all $\beta$.

(C1-3) The range is bounded: there exists $N$ and $L$ *s.t.* if $i \in S$, $|\partial i| \leq N < \infty$ and $d(i,j) \leq L < \infty$ if $j \in \partial i$.

These conditions entail exponential $\alpha$-mixing for $X$ whose covariance is determined uniquely (cf. Theorem 2.3.5).

(C2) $\theta = (\alpha, \beta, \sigma^2)$ belongs to a compact set, and $\alpha \longmapsto m_i(\alpha)$, $\beta \longmapsto c_{ij}(\beta)$ are two equicontinuous families.

Set $\theta_0 = (\alpha_0, \beta_0, \sigma_0^2)$, $M_i(\alpha, \beta) = m_i(\alpha) - \sum_{\partial i} c_{i,j}(\beta) m_j(\alpha)$. Assume the process is observed over $D_n$ with cardinal $d_n$.

(C3-1) *Consistency of the regression:* for each $\beta$,

$$P_0 - \liminf d_n^{-1} \sum_{D_n} [M_i(\alpha, \beta) - M_i(\alpha_0, \beta)]^2 > 0 \text{ if } \alpha \neq \alpha_0.$$

(C3-2) *Consistency of the conditional auto-regression:*

$$P_0 - \liminf d_n^{-1} \sum_{D_n} Var\Big(\sum_{\partial i}(c_{ij}(\beta) - c_{ij}(\beta_0))X_i\Big) > 0 \text{ if } \beta \neq \beta_0.$$

**(4.3.4) Theorem.** [80] *Let $X$ be a Gaussian Markovian field with conditional specification* (4.27) *parametrized by* $\theta = (\alpha,\beta,\sigma^2)$. *Then, under conditions* (C1), (C2) and (C3), *estimation by CLS is consistent.*

The proof is standard. It is based on Theorem 3.4.1 and the fact that $X$ is weakly dependent. Since the $CLS$ contrast is written as

$$U_n(\theta) = d_n^{-1} \sum_{D_n} \log p_i(x_i | \cdot, \theta),$$

the asymptotic normality results in the following two conditions: calling $J_n = Var_{\theta_0}(\sqrt{d_n} U_n^{(1)}(\theta_0))$ and $I_n = E_{\theta_0}(U_n^{(2)}(\theta_0))$, then

- $\liminf J_n > 0$,

•• $(U_n^{(2)}(\theta_0) - I_n) \xrightarrow{P_0} 0$ and $\liminf I_n > 0$.

Under these conditions, $\sqrt{d_n}(\widehat{\theta}_n - \theta_0) \overset{\mathcal{D}}{\sim} \mathcal{N}(0, I_n^{-1} J_n I_n^{-1})$. Reconsider model (4.23):

$$X_i = \alpha \sum_{\partial i} w_{ij} X_j + e_i \quad i = 1, n, \quad e_i \perp X_j \text{ if } i \neq j, \quad Var\ e_i = \sigma^2$$

with $w_{ij} = w_{ji}$, and $w_{ii} = 0$ a matrix $W$ of fixed weights. Setting $n = d_n$,

$$U_n(\alpha) - U_n(\alpha_0) = \frac{2}{n}(\alpha_0 - \alpha) \sum_i e_i v_i + \frac{1}{n}(\alpha_0 - \alpha)^2 \sum_i v_i, \ v_i = \sum_{\partial i} w_{i,j} x_j.$$

In this case, $\liminf(U_n(\alpha) - U_n(\alpha_0)) = (\alpha_0 - \alpha)^2 \liminf\left(\frac{1}{n}\sum_i v_i^2\right)$. Also assume

$$\text{(C1(2))} \qquad 0 < \alpha_0 < 1, \quad w_{ij} \geq 0 \text{ and } \sup_i \sum_{\partial i} w_{ij} \leq 1.$$

The covariance matrix of $X$ is $\Sigma = \sigma^2(I - \alpha_0 W)^{-1} = \sigma^2 \Sigma_{\ell \geq 0} \alpha_0^\ell W^\ell$. In particular, $E(X_i^2) = \Sigma_{ii} \geq \sigma^2\left(1 + \alpha_0^2 \sum_j w_{ij}^2\right)$. Then since

$$E\ v_i^2 = \alpha_0^{-2}(E(X_i^2) - \sigma^2),$$

we get, $\frac{1}{n}\sum_{D_n} E\ v_i^2 \geq \sigma^2 \sum_{ij \in D_n} w_{ij}^2$. Assume now that

$$\text{(C3(1))} \qquad \liminf\left(\frac{1}{n}\sum_{i \in D_n, j} w_{ij}^2\right) > 0.$$

Since the process is weakly dependent, under (C3(1))

$$P - \lim(U_n(\alpha) - U_n(\alpha_0)) > 0 \quad \text{if } \alpha \neq \alpha_0.$$

These conditions also entail the asymptotic normality of $\widehat{\alpha}_n$ (cf. [80]).

**(4.3.5) Theorem.** *For model* (4.23), *conditions* (C1(2)) *and* (C3(1)) *entail the consistency and asymptotic normality of the Conditional Least Squares estimator* $\widehat{\alpha}_n = \frac{X'WX}{X'W^2X}$.

REMARK: Following standard Gaussian-type arguments, we have the following controls over $(\widehat{\alpha}_n - \alpha)$ [16]: $\widehat{\alpha}_n - \alpha = \frac{X'CWX}{X'W^2X}$, $C = I - \alpha W$,

$$E(X'CWX) = \sigma^2 Tr(W) = 0,$$

$$E(X'W^2X) = \sigma^2 Tr(W^2 B^{-1}) \text{ and } Var(X'W^2X) = 2\sigma^4 Tr(W^2).$$

# 4.4. Estimation of Intrinsic AR Models

We restrict our attention to $\mathbb{Z}^2$. Intrinsic processes were introduced by Matheron [119], and intrinsic auto-regressions ($IAR$) by Künsch [109]. Both allow the modelization of nonstationarity.

## *4.4.1. Intrinsic Processes (IP)*

Consider $d \in \mathbb{N}$ and $X$ a field over $\mathbb{Z}^2$.

**(4.4.1) Definition.** $d$-Increments. $\sum \alpha_i X_i$, *a finite linear combination, is a d-increment if for all* $\alpha_1$,$\alpha_2$ *positive integers*, $\alpha_1 + \alpha_2 \leq d$, *we have*

$$\sum \alpha_i \; i_1^{\alpha_1} i_2^{\alpha_2} = 0 \quad if \; i = (i_1, i_2).$$

Over $\mathbb{Z}$, this condition becomes $\sum \alpha_i i^\alpha = 0$ if $0 \leq \alpha \leq d$. A zero increment is also called a contrast: it is a finite linear combination such that $\sum \alpha_i = 0$. To a such increment, we associate the increment process $X_j(\alpha) = \sum_i \alpha_i X_{i+j}$.

**(4.4.2) Definition.** Intrinsic Processes. *$X$ is a IP of order $d$ if every $d$-increment process is stationary.*

The following result ([77], [119]) shows how the spectra of the different increments are related to each other:

**(4.4.1) Theorem.** *Assume processes $X(\alpha)$ are centered with absolutely continuous spectra $f_\alpha$. Then there exists an even function $f(\lambda) \geq 0$, over $]-\pi,\pi]^2$, such that*

$$\int \|\lambda\|^{2d+2} f(\lambda) d\lambda < \infty \quad \text{and} \quad f_\alpha(\lambda) = |\sum \alpha_j e^{ij \cdot \lambda}|^2 f(\lambda).$$

This function $f$ is called the $IP$'s spectral density.

Remarks:

(1) If $X$ is stationary, it is an $IP$ and its spectral density is the $IP$'s spectral density.

(2) The class of $IP$ is greater than that of the stationary processes:

- It contains processes $X = P + Z$ with $Z$ stationary and $P$ a polynomial of degree $d$. In this case, $f = f_Z$.
- •• It contains the class of nonintegrable spectral densities, which have a pole $\lambda = \lambda_0$, $f(\lambda) \sim \|\lambda - \lambda_0\|^{-s}$, $2 \leq s < 2d + 4$.

(3) An $IP$, even in the Gaussian case, is not completely determined by its spectral density $f$, as it is always possible to add a polynomial of degree $d$: its value can be chosen in $\frac{1}{2}(d+1)(d+2)$ arbitrary points. An intrinsic model must be considered as an equivalence class, which admits $f$ as its spectral density.

(4) $\mathbb{Z}$ and $\mathbb{Z}^2$: *ARIMA and IP.*

As with the $ARIMA$ models proposed by Box and Jenkins, it is natural to ask, does there exist an $IP$ $X$ and an increment $X(\alpha)$ of $X$ which is a $CAR$ Markovian model (cf. 1.3.4)? Over $\mathbb{Z}$, the answer is yes. Over $\mathbb{Z}^2$, the answer is no. Indeed, if $f$ is the spectral density of such a process and $\alpha$ its increment, then we would have

$$f_\alpha(\lambda) = \left|\sum \alpha_j e^{i<\lambda,j>}\right|^2 f(\lambda) = \frac{\sigma^2}{Q(\lambda)},$$

$Q$ not zero. Thus at $\lambda = 0$,

$$f(\lambda) \sim \left|\sum \alpha_j < j \cdot \lambda >\right|^{-2},$$

and $\|\lambda\|^2 f(\lambda)$ would not be integrable at 0: the integrability condition for $f$ would not hold. This leads to the definition of intrinsic $AR$ ($IAR$).

## *4.4.2. Intrinsic Auto-Regressive Processes (IAR)*

**(4.4.3) Definition.** *Let $L$ be a finite, symmetric subset of $\mathbb{Z}^2$ with $0 \notin L$. Assume $a_k = a_{-k}$, $k \in L$ are such that*

$$P(\lambda) = 1 - \sum_L a_k \cos < \lambda, k > \geq C^{te} \|\lambda\|^{2d+2}$$

*over $]-\pi,\pi]^2$. The intrinsic model with density $f(\lambda) = \sigma^2 P(\lambda)^{-1}$ is called an IAR with neighborhood $L$. The order $d$ corresponds to the smallest integer such that $P$ satisfies the above inequality.*

Over $T = ]-\pi,\pi]$, if a trigonometric polynomial $P$ has a zero of order $(2d+2)$ at $\lambda = 0$, then

$$P(\lambda) = (1 - \cos\lambda)^{d+1} Q(\lambda).$$

This factorization property shows that an $IAR$ of order $(d+1)$ is an $ARIMA$ $(p, d+1, 0)$, or in other words, the $(d+1)$-th difference is a stationary $AR$. Over $\mathbb{Z}^2$, this factorization is no longer possible.

We will say $\sum_{j:j\neq i} \mu_j X_j$ is an *intrinsic predictor* of $X_i$ if $\left(X_i - \sum_{j:j\neq i} \mu_j X_j\right)$ is an increment. Hence, we have the following interpretation of an $IAR$ as an intrinsic predictor:

**(4.4.2) Theorem.** *If $X$ is an IAR, $\sum_L a_k X_{i+k}$ is the best intrinsic predictor of $X$ in $L^2$.*

*Estimation for IAR with Order 0.*

Let $X$ be an $IAR$ with order $d = 0$, with nonintegrable spectral density $f = \sigma^2 P^{-1}$ ($\sum_L a_k = 1$) and consider the parametrization $\alpha_k = \sigma^{-2} a_k, k \in L$. If $X$ is observed over the rectangle $[1, n] \times [1, m]$, let $\widehat{\gamma}$ be the empirical *semi-variogram:*

$$\widehat{\gamma}_k = \frac{1}{2}(n - |k_1|)^{-1}(m - |k_2|)^{-1} \sum (X_i - X_{i+k})^2,$$

where the sum is over $i \in D_n$ s.t. $(i + k) \in D_n$. Künsch ([109]) proposes and justifies the estimation of $\alpha$ by minimization of the functional:

$$(4.28) \qquad L(\alpha) = -(2\pi)^{-2} \int \log\Big(\frac{\sum_L \alpha_k(1 - \cos < \lambda, k >)}{1 - \frac{1}{2}(\cos\lambda_1 + \cos\lambda_2)}\Big) + \sum_L \alpha_k \widehat{\gamma}_k.$$

That is,

$$(4.29) \qquad \gamma_k(\widehat{\alpha}) = \widehat{\gamma}_k, \quad k \in L.$$

This procedure is analogous to that used in the Gaussian contrast for a stationary process (cf. § 4.2 and 4.19), replacing the empirical covariances by the empirical variogram. When $n, m \to \infty$, the Hessian of $L$ is positive definite with probability tending to 1, and (4.29) has a unique solution. The consistency and asymptotic normality of $\widehat{\alpha}$ follows from the Large Number Theorem and the $CLT$ which the $(\widehat{\gamma}_k)_{k \in L}$ satisfy, since $\{Y_i = X_{i+k} - X_i\}$ is ergodic, and in the Gaussian case, weakly dependent. Notice that if $X$ is a nonstationary $IAR$ of order 0, its variogram $\gamma(k)$ increases as $\log \|k\|$ [109].

*Test for the Stationarity of an IAR.*

Let $X$ be an $IAR$ of order 0 with spectral density,

$$f(\lambda) = \sigma^2(1 - \rho \sum_L a_j \exp i < \lambda, j >)^{-1}, \quad \sum_L a_j = 1.$$

We want to test

$(H_0)$ $\rho = 1$; $X$ is a nonstationary $IAR$

$(H_1)$ $|\rho| < 1$; $X$ is stationary Markovian.

The idea is the following ([145]): under $(H_0)$, we have the Yule–Walker equations for the semi-variogram

$$\sum_L a_j \gamma(k+j) = \gamma_k \quad \text{if } k \neq 0, \quad \sigma^2 \quad \text{if not.}$$

Under $(H_1)$, $\gamma_j = r_0 - r_j$ so that if $k \neq 0$

$$\sum_L a_j \gamma(k+j) = \gamma_k + r_k \frac{1-\rho}{\rho}, \quad \text{if } \rho \neq 0.$$

The test statistic will be, for a fixed set of $k \neq 0$,

$$T_k = \sqrt{nm}(\widehat{\gamma}_k - \sum_L a_j \widehat{\gamma}_{k+j}) \xrightarrow{\mathcal{L}} \begin{cases} \mathcal{N}(0, V) & \text{under } (H_0), \\ \mathcal{N}\Big(r_k \dfrac{1-\rho}{\rho}, V\Big) & \text{under } (H_1). \end{cases}$$

## 4.5. Nonparametric Estimation of the Spectral Density

We shall limit our presentation to the results given in Rosenblatt ([139]) (cf. also [178]). If $D_n = [1, n]^d$, $\widehat{f}_n(\lambda)$ is given by

$$\begin{cases} \widehat{f}_n(\lambda) = (2\pi)^{-d} \displaystyle\sum_{|j_k| \leq n, k=1,d} \widehat{R}_n(j) w_j^{(n)} e^{-i<j,\lambda>}, \\ w_j^{(n)} = a(j \cdot b_n), \quad a(0) = 1, \end{cases}$$

where $a(x) = a(-x)$ is continuous with bounded support, $b_n \longrightarrow 0$ and $nb_n \longrightarrow \infty$.

Assume $X$ is centered, strictly stationary, strongly mixing and with summable eighth-order cumulants (which will be the case if the mixing is exponential). Then

$$(nb_n)^{\frac{d}{2}} (\widehat{f}_n(\lambda) - E\widehat{f}_n(\lambda)) \xrightarrow{\mathcal{D}} \mathcal{N}(0, \sigma^2),$$

where

$$\sigma^2 = e \cdot (2\pi)^{-d} [1 + \eta(2\lambda)] f^2(\lambda) \int_{T^d} W^2(\mu) d\mu,$$

$$\begin{cases} W(\mu) = \displaystyle\int_{T^d} a(u) e^{-i<u,\mu>} d\mu, \\ \eta(\lambda) = \begin{cases} 0 & \text{if } \lambda_1, \lambda_2, \ldots, \lambda_d \text{ are multiples of } 2\pi, \\ 1 & \text{if not.} \end{cases} \end{cases}$$

Here, $e = 1$ if the estimator $\widehat{R}_n$ is the usual one or $R_n^*$, and $e = e(h)$ if $\widehat{R}_n = \widehat{R}_n^h$.
The optimal choice of $b_n$ will depend on the importance of the bias $(E\widehat{f}_n(\lambda) - f(\lambda))$, which in turn will depend on the estimation procedure and on the regularity of $f$ and $W$.

## 4.6. Appendix

**Appendix 1: Proof of Theorem 4.1.3.**

Consider the unidimensional kernel, $k_m(\mu) = (2\pi h_m)^{-1}|H_m(\mu)|^2$, $\mu \in T$ with $h_m = \sum_{j=1}^{m} h^2\left(\frac{j-1/2}{m}\right)$ and $H_m(\mu) = \sum_{j=1}^{m} h\left(\frac{j-1/2}{m}\right)e^{-ij\mu}$. Define

$$K_n(\lambda) = \prod_{i=1,d} k_{n_i}(\lambda_i) = ((2\pi)^d H_{2,n}(0))^{-1} H_{2,n}(\lambda).$$

Thus, $K_n(\lambda)$ is a positive, even kernel with total mass equal to 1. By direct calculation, we find that

$$E\ I_n^h(\lambda) = \int_{T^d} K_n(\alpha) f(\lambda + \alpha) d\alpha.$$

We have the following Lemma, regarding kernel $k_m$ ([44], [171]):

**(4.6.1) Lemma.**

(1) $\int_T k_m(\mu)\sin^2\frac{\mu}{2}d\mu = \frac{1}{4}m^{-2}\frac{\int_0^1 h'(u)^2du}{\int_0^1 h(u)^2du}(1+o(1))$. *$\rho$ can be a function of $m$ if $\rho^{-1} = o(m^{1/3})$. The term $o(1)$ is uniform in $\rho$.*

(2) *Let $0 < \eta \leq \pi$; $\int_{|\mu|>\eta} k_m(\mu)d\mu = 0(m^{-3})\eta^{-3}\rho^{-4}$. The term 0 is uniform in $\rho$.*

*Proof:*

($a$) Letting $h(u) = 0$ if $u \notin [0,1]$ and summing by parts yields

$$\sin\left(\frac{\mu}{2}\right) \cdot H_m(\mu) = -\frac{1}{2}ie^{-i\frac{\mu}{2}}\sum_{j=0}^{m}\delta_m(j)e^{-ij\mu},$$

where

$$\delta_m(j) = h\left(\frac{j+\frac{1}{2}}{m}\right) - h\left(\frac{j-\frac{1}{2}}{m}\right).$$

So that from Parseval's identity, $2\pi\int k_m(\mu)\sin^2\frac{\mu}{2}d\mu = \frac{1}{4}\sum_{j=0}^{m}\delta_m(j)^2$. If $A_j = \left[\frac{j-\frac{1}{2}}{m}, \frac{j+\frac{1}{2}}{m}\right]$, call $D$ the set of $j$, $j = 1, m$ (at most, four indexes $j$)

such that $A_j$ contains an eventual discontinuity point of $h'$ (at 0, $\frac{\varrho}{2}$, $1-\frac{\varrho}{2}$ and 1).

$$\delta_n(j)^2 = m^{-1}\int_{A_j} h'(u)^2du + m^{-1}\int_{A_j}(h'(\overline{u}_j)^2 - h'(u)^2)du$$

for a certain $\overline{u}_j \in A_j$. We now use the $\mathcal{C}^2$ hypothesis over the function $w$ which defines the taper. Let $\gamma = \|h'\|_\infty \|h''\|_\infty$. Then

$$\left|\sum_{j\notin D}\int_{A_j}(h'(\overline{u}_j)^2 - h'(u)^2)du\right| \le \gamma m^{-1},$$

where this term is $o(1)$ if $\gamma$ does not depend on $m$ or if $\gamma = o(m)$, which is true if $\rho_m \longrightarrow 0$ and $\rho^{-1} = o(m^{1/3})$ since $|\gamma| \le C\rho^{-3}$.

Finally, noticing that $m^{-1}h_m \longrightarrow \int_0^1 h^2(u)du$ and that the terms $\delta_m(j)^2$, $j \in D$ do not contribute to the sum, we get $(a)$.

$(b)$ Calling $\delta_m(j) = m^{-1}\int_0^1 h'\Big(\frac{j-\frac{1}{2}+x}{m}\Big)dx$ and again summing by parts, we get

$$H_m(\mu) =$$
$$m^{-1}(e^{i\mu}-1)^{-2}\int_0^1\Big(\sum_{j=-1}^{m}\Big(h'\Big(\frac{j+\frac{1}{2}+x}{m}\Big) - h'\Big(\frac{j-\frac{1}{2}+x}{m}\Big)\Big)e^{-ij\mu}\Big)dx,$$

so that

$$|H_m(\mu)| \le C\, m^{-1}(\sin\frac{\mu}{2})^{-2}\rho^{-2},$$

which yields $(b)$. □

We shall continue with the proof of the Theorem:

$$E(J_n^h(\varphi) - J(\varphi)) = \int_{T^d}\varphi(\alpha)\Delta_n(\alpha)d\alpha, \text{ where}$$

$$\Delta_n(\alpha) = \int_{T^d} K_n(\lambda)[f(\alpha+\lambda) - f(\alpha)]d\lambda.$$

Since $f$ belongs to $\mathcal{C}^2$, we have

$$f(\lambda+\alpha) - f(\alpha) = \sum_1^d \alpha_i f'_{\lambda_i}(\alpha) + 2\sum_{i,j=1}^d \sin\frac{\alpha_i}{2}\sin\frac{\alpha_j}{2}f^{(2)}_{\lambda_i\lambda_j}(\alpha) + R(\alpha,\lambda),$$

where $R(\alpha,\lambda) = o\Big(\sum_{i=1,d}\sin^2\frac{\lambda_i}{2}\Big)$, uniformly in $\alpha$. The fact that the kernels $k_m$ are even gives

$$\Delta_n = 2\sum_1^d f''_{\lambda_i^2}(\alpha)\int_T k_{n_i}(\lambda_i)\sin^2\frac{\lambda_i}{2}d\lambda_i + \int_{T^d} R(\alpha,\lambda)K_n(\lambda)d\lambda.$$

Using part $(a)$ of the Lemma, the principal part of the bias is given by the first term of $\Delta_n$. It remains to check that the contribution of the second term is $o(N^{-d/2})$.

Let $\varepsilon > 0$ and $\eta = \eta(\varepsilon)$ be such that if $\|\lambda\| = \sup_i |\lambda_i| < \eta$:

$$|R(\alpha,\lambda)| \le \varepsilon\Big(\sum_{i=1}^d \sin^2\frac{\lambda_i}{2}\Big),$$

$$\Big|\int_{T^d} R(\alpha,\lambda)K_n(\lambda)d\lambda\Big| \le \varepsilon\int_{T^d} K_n(\lambda)\Big(\sum_{i=1}^d \sin^2\frac{\lambda_i}{2}\Big)d\lambda + C\int_{\|\lambda\|>\eta} K_n(\lambda)d\lambda$$

The first term is $o(N^{-d/2})$, and the second is $O(N^{3/d})\eta^{-3}\rho^{-4}$ because of part $(b)$ of the Lemma. Since $\rho^{-1} = o(N^{1/4d})$, it is possible to choose $\varepsilon = \varepsilon_n \to 0$ so that $\eta_n^{-3}\rho^{-4} = o(N^{1/d})$. In this case, the second term is also $o(N^{-2/d})$. □

**Appendix 2: Proof of Theorem 4.1.4.**

Since $D_n$ is rectangular and the taper factorizes, the results obtained in dimension one (Dahlhaus) can be generalized directly. We shall recall certain definitions and identities. Here, $f_k$ is the spectral density of the $k$-cumulants.

(1) $cum(d_n(\alpha_1),\dots,d_n(\alpha_k)) =$

$\int_{(T^d)^{k-1}} f_k(\gamma)H_{1,n}(\alpha_1-\gamma_1)\cdot\cdot H_{1,n}(\alpha_{k-1}-\gamma_{k-1})H_{1,n}(\alpha_k+\gamma_1+\cdot+\gamma_{k-1})d\gamma$

(2) Consider the kernel $\phi_{k,n}$ over $(T^d)^{k-1}$ defined by

$$\phi_{k,n}(\alpha_1,\dots,\alpha_{k-1})$$
$$= [(2\pi)^{d(k-1)}H_{k,n}(0)]^{-1}H_{n,1}(\alpha_1)..H_{1,n}\Big(\alpha_{k-1})H_{1,n}(-\sum_1^{k-1}\alpha_i\Big)$$

if $H_{k,n}(0) \ne 0$, 0 if not. Then $(\phi_{k,n})$ is an approximate identity

$$\int_{(T^d)^{k-1}} \phi_{k,n}(y)f(x-y)dy = f(x)[1+o(1)].$$

(3) If $f_k$ exists and is continuous, then with $o$ uniform in $\alpha$:

$$cum(d_n(\alpha_1),\dots,d_n(\alpha_k))$$
$$= (2\pi)^{d(k-1)}H_{k,n}\Big(\sum_1^k \alpha_j\Big)f_k(\alpha_1,\dots,\alpha_{k-1})[1+o(1)].$$

Writing $J_n$ for $J_n^h$: $Cov(J_n(\varphi_1), J_n(\varphi_2)) = [(2\pi)^d H_{2,0}(0)]^{-2}(T_1 + T_2 + T_3)$, where

$$T_1 = \int_{T^{2d}} \varphi_1(\alpha)\varphi_2(\beta) cum(d_n(\alpha), d_n(-\alpha), d_n(\beta), d_n(-\beta)) d\alpha d\beta,$$

$$T_2 = \int_{T^{2d}} \varphi_1(\alpha)\varphi_2(\beta) cum(d_n(\alpha), d_n\beta) cum(d_n(-\alpha), d_n(-\beta)) d\alpha d\beta,$$

and $T_3$ is defined like $T_2$ by changing $\beta$ to $-\beta$. This follows from:

$cov(Y_1Y_2, Y_3Y_4) = cum(Y_1, Y_2, Y_3, Y_4) + E(Y_1Y_3)E(Y_2Y_4) + E(Y_1Y_4)E(Y_2)$.

• $T_1$: Since $f_4$ is continuous, identity (3) yields

$$T_1 = (2\pi)^{3d} H_{4,n}(0) \int_{T^{2d}} \varphi_1(\alpha)\varphi_2(\beta) f_4(\alpha, -\alpha, \beta) d\alpha d\beta \cdot [1 + o(1)].$$

•• $T_2$ and $T_3$: After (1),

$cum(d_n(\alpha), d_n(\beta)) cum(d_n(-\alpha), d_n(-\beta)) =$

$$\int_{T^{2d}} f(\gamma)f(\delta)H_{1,n}(\alpha - \gamma)H_{1,n}(-\beta + \gamma)H_{1,n}(-\alpha - \delta)H_{1,n}(-\beta + \delta) d\gamma d\delta.$$

Noting that the sum of the arguments of the four functions $H$ is 0 and using (2), we have

$$\begin{aligned} T_2 &= (2\pi)^{3d} H_{4,n}(0) \int_{(T^d)^4} \varphi_1(\alpha)\varphi_2(\beta)f(\gamma)f(\delta) \\ &\quad \times \phi_{4,n}(\alpha - \gamma, -\beta + \gamma, -\alpha - \delta, -\beta + \delta) d\alpha d\beta d\gamma d\delta \\ &= (2\pi)^{3d} H_{4,n}(0) \Big[\int_{T^d} \varphi_1(\alpha)\varphi_2(\alpha) f^2(\alpha) d\alpha\Big](1 + o(1)). \end{aligned}$$

Theorem 4.1.4 is a consequence of $H_{k,n}(0) = N(\int_0^1 h(x)^k dx)^d (1 + o(1))$. □

**Appendix 3: Proof of Corollary 4.1.**

Part $(a)$ of Corollary 4.1 follows from the following Lemma:

**(4.6.2) Lemma.** *For a $\delta > 0$, assume $X \in L_{4+\delta}$ is strongly mixing, and satisfies for $\alpha = \alpha_{1,3}$ and $\alpha = \alpha_{2,2}$,*

$$\sum_{m=1}^{\infty} m^{3d-1} \alpha(m)^{\delta/2+\delta} < \infty.$$

*Then $\sum_{u_1,u_2,u_3} |c(u_1,u_2,u_3)| < \infty$, and $f_4$ is continuous.*

*Proof:* Define $A = \{u_0, u_1, u_2, u_3\}$ with eventual repetitions

$$\begin{aligned}
cum(X_{u_0}, X_{u_1}, X_{u_2}, X_{u_3}) &= c(u_1 - u_0, u_2 - u_0, u_3 - u_0) \\
&= E(X_{u_0}X_{u_1}X_{u_2}X_{u_3}) - E(X_{u_0}X_{u_1})E(X_{u_2}X_{u_3}) \\
&- E(X_{u_0}X_{u_2})E(X_{u_1}X_{u_3}) - E(X_{u_0}X_{u_3})E(X_{u_1}X_{u_2}).
\end{aligned}$$

Let: $\text{dist}(u_0, \{u_1, u_2, u_3\}) = m = \inf_{i=0,1,2,3} d(u_i, A\backslash\{u_i\}) = d(u_0, u_1)$.

*First case:* $d(u_2, u_3) > m$, then necessarily, $d(u_1, u_2) \leq 3m$ and $d(u_1, u_3) \leq 3m$ (if $d(u_1, u_2) > 3m$, then $d(u_2, \{u_0, u_1, u_3\}) > m$). Define $\alpha_\delta(m) = \alpha(m)^{\delta/2+\delta}$.

(1.1) $|E(X_{u_0}X_{u_1}X_{u_2}X_{u_3})| \leq C\ \alpha_\delta(m)$ since $u_0$ is at distance $m$ from $\{u_1, u_2, u_3\}$. If $u_0$ is fixed, there are $O(m^{d-1})$ possibilities for $u_1$ and $O(m^{2d})$ possibilities for $(u_2, u_3)$ once $u_1$ is fixed. Hence, in the cumulants' sum, the contribution of this expectation under these conditions is finite since it is bounded by $\sum_{m=1}^{\infty} m^{3d-1}\alpha_\delta(m) < \infty$.

(1.2) Behavior of the crossed covariances.
• $|E(X_{u_0}X_{u_1}) \cdot E(X_{u_2}X_{u_3})|$: as $d(u_2, u_3) > m$, the total contribution of these terms is bounded by $C \cdot \sum_{m=1}^{\infty} m^{d-1}\alpha_\delta(m)\ m^{2d}\alpha_\delta(m) < \infty$.
•• $|E(X_{u_0}X_{u_2})E(X_{u_1}X_{u_3})|$: as $d(u_0, u_2) > m$, the total contribution of these terms is bounded by $\sum m^{3d-1}\alpha_\delta(m) < \infty$, as also is the third product.

*Second case:* $d(u_2, u_3) \leq m$.

(2.1) If $d(u_1, u_2)$ and $d(u_1, u_3) \leq 3m$, we have the same bounds as above except for the second term, where we get $\alpha_\delta(m)$ instead of $(\alpha_\delta(m))^2$, which does not change the result.

(2.2) If not, we have for example $d(u_1, u_2) > 3m$.
Define $\ell = d(\{u_0, u_1\}, \{u_2, u_3\})$. Then $\ell > m$, and thus,

• $|cov(X_{u_0}X_{u_1}, X_{u_2}X_{u_3})| = |E(X_{u_0}X_{u_1}X_{u_2}X_{u_3}) - E(X_{u_0}X_{u_1})E(X_{u_2}X_{u_3})| \leq C \cdot \alpha_\delta(\ell)$ so that the contribution to the cumulants' sum is bounded by

$$C' \sum_{\ell=1}^{\infty} \ell^{d-1}\alpha_\delta(\ell)\Big(\sum_{m\leq \ell-1} m^{d-1}m^d\Big) = C'' \sum_{\ell=1}^{\infty} \ell^{3d-1}\alpha_\delta(\ell) < \infty.$$

•• It remains to control the two covariance products which can be treated analogously. For example, $|E(X_{u_0}X_{u_2})E(X_{u_1}X_{u_3})|$. If we write, $p = d(u_0, u_2)$, then $d(u_1, u_3) \leq (p - 2m)^+$ and the contribution of these terms is bounded by $C\sum_{p=1}^{\infty} p^{d-1}\alpha_\delta(p)p^{2d-1}$ since $p \geq \ell > m$.
□

**(4.6.3) Lemma.** *Let $\{\varepsilon_i, i \in \mathbb{Z}^d\}$ be a sequence of real, i.i.d., centered variables of $L^4$ and let $\kappa_4$ be the fourth-order cumulant of $\varepsilon$. Assume that $X$ is the linear process*

$$X_t = \sum_{s \in \mathbb{Z}^d} a_s \varepsilon_{t-s}, \qquad (a_s) \in \ell^1.$$

*Then the fourth-order cumulants of $X$,*

$$c(u_1, u_2, u_3) = \kappa_4 \sum_{t \in \mathbb{Z}^d} a_t a_{t+u_1} a_{t+u_2} a_{t+u_3},$$

*are summable and $f_4(\alpha, -\alpha, \beta) = (2\pi)^{-d} \cdot \frac{\kappa_4}{\sigma^4} f(\alpha) f(\beta)$.*

*Proof:* The first result is a consequence of the multilinearity of the cumulants and of

$$\sum_{(\mathbb{Z}^d)^3} |c(u_1, u_2, u_3)| \leq \Big(\sum_{\mathbb{Z}^d} |a_t|\Big)^4 \cdot \kappa_4.$$

The fourth-order cumulants' density is given by

$$f_4(\lambda_1, \lambda_2, \lambda_3) = (2\pi)^{-3d} \kappa_4 \sum_{(\mathbb{Z}^d)^4} a_t a_{t+u_1} a_{t+u_2} a_{t+u_3} \exp\Big(-i \sum_{j=1}^{3} < \lambda_j, u_j >\Big)$$

$$= \kappa_4 (2\pi)^{-3d} \prod_{j=1}^{3} \Big(\sum_{\mathbb{Z}_d} a_u \exp -i < \lambda_j, u >\Big) \times \sum_{\mathbb{Z}_d} a_u \exp\Big(i \sum_{j=1}^{3} < \lambda_j, u >\Big).$$

Choosing $\lambda_1 = -\lambda_2 = \alpha$, and $\lambda_3 = \beta$ yields the result. □

## 4.7. Bibliographical Comments

Data tapering and its practical interest were introduced by Tuckey (cf. also Brillinger), and studied by Bentkus, Zurbenko and Dahlhaus.

Tapering is not necessary for asymptotics if $d = 1$ or for results analogous to those of § 4.1 and § 4.2, cf. [5]. If $d = 1$, Dahlhaus remarked that tapering allows a better estimation of the peaks of the spectrum in parametric estimation (leakage effect).

If $d \geq 2$, ordinary estimators have an important bias and it becomes necessary to unbias these estimators (cf. Guyon [78], [81]), or to taper the data: tapering for fields was studied by Dahlhaus and Künsch, and in the multidimensional case by Yao [171].

Gaussian contrasts were introduced by Whittle in his pioneer article of 1954. The general form of difference of contrasts for fields is due to Yao. In the Gaussian case, we obtain the same $\chi^2$ test proposed by Whittle (if $d = 1$ and $X$ is Gaussian, cf. also [41]).

The results of § 4.3 for the estimation of a Gaussian, Markovian field that differentiate $d \leq 2$ and $d \geq 3$ are due to Künsch.

Estimation by $PML$ (or $CLS$) was introduced by Besag (1974) for models of type (4.3), which are commonly used in spatial analysis (cf. Besag [15] and [16]; Cliff and Ord [33]; Cressie [30]). A general estimation procedure for nonstationary $CAR$ is given in [80]. The definition of intrinsic processes is due to Matheron, and that of $IAR$ and their statistics to Künsch. Sébastien proposed a test for checking the nonstationarity of an $IAR$.

The result on nonparametric estimation of the spectral density is due to Rosenblatt; it is totally equivalent to the analogous result in dimension one. The work of Yuan and Subba Rao on this field can also be consulted.

## 4.8. Exercises

(1) *Testing factorization in $s$ and $t$ of the covariance.*
$X$ is a stationary, Markovian, Gaussian field over $\mathbb{Z}^2$, with eight nearest neighbors.

(a) Give the representation of $X$ and that of the submodel with factorized covariance.

(b) Assume $X$ is observed over $\{1, 2, \ldots, n\}^2$. Write the $ML$ and $PML$ equations for both models. Test the subhypothesis of covariance factorization.

(c) What is the distribution of $\Delta^2 = (\widehat{R}_{00}\widehat{R}_{11} - \widehat{R}_{10}\widehat{R}_{01})^2$ under this subhypothesis?

(2) *Classical non-$\sqrt{N}$-consistent estimation.*
Assume $X$, an ergodic, stationary process over $\mathbb{Z}^2$, is observed over the rectangle $D_{n,m} = \{1, 2, \ldots, n\} \times \{1, 2, \ldots, m\}$ with $N = nm$ points; $\widehat{R}$ is the ordinary empirical covariance (cf. (4.1)).

(a) Consider $X$ the independent line-by-line $AR(1)$ process: $X_{st} = \rho X_{s-1,t} + \varepsilon_{s,t}$. Check that $\widehat{\rho} = \widehat{R}_{1,0} \cdot \widehat{R}_{00}^{-1}$ is not $\sqrt{N}$ consistent.

(b) Consider $X$ a causal $AR$ and let $\widetilde{\varepsilon}_t$ be the residuals calculated defining $X_u = 0$ if $u \notin D_{n,m}$. Show that the estimation procedure based on the minimization of $\sum_{D_{n,m}} \widetilde{\varepsilon}_t^2$ has a $\sqrt{N}$ bias, whereas that based on $\sum_{\overset{\circ}{D}_{n,m}} \varepsilon_t^2$ does not, with $\overset{\circ}{D}_{n,m} = \{t \in D_{n,m} \text{ s.t. } \widetilde{\varepsilon}_t = \varepsilon_t\}$.

(3) *Approximation of $\sum^{-1}$ in a Markovian model.*
In the standard Markovian model (4.12), we can approximate the inverse of the covariance matrix $\sum_{n,m}^{-1}$ over $D_{n,m}$ by $\sigma_e^{-2} Q_{n,m}$, where $Q_{n,m}(i,i) = 1$, $Q_{m,n}(i,j) = -c_{i-j}$. From this approximation, we find the following approximation of the log-likelihood:

$$-L_{n,m} = N \log \sigma_e^2 - \log |Q_{n,m}| + \sigma_e^{-2} \quad {}^t X(n,m) Q_{n,m} X(n,m).$$

(a) Find $L_{n,m}$ when $X$ is the isotropic and factorizing Markovian model with eight nearest neighbors (one parameter $\alpha$).

(b) Check that the maximum $L_{n,m}(\alpha)$ estimation is $\sqrt{N}$-biased.

(c) Using the covariance factorization and the fact that $\sum_{m,n}$ is a Kronecker product, find the exact likelihood.

(4) Consider the (causal) $AR$: $X_{st} = \alpha X_{s-1,t} + \beta X_{s,t-1} + \varepsilon_{s,t}$, $|\alpha| + |\beta| < 1$, where $\varepsilon = \{\varepsilon_{st}\}$ is the Gaussian white noise $\mathcal{N}(0, \sigma^2)$.

(a) Find the Yule–Walker equations.

(b) Assume $X$ is observed over $D_n = \{1, 2, \ldots, n\}^2$. Give the $LS$ estimators of $\alpha, \beta$ (which are not $\sqrt{n^2}$-biased). Find the tests for $\alpha = \beta$; $\alpha = \beta = 0$.

(5) *Regression/Auto-regression models.* We consider the two models,

$$(R - SAR) : Y = \rho WY + X\beta + \varepsilon, \quad \varepsilon \sim \mathcal{N}_n(0, \sigma^2 I)$$
$$(R - CAR) : Y = \alpha WY + X\beta + e, \quad W = {}^tW$$

where $e$ is the conditional residual (colored) with constant variance $\sigma^2$.

(a) Give the $ML$ estimators for both models. What can be said of the estimation of parameters $(\rho, \beta)$ by ordinary least square ($OLS$) over $\varepsilon$? And of $(\alpha, \beta)$ by $LS$ over $e$?

(b) Give the $R - CAR$ representation of model $R - SAR$ and deduce another consistent estimation procedure for the first model.

(6) (a) Assume $X$ is a Gaussian, nearest neighbor Markovian model with spectral density $f(\lambda, \mu) = \sigma_e^2(1 - 2\alpha \cos\lambda - 2\beta \cos\mu)^{-1}$, $|\alpha| + |\beta| < 1$. For the tapered $ML$ estimation over $D_n = \{1, 2, \ldots, n\}^2$, give the asymptotic distribution of the estimator of $(\sigma_e^2, \alpha, \beta)$. Test $\alpha = \beta$. Consider the same problem for $PML$ estimation.

(b) Consider the same problem for the $ML$ estimator and the models

(b1) $Y = X + \varepsilon$ where $X$ is given in (a), $\varepsilon$ is a Gaussian white noise $\mathcal{N}(0, \sigma^2)$ independent of $X$.

(b2) $X_{st} = \alpha X_{s-1,t} + \beta X_{s,t-1} + \varepsilon_{st}$, $\varepsilon$ is a Gaussian white noise.

(7) *Experimental design for random fields.*
$p$ treatments with mean levels $\mu = {}^t(\mu_1, \ldots, \mu_p)$ are applied, each $r$-times, over the $n = r \times p$ plots of a field $S = \{1, 2, \ldots, n\}$. We observe $Y$ with mean $D\mu$, where $D$ is the $n \times p$ treatments matrix which satisfies ${}^tDD = rI_p$. We shall assume that the process $Y - D\mu = X$ is a $CAR$: $X = \beta WX + e$.

(a) Estimate $\mu$ by ordinary least squares and find an estimator $\widehat{X}$ of $X$.

(b) Estimate $\beta$ by conditional least squares based on $\widehat{X}$.

(c) Find the generalized least squares estimators, on the basis of $\widehat{\beta}$ and of $\mu$, and interpret this estimator as the first step in the iterative process which finds the solution $\mu$ of the equation $r\mu = {}^tDY - \widehat{\beta}^t DW[Y - D\mu]$.

(8) *Intrinsic AR.*

(a) Show that an increment of order 0 can always be written as a linear combination of elementary increments $\Delta_1 X(i,j) = X_{i+1,j} - X_{i,j}$ and $\Delta_2 X(i,j) = X_{i,j+1} - X_{i,j}$, $(i,j) \in \mathbb{Z}^2$. Show that $X$ is intrinsic of order zero if $\Delta_1 X$ and $\Delta_2 X$ are stationary.

(b) $X$, an $IAR$ of order zero, is observed over $D_n = \{1, 2, \ldots, n\}^2$. As $\Delta_1 X$ is stationary, propose an estimation procedure for the model. What is the asymptotic efficiency of this method with respect to that proposed in (4.29)? What advantages are there in considering method (4.29)?

(c) $X$ is an intrinsic process of order zero with known density $f$. How can we simulate $\Delta_1 X$ over $\{i = 1, n-1, j = 1, n\}$? How can we simulate $\Delta_2 X$ over $\{i = 1, j = 1, n-1\}$? Propose a simulation procedure for $X$ over $D_n$, given $X_{00} = x_{00}$ .

(d) Propose an estimation procedure for an $IAR$ with order $d \geq 1$.

(9) Establish the Yule–Walker equations for the semi-variogram of an $IAR$ of order zero.

# CHAPTER 5
# Estimation of Gibbs Fields

In this chapter, $(D_n)$ will be a sequence of cubes of $\mathbb{Z}^d$, $|D|_n \to \infty$, or a sequence of rectangles with all sides tending to infinity. The real model $\mu = P_\theta$, will be a Gibbs state with specifications $\{\pi_\alpha, \alpha \in \Theta\}$, $\theta$ in the interior of a compact $\Theta$ of $\mathbb{R}^p$: $P_\theta \in \mathcal{G}(\pi_\theta)$. We shall assume that the specification is *Markovian with bounded range* $R$ and that the neighbor graph of $\pi_\alpha$ does not depend on $\alpha$ over $\Theta$. We will call $\phi_\alpha$ the associated Gibbs potential and shall assume that

(5.1) $$\|\phi\| = \sup_{\alpha \in \Theta}(\sup_{i \in S} \|\phi_\alpha\|_i) < \infty.$$

## 5.1. Some Estimation Methods

We describe five estimation methods: among them, the $PML$, Coding, Minimum $\chi^2$, and $CLS$ are easy to work with, as opposed to the $ML$ procedure, which requires specific algorithms (cf. § 6.7).

### *5.1.1. Maximum Likelihood (ML)*

If $x \in \Omega$ is a configuration of $X$ over $S = \mathbb{Z}^d$, the specification $\pi_\alpha$ over $D_n$ is

(5.2)
$$\pi_n(\alpha, x) = Z_n^{-1}(\alpha, x) \exp(-H_n(\alpha, x)), \quad H_n(\alpha, x) = \sum_{A: A \cap D_n \neq \emptyset} \phi_A(\alpha, x).$$

It should be remarked that $Z_n(\alpha, x)$ becomes unmanageable, both numerically and analytically, when the number of configurations over $D_n$ is big. This is a drawback of the $ML$ procedure.

*Specifications with Fixed Boundary Conditions.* Let $(\omega^n)$ be a sequence of configurations of $\Omega$. If $x(n)$ is the observation of $X$ over $D_n$, the likelihood contrast, conditional to $\omega^n$, becomes

$$(5.3) \qquad U_n(\alpha) = |D_n|^{-1}[\log Z_n(\alpha, \omega^n) + H_n(\alpha; x(n),\ \omega^n_{S\setminus D_n})].$$

Notice that if we call $\overset{\circ}{D}_n = \{i \in D_n \text{ such that if } A \ni i,\ A \subseteq D_n\}$ and if $\overline{\Phi}_i$ is the mean energy per site (cf. (2.6) and (2.7)), then

$$H_n(\alpha; x(n), \omega^n_{S\setminus D_n}) = -\sum_{i \in \overset{\circ}{D}_n} \overline{\Phi}_i(\alpha, x(n)) + \varepsilon_n.$$

Here, uniformly in $\alpha$, $x$, and $\omega^n$, we have $|\varepsilon_n| \leq \|\phi\| \cdot |\partial_R D_n|$ with $\partial_R D_n$ the $R$-neighborhood of the boundary of $D_n$.

*Boundary Free Conditions.* This is the case when in the energy $H_n$, we eliminate the contribution of potential $\phi_A$ whenever $A$ extends beyond $D_n$. This is equivalent to considering the energy defined by

$$(5.3') \qquad H_n^{\ell}(\alpha, x(n)) = -\sum_{A \subseteq D_n} \phi_A(\alpha, x(n)).$$

REMARK: Assume that the specifications belong to an exponential family. Dropping the index $n$ and fixing the configuration outside the observation domain, we have

$$\pi(x, \alpha) = Z(\alpha)^{-1} e^{-<\alpha, H(x)>},$$
$$\pi_i(x_i | x_{\partial i}, \alpha) = Z_i(x_{\partial i}; \alpha)^{-1} \cdot e^{-<\alpha, H_i(x_i, x_{\partial i})>},$$
$$\frac{\partial^2}{\partial \theta^2} \log \pi(x, \alpha) = -Var_\alpha(H(X)) \leq 0,$$
$$\frac{\partial^2}{\partial \theta^2} \log \pi_i(x_i | x_{\partial_i}, \alpha) = -Var_\alpha(H_i(X_i, x_{\partial i})) \leq 0.$$

The likelihood, as well as the coding likelihood or the conditional pseudo-likelihood, are concave, which assures the unicity of the associated estimator. It also assures the convergence of the gradient algorithms used in the associated optimization problem. Since the gradient of $\log Z(\alpha)$ is $-E_\alpha(H(X))$, the $ML$ equation can be written as

$$H(x) = E_{\widehat{\theta}_{MV}}(H(X)),$$

which is the basis of the stochastic gradient algorithm (cf. 6.7.1).

## *5.1.2. Conditional Pseudo-Likelihood (CPL, Besag [14])*

We shall be considering the expression

$$(5.4) \qquad U_n(\alpha) = -|\overset{\circ}{D}_n|^{-1} \sum_{i \in \overset{\circ}{D}_n} \log \pi_i(x_i | x_{\partial_i}, \alpha),$$

which defines the *conditional pseudo-likelihood.* This expression only depends on the observation $x(n)$ of $X$ over $D_n$, and is free of exterior conditions. If there is independence, it is just the likelihood contrast. We could also consider a $CPL$ relative to any family of subsets (instead of only the sets $\{i\}$, $i \in S$); for example, over $\mathbb{Z}^2$, the family given by $A_{ij}, (i,j) \in \mathbb{Z}^2$,

$$A_{ij} = \{(i,j),\ (i+1,j),\ (i,j+1),\ (i+1,j+1)\}.$$

## *5.1.3. Coding Methods (C; [14])*

A subset $C$ of $S$ is called a *coding subset* for $X$ if for all $i$, $j$ of $C$, $i \neq j$, $i$ and $j$ are not neighbors for the Markovian structure of $X$. Let $C$ be an infinite coding set, $C_n = \overset{\circ}{D}_n \cap C$. The *coding contrast* is

$$(5.5) \qquad U_n(\alpha) = -|c_n|^{-1} \sum_{i \in C_n} \log \pi_i(x_i | x_{\partial i}, \alpha),$$

with $c_n = |C_n|$. Conditional to $x_{\overline{C}} = \{x_j,\ j \notin C\}$, it is the likelihood contrast for observations over $C_n$.

## *5.1.4. Logit or Minimum $\chi^2$ Estimation (L; Possolo [131])*

Consider a binary state space, $E = \{a, b\}$. The distribution $\pi_i(x_i|x_{\partial i}, \alpha)$ is characterized by the quotient of the distributions conditional to $x_{\partial i}$ in $a$ and $b$. We shall assume we have the following exponential stationary model for this quotient:

$$(5.6) \qquad \log \frac{\pi(a|v,\alpha)}{\pi(b|v,\alpha)} = {}^t\alpha \cdot h(v), \quad \alpha, h \in \mathbb{R}^p$$

Here, $v$ is a neighborhood configuration, $v \in E^m$ if $m$ is the number of neighbors of $i$. Define $\mathcal{V} = \{v(1), \ldots, v(K)\}$ as the $K$ equivalence classes over $E^m$ associated to the relationship $x_{\partial i} \sim \widetilde{x}_{\partial i} \iff h(x_{\partial i}) = h(\widetilde{x}_{\partial i})$. We can always reduce $E^m$ to $\mathcal{V}$. Model (5.6) is then characterized by the $p \times K$ matrix: $H = [h(v(1)), \ldots, h(v(K))]$. The log-linearity of (5.6) induces us to consider logistic estimation which consists of minimizing the following contrast:

$$(5.7) \qquad U_n(\alpha) = {}^t(y(n) - {}^t\alpha H) A (y(n) - {}^t\alpha H).$$

Here, $y(n)$ is an estimator of ${}^t\alpha H$, and $A$ is a $K \times K$ symmetric and positive definite matrix. The natural estimator for $\pi(x|v)$ is the frequency of apparition of $(x|v)$ over $\overset{\circ}{D}_n$: $N(x,v) = |\overset{\circ}{D}_n|^{-1} \Sigma_{\overset{\circ}{D}_n} 1(x_i = x \mid v_i = v)$. There are several posibilities of estimators for $y(n)$:

(1)

$$y_n(v) = \log \frac{N(a,v)}{N(b,v)} \quad \text{if } N(a,v)N(b,v) \neq 0; \quad 0 \quad \text{if not.} \tag{5.8}$$

(2) We have prior information about the model which permits us to estimate ${}^t\alpha \cdot h(v)$ by ${}^t\widetilde{\alpha} \cdot h(v)$. Thus, in the above definition, we put $y_n(v) = \log {}^t\widetilde{\alpha} \cdot h(v)$ when $N(a,v)N(b,v) = 0$.

(3) In order to avoid situations in which neither $(a|v)$ nor $(b|v)$ are observed over $\overset{\circ}{D}_n$, and on the basis of prior information about the model, we shall reduce the set of configurations $v$ to those for which both $(a|v)$ and $(b|v)$ are sufficiently probable (preserving the identificability of $\alpha$ in this model).

REMARKS:

(1) If $E = \{a_1, \ldots, a_N\}$, $N > 2$, the method can be extended by considering the following $(N-1)$ equations (and their estimators):

$$\log \frac{\pi(a_\ell|v,\alpha)}{\pi(a_1|v,\alpha)} = {}^t\alpha \cdot h_\ell(v), \quad \ell = 2, 3, \ldots, N.$$

The problem is that as $N$ increases, configurations $(a|v)$ will become scarce.

(2) An advantage of this method with respect to others is that the $\widehat{\alpha}_n$, which minimizes $U_n(\alpha)$, can be explicitly calculated as the $CLS$ estimator of the regression problem

$$y(n) = {}^t\alpha \cdot H + \varepsilon(n).$$

### *5.1.5. Estimation by Conditional Least Squares (CLS; Lele and Ord [113])*

Assume $X$ takes numerical values. Define the conditional mean as $m_i(\alpha, v) = E_\alpha(X_i \mid X_{\partial i} = v)$; the *CLS contrast* is given by

$$U_n(\alpha) = |\overset{\circ}{D}_n|^{-1} \Sigma_{\overset{\circ}{D}_n} (x_i - m(\alpha, v_i))^2, \quad \text{where } v_i = x_{\partial i}. \tag{5.9}$$

For each of the contrast functionals considered above, we shall call $\widehat{\theta}_n$ a minimum contrast estimator: $\widehat{\theta}_n = \operatorname{Arg}\min_{\alpha \in \Theta} U_n(\alpha)$.

# 5.2. Consistency

## 5.2.1. The Case X Stationary and E Compact

Assume $D_n$ is a cube of $\mathbb{Z}^d$, $|D_n| \to \infty$, and $\{\pi_\alpha, \alpha \in \Theta\}$ is a family of specifications which are invariant under translation and Markovian of range $R$ for all $\alpha \in \Theta$, a compact set of $R^p$. The true distribution is $\mu \in \mathcal{G}_s(\pi_\theta)$ with $\theta \in \overset{\circ}{\Theta}$; we are thus assuming that $\mu$ is *stationary*. We shall assume that the space of states $E$ is compact and that $\lambda$ is the reference measure over $E$.

Call $\phi_{A,\alpha}$, $0 \in A$ the potentials that define $\pi_\alpha$:

(C1) $(\alpha, x) \longrightarrow \phi_{A,\alpha}(x)$ are continuous functions of $(\alpha, x), 0 \in A$.
(C2) If $\alpha \neq \beta, \alpha, \beta \in \Theta$ then $\pi_0(\cdot|\cdot,\alpha) \neq \pi_0(\cdot|\cdot,\beta)$ as elements of $L^1(\lambda^{m+1})$ if $m = |\partial i|$.
(C3) The coding subset $C$ contains the sub-group $a\mathbb{Z}^d$ with $a = (a_1, a_2, \ldots, a_d)$, $|a| = a_1 \times \cdots \times a_d > 0$.
(C4) $\alpha \longrightarrow m(\alpha, v), v \in E^m$ is injective $(m(\alpha, v) = E_\alpha(X_i \mid X_{\partial_i = v}))$.

**(5.2.1) Theorem.** *If* $\mu \in \mathcal{G}_s(\pi_\theta)$, *under conditions* (C1) and (C2) *for the* (*ML*), (*PML*), *and* (*L*) *estimators, conditions* (C1) to (C3) *for the coding estimator* (*C*), *and conditions* (C1), (C2) *and* (C4) *for the* (*C.L.S.*) *estimator, we have* $\lim \widehat{\theta}_n = \theta \quad \mu - a.e.$

*Proof:* Since $\mu$ is stationary, it can be written as a convex combination of extremal elements $\mu^*$ of $\mathcal{G}_s(\pi_\theta)$ (cf. § 2.1.5); hence, if for each $\mu^*$ we have the *a.e.* consistency, we get the $\mu$-consistency, as extremal elements are singular. Thus, it is enough to deduce the consistency for such a $\mu^*$ which is ergodic; in other words, for any $\mu$ that is stationary and ergodic. We are going to use the consistency result for a minimum contrast estimator (Theorem 3.4.3):

(1) in order to identify the contrast function: $K(\mu, \alpha) = \lim_n U_n(\alpha)\ \mu - a.e.$,
(2) to check its continuity in $\alpha$ and that $\alpha = \theta$ is the unique minimum of $K$, and
(3) finally, by checking that the continuity property of the potentials entails the condition for the modulus of continuity of $U_n$.

*Maximum Likelihood.*
Consider $U_n(\alpha) = |D_n|^{-1}(\log Z_n(\alpha, \omega^n) - \sum_{\overset{\circ}{D}_n} \overline{\phi}_\alpha(\tau_i x(n)) + \varepsilon_n)$. The first term tends toward the pressure $p(\alpha) = p(\phi_\alpha)$, independently of $\omega^n$, and the last term tends toward 0 uniformly in $x$ and $\omega^n$ (cf. § 5.1.1). The second term tends toward $-E_\mu(\overline{\phi}_\alpha)$ since $\mu$ is ergodic and $\overline{\phi}_\alpha$ is bounded:

$$U_n(\alpha) \longrightarrow p(\alpha) - E_\mu(\overline{\phi}_\alpha) \quad \mu - a.e.,$$

so that $U_n(\alpha) - U_n(\theta) \longrightarrow K_1(\mu, \alpha) = p(\alpha) - E_\mu(\overline{\phi}_\alpha) + h(\mu) \geq 0$, where $h(\mu)$ is the specific entropy of $\mu$ (cf. § 2.1.4).

It is enough to check that the pressure $p(\alpha)$ is continuous in $\alpha$ in order to assure the continuity in $\alpha$ of $K_1$. If the specification belongs to an exponential family, this is true as the pressure is convex:

$$H_n(\alpha, x) = {}^t\alpha \cdot H_n(x) \quad \text{and} \quad |p_n(\alpha, \omega^n) - p_n(\beta, \omega^n)| \leq \|\alpha - \beta\| \, \|\phi\|,$$

and we have the same bound for $p(\alpha)$. If not, write

$$H_n(\alpha) = H_n(\alpha, x(n), \omega^n_{S \setminus \Lambda_n}), \quad \Lambda_n = \overset{\circ}{D}_n,$$

$$\begin{aligned} p_n(\alpha, \omega^n) &- p_n(\beta, \omega^n) \\ &= |\Lambda_n|^{-1} \log \int_{E^{\Lambda_n}} e^{-H_n(\alpha)} \lambda^n(dx) - \log \int_{E^{\Lambda_n}} e^{-H_n(\beta)} \lambda^n(dx) \\ &= |\Lambda_n|^{-1} \log \int_{E^{\Lambda_n}} \exp\{-H_n(\alpha) + H_n(\beta)\} \pi_{n,\beta}(x(n)|\omega^n) \lambda(dx) \\ &= |\Lambda_n|^{-1} E^{\omega^n}_{\Lambda_n,\beta}(H_n(\beta) - H_n(\alpha)) = |\Lambda_n|^{-1} E^{\omega^n}_{\Lambda_n,\alpha}(H_n(\alpha) - H_n(\beta)). \end{aligned}$$

By Jensen's inequality,

$$|\Lambda_n|^{-1} E^{\omega^n}_{\Lambda_n,\beta}(H_n(\beta) - H_n(\alpha)) \leq p_n(\alpha, \omega^n) - p_n(\beta, \omega^n) \leq |\Lambda_n|^{-1} E^{\omega^n}_{\Lambda_n,\alpha}(H_n(\alpha) - H_n(\beta)).$$

Since potentials $\phi_A$, $A \ni 0$ are uniformly continuous, for all $\varepsilon > 0$, there exists $\eta$ such that for all $\|\alpha - \beta\| \leq \eta$, then, uniformly in $n$, $x$, $\omega^n$,

$$|H_n(\alpha) - H_n(\beta)| \text{ and } |p_n(\alpha, \omega^n) - p_n(\beta, \omega^n)| \leq \varepsilon,$$

which gives the continuity of $p = \lim_n p_n$.

On the other hand, under condition (C2) we have

$$\mathcal{G}(\pi_\alpha) \cap \mathcal{G}(\pi_\theta) = \emptyset \text{ if } \alpha \neq \theta$$

so that the Variational Principle (cf. Property 2.1) shows that $K(\mu, \alpha)$ reaches its only minimum at $\alpha = \theta$.

We have the same consistency result for the likelihood with free boundary conditions.

*Pseudo-Maximum Likelihood.*
For $U_n$ defined by (5.4), we have for $K_2(\mu, \alpha) = E_\mu\left(\log \frac{\pi(X_0|X_{\partial 0}, \theta)}{\pi(X_0|X_{\partial 0}, \alpha)}\right)$,

$$\lim_n (U_n(\alpha) - U_n(\theta)) = K_2(\mu, \alpha) \geq 0, \; = 0 \quad \text{only if } \alpha = \theta.$$

The uniform continuity in $\alpha$ of $\log \pi(x_0|x_{\partial 0}, \alpha)$ entails that of $K_2$.

*Coding.*
$X_\partial = \{X_{\partial i},\ i \in a\mathbb{Z}^d\}$ is stationary Markovian with positive specification $\gamma_\theta$ and ergodic distribution $\mu_\partial$ as $\mu$ is. Thus,

$$|C_n \cap a\mathbb{Z}^d|^{-1} \sum_{i \in C_n \cap a\mathbb{Z}^d} 1(X_{\partial i} = v) \longrightarrow \mu(X_{\partial i} = v) \quad \mu - a.e.$$

for all neighborhood configuration $v$. Writing $C_n(v) = \{i \in C_n,\ x_{\partial i} = v\}$, we have

$$\liminf_{n \to \infty} \frac{|C_n(v)|}{|C_n|} \geq \frac{1}{2} \frac{\mu(X_{\partial i} = v)}{|a|} > 0 \quad \mu - a.e.$$

In particular, $|C_n(v)| \longrightarrow \infty$ $\mu - a.e.$ As $(X_i|v)$, $i \in C_n(v)$ are independent:

$$|C_n(v)|^{-1} \sum_{i \in C_n(v)} \log \frac{\pi(x_i|v,\theta)}{\pi(x_i|v,\alpha)} \longrightarrow E_{\mu,v}\Big(\log \frac{\pi(X_0|v,\theta)}{\pi(X_0|v,\alpha)}\Big)$$

so that for the coding contrast given in (5.5):

$$\liminf_n (U_n(\alpha) - U_n(\theta)) \geq K_3(\mu,\alpha) = \frac{K_2(\mu,\alpha)}{2|a|} \quad \mu - a.e.$$

*Minimum Chi 2.*
Consider the estimator $y(n)$ given in (5.8). We have

$$|\overset{\circ}{D}_n|^{-1} \Sigma_{i \in \overset{\circ}{D}_n} 1(X_i = a,\ X_{\partial_i} = v) \longrightarrow \mu_\partial(v)\pi(a|v,\theta) \quad \mu - a.e.$$

so that for $K_4(\mu,\alpha) = {}^t(\theta - \alpha) H A^t H(\theta - \alpha)$

$$(U_n(\alpha) - U_n(\theta)) \longrightarrow K_4(\mu,\alpha) \geq 0 \quad \mu - a.e.$$

and, as Lemma 5.2.1 shows, $K_4(\mu,\alpha)$ is not zero under (C2), except at $\alpha = \theta$.

*Conditional Least Squares.*
Write $V = X_{\partial 0}$. For the contrast defined in (5.9) and with $K_5(\mu,\alpha) = E_\mu[(m(V,\theta) - m(V,\alpha))^2] \geq 0$,

$$\lim (U_n(\alpha) - U_n(\theta)) = K_5(\mu,\alpha) \quad \mu - a.s.$$

$K_5(\mu,\alpha)$ is continuous and under (C4) is only zero for $\alpha = \theta$. □

An example of the fulfilment of (C2) or (C4) is given by the following Lemma:

**(5.2.1) Lemma.** *Assume $E \subseteq \mathbb{R}$, the conditional density belongs to an exponential family $\pi(x|v,\alpha) = Z^{-1}(\alpha,h(v)) \exp(x^t\alpha \cdot h(v))$, $\alpha \in \mathbb{R}^p$ and $h$ is continuous.*

*(a) $E$ finite: let $v(1),v(2),\dots,v(K)$ be the collection of possible neighborhood configurations and $H = [h(v(k)),k = 1,K] = p \times K$. Then (C2)$\Longleftrightarrow$ $H$ has rank $p$.*

*(b) If $E$ is compact, then $\alpha \longrightarrow m(\,\cdot\,,\alpha) = E_{\alpha},(X_i \mid \cdot)$ is injective under* (C2).

*Proof:* If in the conditional energy $(x_t\ {}^t\alpha \cdot h(v_t))$ we change $x_t$ for $x_t - a$, the conditional distribution remains unchanged. Thus we can assume, for example, that $0 \in E$ and that if $E = \{a,b\}$, codify $a$ by 0 and $b$ by 1.

($a$) Assume $\theta \neq \alpha$. Under (C2) there exists $(x,v)$ such that $x^t(\theta-\alpha)h(v) \neq 0$; that is, ${}^t(\theta-\alpha)H \neq 0$. Since this is true for all $(\theta-\alpha)$ in a nonempty open set, it follows that $H$ has full rank $p$. Conversely if $H$ has rank $p$ and if $\theta \neq \alpha$, there exists $v$ such that ${}^t\alpha \cdot h(v) \neq {}^t\theta \cdot h(v)$, and thus if $x \neq 0$, $\frac{\pi(x|v,\alpha)}{\pi(0|v,\alpha)} \neq \frac{\pi(x|v,\theta)}{\pi(0|v,\theta)}$, and (C2) is satisfied.

($b$) Since $h$ is continuous and $E$ compact, condition (C2) implies that there exists $K$ and $v(1),\dots,v(K)$ neighborhood configurations such that

$$H = [h(v(1)),\dots,h(v(K))] = p \times K$$

has rank $p$. Let $\alpha \neq \theta$: assume that over $\mathcal{V} = \{v(1),\dots,v(K)\}$, we have $m(v,\alpha) = m(v,\theta)$. As $m'_{\alpha}(v,\alpha) = h(v)\ Var_{\alpha,x_{\partial_i}=v}(X_i)$, we deduce that over $\mathcal{V}$, $m(v,\alpha) - m(v,\theta) = {}^t(\alpha-\theta)h(v)\ Var_{\alpha^*,v}(X_i) = 0$; that is, ${}^t(\alpha - \theta)H = 0$, which is impossible. This yields ($b$). □

### *5.2.2. Case of a Translation Invariant Specification and a Compact State Space E*

In this paragraph, we shall assume that the observed distribution $\mu \in \mathcal{G}(\pi_\theta)$ with $\pi_\theta$ a translation invariant specification. We shall assume the same hypothesis as in 5.2.1 *except the stationarity of* $\mu$. The Large Deviations ($LD$) exponential bounds (Theorem 2.1.7) will take the place of the ergodicity condition used in 5.2.1 if $\mu$ is stationary: this approach and the resulting consistency are due to Comets ([35]; cf. also [36]). This consistency has exponential rates.

Let $U_n(\alpha) = U_n(\alpha,x(n))$ be a contrast process and $K : \mathcal{P}(\Omega,\mathcal{F}) \times \Theta \longrightarrow \mathbb{R}$ be the associated continuous contrast function and $R_{n,x}$ the empirical field over $D_n = [0,n-1]^d$ associated to observation $x$:

$$R_{n,x} = \frac{1}{|D_n|} \sum_{i \in D_n} \delta_{\tau_i x^n}$$

with $x^n$ an arbitrary prolongation of $x(n)$ to the exterior of $D_n$.

*First Case*: $\Theta$ compact

(C) (C1) $U_n(\alpha) = K(R_{n,x}, \alpha) + \varepsilon_n$, with $\varepsilon_n \to 0$ uniformly in $x$.
(C2) For all $\nu \in \mathcal{G}_s(\theta)$, $\alpha \longrightarrow K(\nu, \alpha)$ has a unique minimum at $\alpha = \theta$.

REMARK: The usual equicontinuity condition over $U_n$ is not necessary as it is implied by (C1) and $\Theta$ compact.

*Second Case*: $\Theta$ an open convex of $\mathbb{R}^p$, and $U_n$ a convex contrast

(C') (C1')=(C1) uniformly in $\alpha$ over compacts; (C'2)=(C2).
(C3') $\alpha \longmapsto U_n(\alpha)$ is convex for every observation.

**(5.2.2) Theorem.** *Under one of the conditions* (C) *or* (C'), $\widehat{\theta}_n$, *the minimum contrast estimator of* $U_n$, *is strongly consistent if the distribution* $\mu$ *of the observation belongs to* $\mathcal{G}(\theta)$.

*Proof:*
*Under (C):* Let $\varepsilon$, $\delta > 0$. Because of (C1), for $n$ big

$$\{|\widehat{\theta}_n - \theta_0| \geq \delta\} \subseteq \Big\{\inf_{|\theta-\theta_0|\geq\delta} U_n(\theta) \leq U_n(\theta_0)\Big\}$$
$$\subseteq \Big\{\inf_{|\theta-\theta_0|\geq\delta} (K(\theta, R_{n,x}) - K(\theta_0, R_{n,x})) \leq 2\varepsilon\Big\}.$$

Since $\{\theta \in \Theta : |\theta - \theta_0| \geq \delta\}$ is compact and $K$ continuous, the function

$$R \longrightarrow \inf_{|\theta-\theta_0|\geq\delta} (K(\theta, R) - K(\theta_0, R))$$

is continuous. Thus, if $R \in \mathcal{G}_s(\theta_0)$ and $|\theta - \theta_0| \geq \delta$, $K(\theta, R) - K(\theta_0, R) > 0$. This, added to the fact that $\mathcal{G}_s(\theta_0)$ is compact, yields

$$\delta_0 = \inf\{K(\theta, R) - K(\theta_0, R),\ R \in \mathcal{G}_s(\theta_0),\ |\theta - \theta_0| \leq \delta\} > 0.$$

Choose $\varepsilon = \dfrac{\delta_0}{4}$. For $n$ big and using the $LD$ bounds,

$$P_0(|\widehat{\theta}_n - \theta_0| \geq \delta) \leq P_0\Big\{\sup_{|\theta-\theta_0|\geq\delta} K(\theta, R_{n,x}) - K(\theta_0, R_{n,x}) > -\frac{\delta_0}{2}\Big\}$$
$$\leq C\ \exp -\delta_1 |\Lambda_n| \quad \text{for } \delta_1 > 0.$$

*Under (C'):* Write $\zeta(\theta, \varepsilon) = \{\alpha : \|\alpha - \theta\| = \varepsilon\}$, and choose $\varepsilon > 0$, $\delta > 0$. Since $\alpha \longmapsto U_n(\alpha)$ is convex,

$$\{|\widehat{\theta}_n - \theta| \geq \varepsilon\} \subset \{\exists \alpha \in \zeta(\theta, \varepsilon) : U_n(\alpha) \leq U_n(\theta)\}$$
$$\subset \{\inf_{\alpha\in\zeta(\theta,\varepsilon)} K(R_{n,x}, \alpha) \leq K(R_{n,x}, \theta) + \delta\}$$

for big enough $n$ because of (C1'), the continuity of $K$ and the compacity of $\zeta(\theta,\varepsilon)$. Also because of this continuity and the compacity of $\zeta(\theta,\varepsilon)$, the mapping of $\mathcal{P}_s(\Omega)$ in $\mathbb{R}$ given by

$$F(\nu) = \Big(\inf_{\alpha\in\zeta(\theta,\varepsilon)} K(\nu,\alpha)\Big) - K(\nu,\theta)$$

is continuous and $\varepsilon_0 = \inf_{\nu\in\mathcal{G}_s(\theta)} F(\nu) > 0$. Choose $\delta_0 = \dfrac{\varepsilon_0}{2}$; then

$$\{|\widehat{\theta}_n-\theta| \geq \varepsilon\} \subset \Big\{F(R_{n,x}) \leq \frac{\varepsilon_0}{2}\Big\} \subset \Big\{R_{n,x} : \text{dist}(F(R_{n,x}), F(\mathcal{G}_s(\theta))) \geq \frac{\varepsilon_0}{2}\Big\}$$

so that the $LD$ exponential bound yields, for some $\delta_1 > 0$ and $n$ big enough,

$$|\Lambda_n|^{-1} \log \mu\{|\widehat{\theta}_n - \theta| \geq \varepsilon\} \leq C \ \exp -\delta_1|\Lambda_n|.$$

□

*Examples of Applications.*

(*a*) *Maximum Likelihood:* (C1) and (C1') are satisfied (cf. 5.2.1). If the representation in $\alpha$ of the specification is injective, the variational principle entails (C2). If the specification model is exponential, the contrast is convex. The equicontinuity condition (C3) is assured whenever the potentials $\phi_A$, $A \ni 0$, are continuous.

(*b*) *PML:* Under the conditions of (5.2.1), (C) can be readily checked; if the conditional density belongs to an exponential model, then $-\log \pi(x_i|x_{\partial i},\alpha)$ is convex in $\alpha$, which entails the convexity in $\alpha$ of the contrast.

### *5.2.3. The Case X Markovian with Bounded Range*

In this paragraph, $S$ is a numerable infinite set of sites, without any particular structure, and the field $X = \{X_i, i \in S\}$ takes values over a compact set $\Omega = \prod_{i\in S} E_i$. Let $\mu = P_\theta \in \mathcal{G}(\pi_\theta)$, where $\theta$ belongs to the parameter set $\Theta$, a compact of $\mathbb{R}^p$, for the Markov specifications family $\{\pi_\alpha, \alpha \in \Theta\}$. We shall assume that the Markovian graph is independent of $\alpha$. In what follows, $C \subseteq S$ is an infinite coding subset, $(D_n)$ is a sequence of strictly increasing finite subsets of $S$, $d_n = |\overset{\circ}{D}_n|$, and $c_n = |C_n|$ where $C_n = \overset{\circ}{D}_n \cap C$. We shall use the following three conditions. The first one refers to the Markov graph and the other two to the specification:

(M1) $C$ is the union of disjoint sets $C_k$, $k = 1, K$ such that

(1) Each $C_k$ is a *strong coding* set in the sense that for all $i$, $j \in C_k$ $i \neq j$, $\partial i$ and $\partial j$ are not neighbors.

(2) For $i \in C_1$, $\Omega_{\partial i} = \prod_{j \in \partial i} E_j = \mathcal{X}$ is a fixed space, and if $c_{n,1}$ is the cardinal of $(C_1 \cap \overset{\circ}{D}_n)$, then $\liminf_n c_{n,1} c_n^{-1} = \tau > 0$.

(3) $\liminf_n (c_n d_n^{-1}) > 0$.

(M2) $\pi_i(x_i | x_{\partial i}, \alpha)$ and $\pi_{\partial i}(x_{\partial i} | x_{\partial\partial i}, \alpha) \geq c > 0$, with $\pi_i$ a continuous function of its three arguments, uniformly in $i$.

Define

$$m_i(\theta, \alpha; x_{\partial i}) = -E_\theta \Big( \log \frac{\pi_i(X_i | x_{\partial i}, \alpha)}{\pi_i(X_i | x_{\partial i}, \theta)} \Big) \geq 0,$$

and for each $i \in C_1$, $\lambda$ is the reference measure over $\mathcal{X}$.

(M3) There exists $m(\theta, \alpha; z) \geq 0$, $(\alpha, z) \in \Theta \times \mathcal{X}$, continuous in $\alpha$ and $\lambda$-integrable for all $\alpha$, such that

(1) $m_i(\theta, \alpha; z) \geq m(\theta, \alpha; z)$ if $i \in C_1$.
(2) $\alpha \longmapsto K(\theta, \alpha) = \int_{\mathcal{X}} m(\theta, \alpha; z) \lambda(dz)$ is continuous with a unique minimum at $\alpha = \theta$.

**(5.2.3) Theorem.** *Under conditions* (M1) *to* (M3), *the PML and Coding estimators are consistent.*

*Proof:* First we will show the consistency of the Coding estimator $\widehat{\theta}_n^c$. The basic tool is the conditional independence of $X_i$, $i \in C$ given $X_j$, $j \notin C$. In order to prove consistency, we will need the following Lemma, which gives the fundamental subergodicity property (Geman and Graffigne, [70]; [79], [99]).

**(5.2.2) Lemma.** *Let $A$ be a measurable set of $\mathcal{X}$ and $F_n(C_1, A)$ the empirical frequency of $A$ over $C_1$: $F_n(C_1, A) = c_{n,1}^{-1} \sum_{C_{n,1}} \mathbb{1}(X_{\partial i} \in A)$. Then*

$$\liminf_n F_n(C_1, A) \geq \frac{c}{2} \lambda(A) \quad P_\theta - a.e.$$

*Proof:* Because of (M2), variables $\mathbb{1}(X_{\partial i} \in A)$, $i \in C_1$ have a mean which is greater than $c\lambda(A)$, a variance which is bounded by 1, and are independent conditional to $x_{\overline{\partial C_1}}$. Thus, following the Strong Law of Large Numbers for independent variables of $L^2$, we have [22]

$$\liminf_n F_n(C_1, A) \geq \frac{c}{2} \lambda(A) \quad P_{\theta, x_{\overline{\partial C_1}}} - a.e.$$

Since neither of the above terms depends on $x_{\overline{\partial C_1}}$, the inequality remains true $P_\theta - a.e.$ □

Consider the $C_1$-contrast: $U_n^{C_1}(\alpha) = -c_{n,1}^{-1} \sum_{C_{n,1}} \log \pi_i(x_i|x_{\partial i}, \alpha)$. Write: $U_n^{C_1}(\alpha) - U_n^{C_1}(\theta) = A_n + B_n$, with

$$A_n = -c_{n,1}^{-1} \sum_{C_{n,1}} \Big( \log \frac{\pi_i(x_i|x_{\partial i}, \alpha)}{\pi_i(x_i|x_{\partial i}, \theta)} + m_i(\theta, \alpha; x_{\partial i}) \Big).$$

$A_n$ is a sum of centered variables with bounded variances (M2) and independent conditional to $x_{\overline{C}_1}$. Hence, proceeding as in the Lemma,

$$\lim_n A_n = 0 \quad P_\theta - a.e.,$$

and $P_\theta - a.e.$, we have the following:

(5.10)

$$\begin{aligned}
\liminf_n \quad (U_n^{C_1}(\alpha) - U_n^{C_1}(\theta)) &= \liminf_n B_n \\
&\geq \liminf \int_{\mathcal{X}} m(\theta, \alpha; z) F_n(C_1, dz) && \text{(because of (M3(1))} \\
&\geq \int_{\mathcal{X}} m(\theta, \alpha; z) \liminf F_n(C_1, dz) && \text{(since } m \geq 0) \\
&\geq \frac{c}{2} \int_{\mathcal{X}} m(\theta, \alpha; z) \lambda(dz) := \frac{c}{2} K(\theta, \alpha) && \text{(Lemma 5.2.3)}
\end{aligned}$$

*Consistency of the Coding Estimator.*
$U_n^C(\alpha) = \sum_{k=1,K} c_{n,k} c_n^{-1} U_n^{C_k}(\alpha)$ where $U_n^{C_k}$ is the $C_k$-Coding contrast and $c_{n,k} = |C_k \cap \overset{\circ}{D}_n|$. The consistency of $\widehat{\theta}_n^c = \operatorname{Arg\,min}_{\Theta} U_n^c(\alpha)$ follows from (5.10), from (M1(2)), from (M2) and from Corollary 3.2.

*The consistency of the PML* estimator follows from (M1(3)) and

$$U_n = c_n d_n^{-1} U_n^C + (d_n - c_n) d_n^{-1} U_n^{\overline{C}}.$$

□

**(5.2.1) Example.** $S = \mathbb{Z}^d$, $\pi_\alpha$ is invariant under translation, $\pi_0(\cdot|\cdot, \alpha)$ and $\pi_{\partial 0}(\cdot|\cdot, \alpha)$ are continuous functions of the three arguments, and $\alpha \longmapsto \pi_\alpha$ is injective:

if $\alpha \neq \beta$, there exists $(x, z) \in E \times E^{\partial 0}$ such that $\pi(x|z, \alpha) \neq \pi(x|z, \beta)$.

Then if there exists $a = (a_1, \ldots, a_d)$, $a_i > 0$, $i = 1, d$ with $C \supseteq a\mathbb{Z}^d$, conditions (M1) to (M3) are satisfied.

**(5.2.2) Example.** $S = \mathbb{Z}^2$ and consider the noninvariant Ising model with four nearest neighbors,

$$E_i = \{+1, -1\}, \quad \partial i = \{j : |j - i| = 1\},$$

$$C = \{(i,j) \in \mathbb{Z}^2 : i + j \text{ is even}\}, \quad \pi_i(x_i | x_{\partial i}, \alpha) = \frac{\exp x_i v_i(x_{\partial i}, \alpha)}{2ch(v_i(x_{\partial i}, \alpha))},$$

such that

*(i)* The $v_i$ are continuous functions of $\alpha$, uniformly in $i$, and bounded by $A < \infty$.

*(ii)* If $\alpha \neq \theta$, there exists continuous $\eta_\theta(\alpha) > 0$ *s.t.* for all $i$ in $C$, $x^*_{\partial i} = x^*_{\partial i}(\alpha)$

$$|v_i(x^*_{\partial i}, \alpha) - v_i(x^*_{\partial i}, \theta)| \geq \eta_\theta(\alpha).$$

*(iii)* $\pi_i(x_{\partial i} \mid x_{\partial \partial i}, \alpha) \geq c > 0$.

Then (M1) and (M2) are satisfied. On the other hand,

$$m_i(\theta, \alpha; x_{\partial i}) = m(v_i(x_{\partial i}, \theta), v_i(x_{\partial i}, \alpha)),$$

where $m(a,b) = (a - b)th\ a - \log \frac{ch\ a}{ch\ b} \geq 0$, is continuous.

Since the set of neighborhood configurations is finite, there exists $z^*(\alpha)$ in $\{-1,1\}^4$, $C^* \subseteq C$ *s.t.* $\liminf c_n^{-1} c_n^* > 0$, $c_n^* = |c^* \cap \overset{\circ}{D}_n|$ and over $C^* : |a_i(z^*, \theta) - a_i(z^*, \alpha)| \geq \eta_\theta(\alpha)$. Call

$$M_\theta(\alpha) = \inf\{m(a,b), 0 \leq |a|, |b| \leq A, |a - b| \leq \eta_\theta(\alpha)\}.$$

$\alpha \longmapsto M_\theta(\alpha) \geq 0$ is continuous and is zero only at $\alpha = \theta$. (M3) is satisfied restricting $C$ to $C^*$ and for

$$m(\theta, \alpha, z) = 0 \quad \text{if } z \neq z^*(\alpha), \quad M_\theta(\alpha) \quad \text{if not.}$$

We will further specify the model by setting

$$v_i(x_{\partial i}, \alpha) = \alpha_1 a_i + \alpha_2 \textstyle\sum_{j \in \partial i} w_{ij} x_j, \quad \alpha = (\alpha_1, \alpha_2)$$

$$0 < d \leq a_i \leq A, \quad 0 \leq w_{ij} \leq A, \quad \textstyle\sum_{\partial i} w_{ij} \geq d.$$

Condition *(ii)* is satisfied. Indeed,

$$|v_i(x_{\partial i}, \alpha) - v_i(x_{\partial i}, \theta)| = |(\alpha_1 - \theta_1)a_i + (\alpha_2 - \theta_2) \sum_{\partial i} w_{ij} x_j|.$$

If $(\alpha_1 - \theta_1) \times (\alpha_2 - \theta_2) \geq 0$, choose $x^*_j = 1$ if $j \in \partial i$. If not choose $x^*_j = -1$, $j \in \partial i$ so that for all $i \in C$ (here $C \equiv C^*$),

$$|v_i(x^*, \alpha) - v_i(x^*, \theta)| \geq \eta_\theta(\alpha) = d\|\theta - \alpha\|.$$

## 5.3. Asymptotic Distributions and Tests

We shall examine the following three situations:

(1) $X$ is a Gibbs field that belongs to an exponential parametric family. We shall also assume $X$ is *stationary and ergodic.* We shall establish the asymptotic normality of the $PML$, Coding, Minimum $\chi^2$ (binary data) and $CLS$ estimators. Asymptotic tests for subhypothesis are distributed as a mixture of independent $\chi_1^2$.

(2) $X$ is *weakly dependent*: For all estimators, including the $ML$, we get asymptotic normality by standard methods. The results do not require invariance under translation or lattice regularity.

(3) $X$ is *Markovian* over a nonnecessarily regular lattice. Then, without any other hypothesis, we show that the *coding* estimator is conditionally asymptotically normal and construct an explicit $\chi^2_{p-q}$ subhypothesis difference of coding test.

### *5.3.1. The Case X, a Stationary Ergodic Field*

$X$ is a field over $S = \mathbb{Z}^d$, with compact state space $E$ and whose specification $\pi_\theta$ belongs to an exponential family. In particular, its conditional distribution at site $i$ is

$$\begin{cases} \pi(X_i = x_i \mid V_i = v_i, \alpha) = \dfrac{\exp {}^t\alpha \cdot \phi(x_i, v_i)}{Z(\alpha, v_i)}, \\ Z(\alpha(v_i)) = \displaystyle\int_E \exp({}^t\alpha \cdot \phi(x, v_i))\ \lambda(dx), \quad \alpha, \phi(\cdot,\cdot) \in \mathbb{R}^p, \end{cases}$$

with $v_i$, the neighbor configuration of $i$, belonging to a compact $\mathbb{V}$.

Assume that the true value of $\theta$ is in the interior of a parameter set $\Theta$. Then, if the minimum contrast estimator $\widehat{\theta}_n$ is consistent, with tending to 1 probability and $\Delta_n$ defined as in § 3.4.2, we have

$$U_n^{(1)}(\widehat{\theta}_n) = 0 = d_n^{1/2} U_n^{(1)}(\theta) + \Delta_n(\theta, \widehat{\theta}_n) d_n^{1/2}(\widehat{\theta}_n - \theta). \tag{5.10}$$

The asymptotic normality of $d_n^{1/2}(\widehat{\theta}_n - \theta)$ will be proved

(a) by checking that $\Delta_n(\theta, \widehat{\theta}_n) \to I(\theta)$ in $P_\theta$-probability (ergodicity).

(b) by showing that $d_n^{1/2} U_n^{(1)}(\theta) = d_n^{-1/2} \sum_{\overset{\circ}{D}_n} Z_t$ converges weakly to a $\mathcal{N}_p(0, J(\theta))$; here, we use the *Markovian property* of $X$ via the $CLT$ given in Theorem 3.3.3, with

$$J(\theta) = \sum_{t \in \mathcal{V}(0)} cov(Z_0, Z_t), \quad \text{with } \mathcal{V}(0) = \{0\} \cup \partial\{0\}. \tag{5.11}$$

We shall respectively write (1), (2), (3), and (4) for the $PML$, Coding, minimum $\chi^2$, and $CLS$ methods. If $E \subseteq \mathbb{R}$ et $\phi(x,v) = x \cdot h(v)$, we define

$$\mu(\theta, V_t) = E_\theta^{V_t} X_t, \quad \sigma^2(\theta, V_t) = Var_\theta^{V_t}(X_t), \quad P_\theta(v) = P_\theta(V_t = v).$$

**(5.3.1) Theorem.** *Assume $\phi(x,v)$ is continuous over $E \times \mathbb{V}$, and can be written as $x \cdot h(v)$ in the context of (3) and (4). If $\widehat{\theta}_n$ is consistent and if $X$ is stationary and ergodic, then, whenever the variance covariance matrices defined below are positive definite,*

- $i = 1,3,4$: *for* $\Delta_i = I_i(\theta)^{-1} J_i(\theta) I_i(\theta)^{-1}$

$$d_n^{1/2}(\widehat{\theta}_n(i) - \theta) \xrightarrow{d} \mathcal{N}_p(0, \Delta_i)$$

•• $i = 2$, $c_n = |C_n|$, $C_n = a\mathbb{Z}^d \cap D_n$ *the coding set*

$$c_n^{1/2}(\widehat{\theta}_n^c(2) - \theta) \xrightarrow{d} \mathcal{N}_p(0, I_1(\theta)^{-1}).$$

| Method | $Z_t$ ($J$ as in (5.11)) | $I(\theta)$ |
|---|---|---|
| $i = 1 : PML$ | $\phi(X_t, V_t) - E_\theta^{V_t} \phi(X_t, V_t)$ | $E_\theta(Var_\theta(Z_t \mid V_t))$ |
| $i = 3$ : Min. $\chi^2$ | $(HA)(V_t)\sigma^{-2}(\theta, V_t) P_\theta^{-1}(V_t)(X_t - \mu(\theta, V_t))$ | $HA^t H$ |
| $i = 4 : CLS$ | $\sigma^2(\theta, V_t) h(V_t)(X_t - \mu(\theta, V_t))$ | $E_\theta(\sigma^4(\theta, V_t)(h^t h)(V_t))$ |

*Proof:*

*PML Estimator.*

*Analysis of $\Delta_n$*: since $\phi(x,v)$ is continuous and $\Theta$ and the state space are compact, $U_n^{(2)}$ is uniformly continuous, uniformly in $n$. It follows easily from the ergodicity that $\Delta_n(\theta, \widehat{\theta}_n) \xrightarrow{P_0} I_1(\theta)$.

*Normality of $d_n^{1/2} U_n^{(1)}(\theta)$*: $d_n^{1/2} U_n^{(1)}(\theta) = d_n^{1/2} \sum_{\overset{\circ}{D}_n} Z_t$, where

$$Z_t = \phi(X_t, V_t) - E_\theta^{V_t}(\phi(X_t, V_t)) \text{ satisfies } E_\theta^{V_t}(Z_t) = 0.$$

Hence, by applying Theorem 3.3.3, we have $d_n^{1/2} U_n^{(1)}(\theta) \xrightarrow{d} \mathcal{N}_p(0, J_1(\theta))$, $J_1(\theta) = \sum_{t \in \mathcal{V}} \Gamma_t(\theta, \phi)$, $\mathcal{V} = \{0\} \cup \partial\{0\}$, and $\Gamma_t(\theta, \phi) = E(Z_0 Z_t)$.

*Coding Estimator:* $a = (a_i, i = 1, d)$, $a_i > 0$, $i = 1, d$, $C = a\mathbb{Z}^d$. $d_n$, $\overset{\circ}{D}_n$ are replaced by $c_n = |C_n|$, $C_n = C \cap D_n$ in the expressions for $U_n^{(\ell)}$, $\ell = 0, 1, 2$. As $(X_t, V_t)$ is ergodic over the lattice $C$, we have $\Delta_n(\theta, \widehat{\theta}_n) \longrightarrow I_1(\theta) = I_2(\theta)$.

On the other hand, if $t$ isn't a neighbor of 0, noticing that $Z_0$ and $Z_t$ are independent conditional to $\{X_s,\ s \neq 0 \text{ and } s \neq t\}$,

$$d_n^{1/2} U_n^{(1)}(\theta) \xrightarrow{d} \mathcal{N}_p(0, I_1(\theta)) \quad (J_2(\theta) = I_2(\theta)).$$

*Minimum* $\chi^2$ *Estimator:* Considering $y(v)$ given by (5.8) and developing the Taylor series of $\log u$ around $u = 1$, $u = P_\theta(x,v)^{-1}N(x,v)$, $x = a, b$, yields

$$y(v) = {}^t\theta \cdot h(v) + \Big(\frac{N(a,v)}{P_\theta(a,v)} - \frac{N(b,v)}{P_\theta(b,v)}\Big)(1 + o_p(1)).$$

The parentheses can be yet written as (the sum is over $\overset{\circ}{D}_n$):

$$(\pi_\theta(a|v)\pi_\theta(b|v)P_\theta(v))^{-1} d_n^{-1} \sum (1(X_t{=}a)\pi_\theta(b|v) - 1(X_t{=}b)\pi_\theta(a|v)) \cdot 1(V_t{=}v).$$

Coding $a, b$, respectively, by $1, 0$, we notice that the parentheses that appears in the sum is $X_t - \mu(\theta, V_t)$. That is,

$$y(v) = {}^t\theta \cdot h(v) + [\pi_\theta(1|v)\pi_\theta(0|v)P_\theta(v)]^{-1} d_n^{-1} \sum (X_t - \mu(\theta,V_t)) 1(V_t{=}v) \times (1 + o_p(1)).$$

For the $A$-least squares contrast associated to (5.7), we have

$$U_n^{(1)}(\theta) = -2HA(y - {}^t\theta \cdot H), \quad U_n^{(2)}(\alpha) \equiv -2HA\,{}^tH,$$
$$-\tfrac{1}{2}U_n^{(1)}(\theta) = d_n^{1/2} \Sigma_{\overset{\circ}{D}_n} Z_t,$$

and if $HA(v)$ is the $v$-th column of $HA$,

$$Z_t = HA(V_t)\sigma^{-2}(\theta, V_t)P_\theta^{-1}(V_t)(X_t - \mu(\theta, V_t)).$$

*CLS Estimator:* $E \subset \mathbb{R}$, $\phi(X_t, V_t) = X_t h(V_t)$. A straightforward calculation shows that

$$Z_t = (X_t - \mu(\theta, V_t))h(V_t)\sigma^2(\theta, V_t), \; U_n^{(1)}(\theta) = 2d_n^{-1} \sum Z_t,$$
$$U_n^{(2)}(\alpha) = -2\sigma^4(\alpha, V_t)(h \cdot {}^th)(V_t) + 2\mu^{(2)}(\alpha, V_t)(X_t - \mu(\alpha, V_t)).$$

Using $X$'s ergodicity, the consistency of $\widehat{\theta}_n$ and the fact that $(X_t - \mu(\theta, V_t))$ is centered, we get

$$-\tfrac{1}{2}U_n^{(2)}(\theta_n^*) \longrightarrow I_4(\theta) = E_\theta(\sigma^4(\theta, V_t)(h\,{}^th)(V_t)) \text{ in } P_\theta - \text{probability}$$
$$\tfrac{1}{2}d_n^{1/2}U_n^{(1)}(\theta) \xrightarrow{d} \mathcal{N}_p(0, J_4(\theta)).$$

□

REMARKS:

(1) If the parametrization is injective, matrices $I(\theta)$ are nonsingular.

(2) For estimators (1) and (2) under a model with quantitative states $\phi(x_t, v_t) = x_t h(v_t)$, we have

$$Z_t(1) = (X_t - \mu(\theta, V_t))h(V_t),\ I_1(\theta) = E_\theta(\sigma^2(\theta, V_t)(h \cdot h^T)(V_t)).$$

(3) $I$ and $J$ cannot, in general, be obtained analytically, which makes it difficult to check the nonsingularity of $J$. Numerical calculus of $I$ and $J$ will be done either by Monte Carlo methods, using the Gibbs sampler under $\pi_{\widehat{\theta}_n}$ (cf. 6.2), or as a consequence of $X$'s ergodicity, by empirical estimation.

If the state space $E$ is binary, the following result shows that the $PML$ and the $CLS$ estimators are equivalent to minimum $\chi^2$ estimation for an adequate choice of the metric $A$ [83].

**(5.3.1) Corollary.** *If the state space $E$ is binary,*

*(a) PML estimation is asymptotically equivalent to minimum $\chi^2$ if we choose diagonal matrix $A$:$a(v) = P_\theta(v)\sigma^2(\theta,v)$*

*(b) CLS estimation is asymptotically equivalent to minimum $\chi^2$ if we choose the diagonal matrix $A$: $a(v) = P_\theta(v)\sigma^4(\theta,v)$.*

*Proof:*

($a$) For the first choice of $a(v)$,

$$I_3(\theta) = HA_\theta{}^tH = \sum_v P_\theta(v)\sigma^2(\theta, v)h(v)^t h(v) = I_1(\theta),$$

$$\begin{aligned} Z_t(3) &= (HA_\theta)(V_t)\sigma^{-2}(\theta, V_t)P_\theta(V_t)^{-1}(X_t - \mu(\theta, V_t)) \\ &= P_\theta(V_t)\sigma^2(\theta, V_t)h(V_t)\sigma^{-2}(\theta, V_t)P_\theta(V_t)^{-1}(X_t - \mu(\theta, V_t)) = Z_t(1). \end{aligned}$$

($b$) For the second choice of $a(v)$, we check likewise: $I_3(\theta) = I_4(\theta)$, $Z_t(3) = Z_t(4)$. □

REMARKS:

(1) For minimum $\chi^2$, the best choice for $A$ is the inverse of the asymptotic variance of $d_n^{1/2}\ {}^t(y(v), v \in \mathcal{V})$. In this case, $a(v, v') =$

$$[P_\theta(v)P_\theta(v')\sigma^2(\theta, v)\sigma^2(\theta, v')]^{-1} \sum_{t \in \mathcal{V}(0)} E_\theta[(X_0 - \mu(\theta,v))(X_t - \mu(\theta,v'))] \mathbb{1}(V_0 = v, V_t = v').$$

This choice is different from that considered in the Corollary; this leads to believing that a good metric choice for the minimum $\chi^2$ estimator gives better estimators than $PML$ or $CLS$ methods. Simulations carried out by Possolo [131] follow this direction, but we do not know of any explicit analytical examples that exhibit this superiority; in the case of a nearest neighbor Ising model over $\mathbb{Z}$, all expressions can be explicitly calculated and the $PML$ (as the $CLS$) is equivalent to the minimum $\chi^2$ for ordinary least squares, which here coincide with $CLS$.

(2) If $A_n \to A$ in probability, Theorem (5.3.1) is still true for minimum $\chi^2$ estimation. In particular, the limiting covariance matrix for $((y(v), v \in \mathcal{V})$ can be estimated either parametrically (cf. above formula), or non-parametrically using resampling methods (cf. [131]).

*Asymptotic Tests.*

Under the hypothesis of Theorem (5.3.1), and for each of the proposed methods, an asymptotic test can be considered as a result of Property (3.8) in its stationary version: if $\varphi \in \Lambda$ defines $(H_q)$, subhypothesis of $(H_p)$,

$$2d_n(U_n(\widehat{\varphi}_n) - U_n(\widehat{\theta}_n)) \xrightarrow[H_q]{d} \sum_{i=1}^{p-q} \lambda_i \chi_{i1}^2.$$

The eigenvalues $\lambda_i$ of matrix $A_0$ (cf. Remark 3 of Property 3.8) can be obtained for methods (1), (3) and (4) in terms of the estimated $I_0$, $J_0$, and $\overline{I}_0$. For the coding estimator, the *statistic is explicit*:

$$2c_n(U_n(\widehat{\varphi}_n) - U_n(\widehat{\theta}_n)) \xrightarrow[H_q]{\mathcal{D}} \chi_{p-q}^2.$$

As we shall see in the two following paragraphs, this property of the *coding test* is general and doesn't require invariance of the specification, ergodicity, or weak dependence of the field, nor regularity of the lattice.

## *5.3.2. The Case of a Field that Satisfies the Dobrushin–Simon Condition (cf. (2.10))*

Assume the infinite numerable set of sites $S$ is equipped with a metric $d$ and that the state space, $(E, \mathcal{E}, \lambda)$, the same for each site, is a polish space. We observe a realization of $\mu \in \mathcal{G}(\pi_\theta)$. The Dobruschin–Simon $(DS)$ condition assures the uniqueness of $\mathcal{G}(\pi_\theta)$. This condition entails several consequences:

(1) $\mu$ is uniformly mixing (Theorem 2.3): if $A \in \mathcal{S}$ and $B \subset S$,

$$\varphi_\mu(A, B) \le C\ |A| \exp\{-cd(A, B)\}, \quad C < \infty,\ c > 0.$$

(2) If $f$ is quasi-locally bounded (cf.§ 2.1.3; if $E$ is compact, $f$ continuous is quasi-local), if $(\omega^n)$ is a sequence of $\Omega$, and if $\pi_{\theta,n}^{\omega^n}$ is the specification over $D_n$, then ([71], Theorem 8.23)

$$E_{\pi_{\theta,n}^{\omega^n}}(f) \longrightarrow E_\mu(f), \quad \text{uniformly.}$$

(3) If the $DS$ condition is satisfied in a neighborhood of $\theta$ and if $\mathcal{G}(\pi_\alpha) = \{\mu_\alpha\}$ over this neighborhood, then $\alpha \longmapsto \mu_\alpha(f)$ belongs to $\mathcal{C}^1$ (Theorem 2.1.6).

*Hypothesis over the Gibbs Model.*

(H1) $\mu_\alpha$ is associated to a family of potentials $\{\phi_A(\alpha)\}$ of range $m$, $\alpha \in \Theta$ compact of $\mathbb{R}^p$, and each $\phi_A$ is of class $\mathcal{C}^2$ over $\overset{\circ}{\Theta}$ with

$$\sup_A \sup_{\alpha\in\Theta} \sup_{\ell=0,1,2} \|\phi_A^{(\ell)}(\alpha)\| \le C < \infty.$$

(H2) The uniqueness condition $(DS)$ is satisfied at $\theta \in \overset{\circ}{\Theta}$, the true value of the parameter

$$|||\phi_\theta||| = \sup_{i\in S} \sum_{A\in\mathcal{S}:A\ni i} e^{d(A)}(|A|-1)\|\phi_A(\theta)\| < 1,$$

where $d(A)$ is the diameter of $A$ with respect to the metric $d$.

(H3) $D_n \nearrow S$ and $|D_n \setminus \overset{\circ}{D}_n| = o(|D_n|)$
Here, $\overset{\circ}{D}_n = \{i \in D_n$, such that if $\phi_A \not\equiv 0$, $A \ni i$, then $A \subseteq D_n\}$.

(H4) Define $\overline{\phi}_i(\theta) = \sum_{A\ni i} |A|^{-1}\phi_A(\theta)$ and for $\mu = \mu_\theta$, $d_n = |D_n|$

$$I_n(\mu) = d_n^{-1} \sum_{i,j\in D_n} Cov_\mu(\overline{\phi}_i^{(1)}(\theta), \overline{\phi}_j^{(1)}(\theta)).$$

Then there exists $I_0$, symmetric and positive definite, *s.t.* for $n$ big enough, $I_n(\mu) \ge I_0 > 0$.

Observation: Assume $S \subseteq \mathbb{R}^d$, $\delta = \Delta d$ with $d$ the Euclidean metric. If

$$\sup_{i\in S} \sum_{\substack{A\in\mathcal{S}\\ A\ni i}} (|A|-1)\|\phi_A(\theta)\| < 1,$$

we can deduce (H2) choosing $\Delta$ small enough whenever the potentials satisfy (H1). On the other hand, because of (H1), there exists $\varepsilon > 0$ such that $\overline{B}(\theta,\varepsilon) \subseteq \Theta$, and over this closed ball, and for a certain $c < 1$,

$$\sup_{\alpha\in\overline{B}(\theta,\varepsilon)} \left( \sup_{i\in S} \sum_{A:A\ni i} e^{\delta(A)}(|A|-1)\|\phi_A(\alpha)\| \right) \le c < 1.$$

In particular, we can deduce that $\mu_\alpha$ is mixing uniformly in $\alpha \in \overline{B}(\theta,\varepsilon)$

*5.3.2.1. Estimation by Maximum Likelihood*

The consistency of the $ML$ estimator was established for the case $\mu$ stationary (§ 5.2.1) and for the case $\mu$, a translation invariant specification (§ 5.2.2). For the more general case where $S$ is arbitrary but the potentials have bounded range, weak consistency can be deduced easily from Theorem (3.4.1); indeed, the energy $H_n$ can be expressed additively in terms of the $\overline{\phi}_i$ and condition (3.13) follows directly from the mixing assumption over $\mu$ and from a well-defined parametrization hypothesis. Hence, in what follows, we shall assume that the M.L. estimation is consistent. Let $x \in \Omega$ be the realization of $\mu$ observed over a sequence of domains $D_n \nearrow S$, $d_n = |D_n|$ and $\pi_n(x,\alpha)$ the distribution $\mu_\alpha$ over $D_n$ conditional to $x^n$. Let $U_n$ be the opposite of the log-likelihood.

**(5.3.2) Theorem.** *Assume that* $\widehat{\theta}_n$, *the ML estimator, is consistent and that conditions* (H1) *to* (H4) *are satisfied. Then*

*(a)* $d_n^{1/2} I_n^{1/2}(\mu)(\widehat{\theta}_n - \theta) \xrightarrow{\mathcal{D}} \mathcal{N}_p(0,I)$.

*(b) If* $\varphi \in \Lambda \subseteq \mathbb{R}^q$ *is a regular subhypothesis* $(H_q)$ *of class* $\mathcal{C}^2$ *of dimension* $q$ *and* $\widehat{\varphi}_n$, *the ML estimator of* $\varphi$ *under* $(H_q)$, *then*

$$2d_n(U_n(\widehat{\varphi}_n) - U_n(\widehat{\theta}_n)) \xrightarrow[(H_q)]{\mathcal{D}} \chi^2_{p-q}.$$

*Proof:* The likelihood contrast is

$$U_n(\alpha) = d_n^{-1}\{\log Z_n(\alpha, x^n) + H_n(x,\alpha)\}, \text{ with } H_n(x,\alpha) = \sum_{A\cap D_n \neq \emptyset} \phi_A(x,\alpha).$$

Under (H1), we have

$$U_n^{(1)}(\theta) = d_n^{-1}\{H_n^{(1)}(x,\theta) - E_\alpha^{x^n}(H_n^{(1)}(X_n,\theta))\},$$

$$U_n^{(2)}(\alpha) = d_n^{-1}\{[H_n^{(2)}(\alpha) - E_\alpha^{x^n}(H_n^{(2)}(\alpha))] - Var_\alpha^{x^n}(H_n^{(1)}(\alpha))\}.$$

*(a1) Behavior of* $d_n^{1/2}U_n^{(1)}(\theta)$ *under* $\mu$: Following (2.6) for $D_n = \Lambda$, we get

$$d_n^{1/2}U_n^{(1)}(\theta) = C_n + R_n, \quad \text{with}$$

$$C_n(\theta) = d_n^{-1/2} \sum_{i\in D_n} \{\overline{\phi}_i^{(1)}(\theta) - E_\theta^{x^n}(\overline{\phi}_i^{(1)}(\theta))\},$$

$$R_n(\theta) = d_n^{-1/2} \sum_{\substack{A_n\cap D_n \neq \emptyset \\ A \not\subseteq D_n}} \left(1 - \frac{|A\cap D_n|}{|A|}\right)\{\phi_A^{(1)}(\theta) - E_\theta^{x^n}(\phi_A^{(1)}(\theta)\}.$$

Under $\mu$, $C_n$, as well as the residual term $R_n$, is centered. We will now show the normal convergence of $C_n$ and the noncontribution of $R_n$.

*Normal Convergence of* $C_n$. Since the potentials have bounded range, $\{Y_i = \overline{\phi}_i(\theta) - E_\theta^{x^n}(\overline{\phi}_i)\}$ is entailed with the same mixing property as $\mu$. Condition (H4) and Theorem 3.3.1 yield the asymptotic normality:

$$I_n^{-1/2}(\mu)C_n(\theta) \xrightarrow{\mathcal{D}} \mathcal{N}_p(0,I).$$

*Noncontribution of* $R_n$. As $R_n$ is centered, we will check that $Var(R_n) \to 0$.

$$Var_\mu(R_n) = d_n^{-1} \sum\sum cov_\mu(\phi_A^{(1)}(\theta),\ \phi_{A'}^{(1)}(\theta)),$$

where the sum in $A$, $A'$ is over all sets, for example, $A$, *s.t.* $A \cap D_n \neq \emptyset$, $A \not\subseteq D_n$. However, for all $A$: $\sum_{A'} |cov_\mu(\phi_A^{(1)}, \phi_{A'}^{(1)})| \leq C < \infty$. Condition (H3) over the boundary of $(D_n)$ allows us to conclude that $Var(R_n) \to 0$.

$(a2)$ *Behavior of* $U_n^{(2)}(\theta)$: As the potentials belong to $\mathcal{C}^2$, we have

$$U_n^{(2)}(\theta) = T_1 + T_2, \text{ with } T_1 = d_n^{-1}(H_n^{(2)}(\theta) - E_{\theta_n}^{x^n}(H^{(2)}(\theta))),$$

$$\text{and } T_2 = -d_n^{-1} Var_\theta^{x^n}(H_n^{(1)}(\theta)).$$

$Var\ T_1 \longrightarrow 0$; hence, $T_1 \longrightarrow 0$ in probability. On the other hand, as in $(a1)$, $H_n^{(1)}(\theta) = C_n^* + R_n^*$ where only $C_n^*$ contributes and has the same asymptotic variance as $C_n$. It follows that $U_n^{(2)}(\theta) - I_n(\mu) \xrightarrow{P_0} 0$. $(a)$ now results from Theorem 3.4.5.

(b) is obtained in a standard fashion. □

### 5.3.2.2. *Estimation by PML or by Coding*

As the conditional energy at site $i$ under $\alpha$ is equal to $H_i(\alpha) = H_i(x_i \mid x_{\partial i}, \alpha) = \sum_{A: A \ni i} \phi_A(\alpha)$, we shall use the *pseudoinformation* matrices $J_n$, and $I_n$ to describe the behavior of the $PML$ estimator $\widehat{\theta}_n$ and the $C$ *information* matrix $I_n^C = J_n^C$ for the coding estimator $\widehat{\theta}_n^C$. They are defined as follows:

$$J_n(\mu) = d_n^{-1} \sum_{i,j \in D_n} Cov_\mu(H_i^{(1)}(\theta), H_j^{(1)}(\theta)),$$

$$I_n(\mu) = d_n^{-1} \sum_{i \in D_n} E_\mu(Var_\theta^{X_{\partial i}}(H_i^{(1)}(X_i \mid X_{\partial i}, \theta))),$$

$$I_n^C(\mu) = c_n^{-1} \sum_{C_n} E_\mu(Var_\theta^{X_{\partial i}}(H_i^{(1)}(X_i \mid X_{\partial i}, \theta))).$$

According to whether we consider the $PML$ estimator or the coding estimator, hypothesis (H4) shall be replaced by (H'4) or (H"4):

(H'4) There exist two $p \times p$ symmetric, positive definite matrices $J_0$ and $I_0$ such that from a certain point on, $J_n(\mu) \geq J_0 > 0, \quad I_n(\mu) \geq I_0 > 0$.

(H"4) There exists a $p \times p$ symmetric, positive definite matrix $I_0$ such that from a certain point on, $I_n^C(\mu) \geq I_0 > 0$.

**(5.3.3) Theorem.**

*(a) Assume that* $\widehat{\theta}_n$*, the PML estimator, is consistent and that* (H1) *to* (H3) *and* (H'4) *are satisfied. Then*

$$d_n^{1/2} J_n^{-1/2}(\mu) I_n(\mu)(\widehat{\theta}_n - \theta) \xrightarrow{\mathcal{D}} \mathcal{N}_p(0, I).$$

*(b) If the coding estimator* $\widehat{\theta}_n^C$ *is consistent, and if* (H1) *to* (H3) *and* (H"4) *are satisfied, then* $c_n^{1/2} I_n^C(\mu)^{1/2}(\widehat{\theta}_n^C - \theta) \xrightarrow{\mathcal{D}} \mathcal{N}_p(0,I)$.

The proof of this Theorem is analogous to that for $ML$ estimation, without the additional difficulty of having to examine the complementary term $R_n$. When studying the behavior of the $PML$ estimator, we are in a slightly different situation: contrary to what happens with $ML$ estimation or coding estimation, the matrix, which is asymptotically equivalent to $-U_n^{(2)}(\theta_n^*)$, that is, $J_n(\mu)$, is different from the asymptotic variance of $d_n^{1/2} U_n^{(1)}(\theta)$. This is what explains the particular normalization factor in terms of $I_n(\mu)$ and $J_n(\mu)$ that appears for the $PML$ estimator $(a)$. It is this same difference that is responsible for the fact that for $PML$, the subhypothesis test is a weighted sum of $\chi^2_{i,1}$ (cf. § 3.2) and not directly a $\chi^2_{p-q}$ test, as is the case for tests based on coding or $ML$ estimation.

*5.3.2.3. Results for the Stationary and Ergodic Case,* $S = \mathbb{Z}^d$

The potentials are generated by $K$ base potentials $\phi_{A_k}$, $0 \in A_k$, $k = 1, K$ and $\phi_{A_k+i}(x) = \phi_{A_k}(\tau_i x)$:

$$(H2) \qquad \sum_{k=1}^{K} (|A_k| - 1)\|\phi_{A_k}(\theta)\| < 1.$$

*Estimation by ML:* $\overline{\phi}_0 = \sum_{A \ni 0} |A| \phi_A(\theta)$, $\overline{\phi}_i(x) = \overline{\phi}_0(\tau_i x)$.

$$I_n(\mu) \longrightarrow I_0(\mu) = \sum_{\|s\| \le 2m} cov_\mu(\overline{\phi}_0^{(1)}(\theta), \overline{\phi}_s^{(1)}(\theta)),$$

where $m$ is the Markov range. In the case of an exponential family,

$$H_\Lambda(x, \theta) = \sum_{\ell=1,p} \theta_\ell \cdot H_{\Lambda,\ell}(x),$$

we can find $I_0(\mu)$ explicitly. If $H_{\Lambda,\ell}(x)$ is defined by a potential $\phi_{A_\ell}$, $0 \in A_\ell$

$$H_{\Lambda,\ell}(x) = \sum_{i:(A_\ell+i)\cap\Lambda \neq \emptyset} \phi_{A_\ell}(\tau_i x)$$

and if we write $\varphi_\ell = |A_\ell|^{-1} \phi_{A_\ell}$, $\ell = 1, p$, then $\overline{\phi}_0 = {}^t(\varphi_\ell,\ \ell = 1, p)$

$$I_0(\mu) = \textstyle\sum_{\|s\| \le 2m} Cov_\mu(\overline{\phi}_0, \overline{\phi}_s).$$

$I_0(\mu)$ must be evaluated at $\widehat{\mu} = \mu_{\widehat{\theta}}$ by simulation over the finite domain $D_n$, for example, by means of the Gibbs sampler (cf.§ 6.2) or, in the ergodic case, by empirical estimation.

*Estimation by PML:* We shall limit ourselves to the case of an exponential family

$$H_i(\theta) = {}^t\theta \cdot h(x_i, x_{\partial i}), \quad \theta, h \in \mathbb{R}^p,$$

$$I_n(\mu) \text{ and } I_n^C(\mu) \longrightarrow I_0(\mu) = E_\mu(Var_\theta^{X_{\partial 0}}(h(X_0, X_{\partial 0}))),$$

$$J_n(\mu) \longrightarrow J_0(\mu) = \sum_{\|s\| \leq 2m} Cov_\mu(h(X_0, X_{\partial 0}), h(X_s, X_{\partial s})).$$

If the parametrization in $\theta$ is well defined, $I_0(\mu)$ is regular.

### 5.3.3. X Markovian: Conditional Normality of the Coding Estimator, $\chi^2$ Difference of Coding Test, and Identification

This paragraph is the logical continuation of § 5.2.3. Here we deal with the consistency of $PML$ and Coding estimators for a Markov field: $S$ is a general collection of sites, the specifications are not assumed invariant, and the distribution $\mu$ of the field is not assumed weakly dependent. $C \subseteq S$ is an infinite coding subset and we shall assume that conditions (M1) to (M3) of § 5.2.3 are satisfied. We shall assume the following conditions are true for the conditional distributions: $\pi_i : \Omega_i \times \Omega_{\partial i} \times \Theta \longmapsto \mathbb{R}^+$, $C_1$ and $\chi$ are defined in (M1):

(N1) For all $i \in S$, $\pi_i$ is of class $\mathcal{C}^3$ in $\alpha$ and $\pi_i^{-1}$, $\pi_i^{(k)}$, $k = 1, 2, 3$ are uniformly bounded in $i$, $x_i$, $x_{\partial i}$, $\alpha$.

(N2) There exists $V : \mathcal{X} \longrightarrow \mathbb{R}^p$ such that for all $i \in C_1$

(1) $E_{\theta, x_{\partial i}}[\pi_i^{(1)} \cdot \pi_i^{(1)}(X_i, x_{\partial i}, \theta)] \geq V^t V(x_{\partial i})$.
(2) $\int_{\mathcal{X}} V^t V(y) \lambda(dy)$ is regular.

Let $x = (x_C, x_{\overline{C}})$ be the realization of a field over $S$ with $x_C$ its component over $C$ and $x_{\overline{C}}$ its component over $S \backslash C = \overline{C}$. The $C$ coding estimator $\widehat{\theta}_n = \widehat{\theta}_n(x_C, x_{\overline{C}})$ is a r.v. which depends on both the realization of $x$ over $C$, and over $\overline{C}$. Define

$$Z_i = -\frac{\partial}{\partial \theta} \log \pi_i(X_i, x_{\partial i}, \theta),$$

$$I_i(\theta, x_{\partial i}) = Var_{\theta, x_{\partial i}}(Z_i), \quad I_n(\theta, x_{\overline{C}}) = c_n^{-1} \textstyle\sum_{i \in C_n} I_i(\theta, x_{\partial i}).$$

**(5.3.4) Theorem.** *Assume that* (M1) *to* (M3) *and* (N1) *and* (N2) *are satisfied; then*

(a) *Conditional to* $x_{\overline{C}}$, $I_n(\theta, x_{\overline{C}})^{1/2} \cdot c_n^{1/2}(\widehat{\theta}_n - \theta) \xrightarrow{d} \mathcal{N}(0,I)$.

(b) *Let* $(H_q)$: $\varphi \in \Lambda \subseteq \mathbb{R}^q$ *be a regular subhypothesis of* $(H_p)$: $\theta \in \Theta \subseteq \mathbb{R}^p$. *Then, unconditionally*

$$2c_n(U_n(\widehat{\varphi}_n) - U_n(\widehat{\theta}_n)) \xrightarrow[(H_q)]{d} \chi^2_{p-q}.$$

*($U_n$ is the $C$ coding contrast under $(H_p)$, and $\widehat{\varphi}_n$ is the estimator under $(H_q)$ for this same contrast.)*

*Proof:*

(a) Fix $x_{\overline{C}}$ the configuration of $x$ over $S \backslash C$ and *work under* $\overline{P} = P_{\theta, x_{\overline{C}}}$, the probability of $\{X_i, i \in C\}$ conditional to $x_{\overline{C}}$. With a probability tending to 1, we have (cf. Theorem 3.4.5)

$$0 = \sqrt{c_n} U_n^{(1)}(\theta) + \Delta_n(\theta, \widehat{\theta}_n)\sqrt{c_n}(\widehat{\theta}_n - \theta).$$

(i) *Asymptotic Normality of* $\sqrt{c_n} U_n^{(1)}(\theta)$: $\sqrt{c_n} U_n^{(1)}(\theta) = c_n^{-1/2} \sum_{C_n} Z_i$; variables $Z_i$, $i \in C$ are centered, independent but not identically distributed. Hence, we can use a *CLT* for independent but not identically distributed variables [22], if we check that for a symmetric, positive definite matrix $I_0$, $\liminf\ c_n^{-1} Var\left(\sum_{C_n} Z_i\right) \geq I_0 > 0$. This condition follows from (N2) and Lemma 5.2.3.

$$\liminf_{n\to\infty}\left(c_n^{-1} Var\left(\sum_{C_n} Z_i\right)\right) \geq \frac{\tau}{2} \liminf_{n\to\infty} \left(c_{n,1}^{-1} Var \sum_{C_{n,1}} Z_i\right) \quad (\text{M1}-2)$$

$$\geq \frac{\tau}{2c^2} \liminf_{n\to\infty} \left(c_{n,1}^{-1} \sum_{C_{n,1}} E_\theta[\pi_i^{(1)t} \pi_i^{(1)}]\right) \geq \frac{\tau}{4c} \int_{\mathcal{X}} V^t V(y)\lambda(dy) = I_0.$$

(ii) On the other hand, we can check directly that

$$(\Delta_n(\theta, \widehat{\theta}_n) + I_n(\theta, x_{\bar{c}})) \xrightarrow{\overline{P}} 0.$$

(b) Call $\overline{P}_\varphi = P_{\varphi, x_{\overline{C}}}$ the probability conditional to $x_{\overline{C}}$ under $\varphi \in (H_q)$. We are going to check that for each $x_{\overline{C}} \in \Omega_{\overline{C}}$,

$$2c_n(U_n(\widehat{\varphi}_n) - U_n(\widehat{\theta}_n)) \xrightarrow[\overline{P}_\varphi]{\mathcal{D}} \chi^2_{p-q}. \tag{5.12}$$

As the limiting distribution doesn't depend on $x_{\overline{C}}$, the result is still true unconditional to $x_{\overline{C}}$. Defining $\theta = r(\varphi)$ the parametrization of $(H_q)$, we have

$$I_n(\varphi) = {}^t R\, I_n(\theta) R, \text{ where } R = \frac{\partial r}{\partial \varphi}(\varphi)$$

$$\Delta_n(\theta, \widehat{\theta}_n) + I_n(\theta) \text{ and } \Delta_n(\varphi, \widehat{\varphi}_n) + I_n(\varphi) \to 0 \text{ in } \overline{P}_\varphi \text{ probability.}$$

The asymptotic normality result $(a)$, as well as Property 3.8, yields (5.12) as $I_n(\theta) = J_n(\theta)$. □

REMARK: There are as many coding tests as ways to code $S$. These tests are correlated among them and in general it isn't possible to obtain a global statistic from these different tests.

*The case $S = \mathbb{Z}^d$, $\pi$ translation invariant, and $E$ compact:*

(M)

(M1') $C = a\mathbb{Z}^d$, $a = (a_i, i = 1, d)$, $a_i > 0$, $i = 1, d$.

(M2') $\pi_0(x \mid v, \alpha)$, $\pi_{\partial 0}(v \mid w, \alpha)$ are positive and continuous in $x$, $v$, $w$, $\alpha$.

(M3') $\alpha \longrightarrow \pi_0(\cdot \mid \cdot, \alpha)$ is injective.

(N)

(N1') $\pi_0^{(k)}$, $k = 1, 3$ (derivatives in $\alpha$) exist and are continuous.

(N2') $\int E_{\theta,v}[\pi^{(1)t}\pi^{(1)}(X \mid v, \theta)]\lambda(dv)$ is regular.

(M) and (N) are sufficient to assure (M1) to (M3) as follows from § 2.3 and conditions (N1) and (N2). If the conditional distribution in 0 is originated by an exponential family with conditional energy $H_s(x_s, v_s) = {}^t\alpha \cdot h(x_s, v_s)$, and if the representation in $\theta$ is injective, then (M3') and (N2') are satisfied.

**(5.3.1) Example.** *Isotropy Test for the Nearest Neighbor Ising Model.* $S = \mathbb{Z}^2$, $x_s \in \{-1, +1\}$, and if $s = (i, j)$, we have the model $(H_3)$:

$$H_s(x_s, v_s, \theta) = (h + \beta_1(x_{i-1,j} + x_{i+1,j}) + \beta_2(x_{i,j-1} + x_{i,j+1}))x_{ij}.$$

Let $C = \{(i, j),\ i + j \text{ pair}\}$ be a coding set under $(H_3)$ and $(H_2)$ the isotropy hypothesis $\beta_1 = \beta_2$ ($C$ is also an optimal coding set under $(H_2)$). The difference of coding statistic will be, under $(H_2)$, a $\chi_1^2$.

**(5.3.2) Example.** *Exchangeability Test in a K Color Model.*
Let $E = \{1, 2, \ldots, K\}$ be a qualitative state space; $S = \mathbb{Z}^2$ and we consider the translation invariant conditional distribution associated to the conditional energy at site $i$:

$$\begin{cases} H(X_i = k \mid x_{\partial i}, \theta) = \alpha_k + \sum_{\ell \neq k} \beta_{k\ell} n_i(\ell), \quad \text{with} \quad \alpha_K = 0, \\ n_i(\ell) = \sum_{j \in \partial i} 1(X_j = \ell), \quad \beta_{k\ell} = \beta_{\ell k}, \quad \ell \neq k, \quad \ell, k = 1, K. \end{cases}$$

$\theta = (\alpha_1, \ldots, \alpha_{K-1}; \beta_{k,\ell}, 1 \le k < \ell \le K)$ is a well-defined parametrization. The exchangeability of color labels translates into $\alpha_1 = \cdots = \alpha_{K-1} = 0$, $\beta_{k,\ell} = \beta$. We can thus use the difference of coding statistic which, under this hypothesis, is asymptotically a $\chi^2\left(\frac{1}{2}(K^2 + K - 4)\right)$.

**(5.3.3) Example.** *Parametric Specification Test given by a Markovian Specification.*
Assume $S$ is equal to $\mathbb{Z}^2$, $V$ is a symmetric, finite subset of $\mathbb{Z}^2$ which contains the origin, and $\pi$ is an invariant under translation specification which is $V$-Markovian over $E^S$, $E = \{0, 1, \ldots, n-1\}$. Several families of conditional models that fit this description can be considered.

*(a) $\mathcal{C}(V)$: General conditional distributions $p(x_i \mid x_{\partial i})$.*
Without constraints, this parametric family has dimension $n^K(n-1)$ for $K = |\partial i|$. We consider the natural parametrization $p(\ell \mid x_{\partial i})$, $\ell = 1, n-1$, $x_{\partial i}$ the neighborhood configuration. The probability estimators will be the observed empirical frequencies. In general, we can't guarantee that these conditional distributions are coherent, that is, that exists a $V$-Markovian model with these conditional distributions.

*(b) $\mathcal{M}(V)$: Conditional distributions based on a general $V$-Markovian model.*
This family has the same parametric dimension as the joint Markovian model. Parameter description and the dimension depend on the description of the family of cliques; for example, for the eight nearest neighbor model, there is one kind of clique with 1 point, 4 with 2 points, 4 with 3 points and 1 with 4 points, which gives $(n-1) + 4(n-1)^2 + 4(n-1)^3 + (n-1)^4$ as the dimension of $\mathcal{M}(V)$.

*(c) Submodels of $\mathcal{M}(V)$.*
For example, $\mathcal{M}_2(V)$ = cliques with, at most, 2 points; $\mathcal{MR}_2(V)$ = cliques with, at most, 2 points and reversible potentials (cf. 2.2.4); $\mathcal{AB}(V)$ = auto-binomial; $\mathcal{ABI}(V)$ = auto-binomial and isotropic.

These models are nested, $\mathcal{M}(V) \supset \mathcal{M}_2(V) \supset \cdots \supset \mathcal{ABI}(V)$. In terms of the respective dimensions, and for the eight nearest neighbor relationship, starting by $\mathcal{M}_2(V)$, we have $(n-1) + 4(n-1)^2$; $(n-1) + 2n(n-1), 5, 3$.

REMARK: Contrary to model $\mathcal{C}(V)$, *parameters from other models will never be* conditional probabilities: estimation must be done directly based on the coding likelihood expressed in terms of the unique parametrization of the potentials. For example, for $\mathcal{MR}_2$ with four nearest neighbors and isotropy, and $n = 3$, $\phi_1$, and $\phi_2$ potentials of order 1 and 2, the parameters are given by $\varphi_1 = \phi_1(1)$, $\varphi_2 = \phi_1(2)$; $\varphi_{11} = \phi_2(1,1)$, $\varphi_{22} = \phi_2(2,2)$, and $\varphi_{12} = \varphi_{21} = \phi_2(1,2) = \phi_2(2,1)$. The coding likelihood turns out to be

$$L_n^C(\varphi) = \prod_{i \in C_n} \frac{\exp[\phi_1(x_i) + \sum_{j:|j-i|=1} \phi_2(x_i, x_j)]}{\sum_{x \in \{0,1,2\}} \exp[\phi_1(x) + \sum_{j:|j-i|=1} \phi_2(x, x_j)]}.$$

(*d*) *Markovian Range Test.*
In the context of the previous example, let $W \subset V$ be a Markov neighborhood submodel. The test for $\mathcal{M}(W)$ in $\mathcal{M}(V)$ will be standard *based on a V-coding.* For example, if $V$ is given by the eight nearest neighbors, and $W$ by the four nearest neighbors, the test statistic will be under $\mathcal{M}(W)$ a $\chi^2$ with $[2(n-1)^2 + 4(n-1)^3 + (n-1)^4]$ d.f.

*Model Identification by Penalized Coding Contrasts.*
For a coding contrast, replacing $d_n$ by $c_n$, we use Corollary 3.10 which deals with the identification of a (conditionally) independent field model: the rate $v(n)$-Penalized Coding contrast leads to the choice

$$\widehat{\delta}_n = \underset{\delta \in \mathcal{E}}{\text{ArgMin}} \, \{U_n(\widehat{\alpha}_{n,\delta}) + \frac{v(n)}{c_n} |\delta(\alpha)|\}$$

for a family $\mathcal{E}$ of models $\delta$ of parametric dimension $|\delta|$.

**(5.3.5) Theorem.** *Under conditions* (M1) *to* (M3), (N1) *and* (N2), *and if the coding estimator* $\widehat{\theta}_n^C$ *is strongly consistent, the identification of the true model by penalized coding will occur a.s. if*

$$\lim_n \frac{v(n)}{c_n} = 0, \quad \liminf_n \frac{v(n)}{\log \log c_n} > A,$$

*where A is the trace of matrix* $I_0$ *defined in* (N2(2)).

## *5.3.4. Comparing Methods*

Assume $X$ is defined over a regular lattice and is stationary and ergodic. The estimation precision increases from Coding (C) to $PML$, from $PML$ to $ML$. The advantage of the first two methods is a greater numerical simplicity. The advantage of Coding estimation is that it provides simple subhypothesis tests. We compare the precision of the different methods in terms of the asymptotic variance ([15], [16], [83]).

### *5.3.4.1. The Gaussian Case*

We shall consider the case of a Gaussian isotropic model with $\nu$ nearest neighbors over a regular lattice

$$X_s = \beta \sum_{t:<t,s>} X_t + e_s.$$

If $\rho$ is the correlation of two nearest neighbors and $\tau$ is the coding rate, the precision is given by

$$\lim_n (n\ Var\widehat{\beta}_C) = \frac{\beta(1-\nu\beta\rho)}{\tau\nu\rho} \quad \text{if} \quad \beta \neq 0, \quad = \frac{1}{\tau\nu} \quad \text{if } \beta = 0;$$

$$\lim_n (n\ Var\widehat{\beta}_{PMV}) = \frac{2\beta^2(1-\nu\beta\rho)^2}{\nu\rho^2} \quad \text{if} \quad \beta \neq 0, \quad = \frac{2}{\nu} \quad \text{if } \beta = 0.$$

If the lattice is $\mathbb{Z}^d$ and we consider the $\nu = 2d$ nearest neighbors model, we have

$$\lim_n\ (n\ Var\widehat{\beta}_{MV}) = \tfrac{1}{2} \jmath(\beta)^{-1},$$

$$\jmath(\beta) = (2\pi)^{-d} \int_{T^d} \frac{(\sum_1^d \cos\lambda_i)^2}{(1-2\beta\sum_1^d \cos\lambda_i)^2} d\lambda_1 \cdots d\lambda_d.$$

This expression equals $\frac{1}{d} = \frac{2}{\nu}$ if $\beta = 0$. We remark that under $\beta = 0$, $\frac{2}{\nu}$ is the precision of the $ML$ estimator for a regular lattice.

*Under the hypothesis of independence*, the $ML$ and $PML$ procedures are equivalent. Coding will be equivalent if the lattice is $\mathbb{Z}^d$ with $2d$-neighbors or in the case of a triangular lattice in the plane. However, for $\mathbb{Z}^2$ and the isotropic model with eight nearest neighbors, the efficiency is $\frac{1}{2}$ since the coding rate is $\frac{1}{4}$.

In the case of the four nearest neighbor model over $\mathbb{Z}^d$, the relative efficiency $e_1$ of the $ML$ estimator with respect to the Coding estimator and $e_2$ of the $ML$ with respect to the $PML$ was calculated by Besag [15], as well as the correlation $\rho$ of the nearest neighbors.

| $\pm 4\beta$ | 0.0 | 0.1 | 0.2 | 0.3 | 0.4 | 0.5 | 0.6 | 0.7 | 0.8 | 0.9 |
|---|---|---|---|---|---|---|---|---|---|---|
| $\pm\rho$ | 0.0 | 0.03 | 0.05 | 0.08 | 0.11 | 0.14 | 0.17 | 0.21 | 0.27 | 0.35 |
| $e_1$ | 1.00 | 0.99 | 0.97 | 0.92 | 0.86 | 0.78 | 0.68 | 0.56 | 0.42 | 0.25 |
| $e_2$ | 1.00 | 1.00 | 0.99 | 0.97 | 0.95 | 0.91 | 0.87 | 0.80 | 0.71 | 0.56 |

*Efficiency ML/Coding* ($e_1$), *ML/PML* ($e_2$) *for the Gaussian field with four nearest neighbors*

The slow growth of $\rho$ toward 1 when $4\beta \to 1$ is explained in Exercise 10 of Chapter 1.

It is also interesting to compare the efficiency of the three methods when $\delta = 1 - 4\beta \to 0_+$. An elementary calculation of the order of the integrals that yield $\rho(\beta)$ and $j(\beta)$ leads to the following Table, which gives the order of $\lim_n (n\ Var\widehat{\beta})$, up to a multiplicative constant, in terms of $\delta$ for the lattice $\mathbb{Z}^d$ and the $2d$ nearest neighbors relationship [83]:

| Method | $d = 1$ | 2 | 3 | 4 | $\geq 5$ |
|---|---|---|---|---|---|
| Coding | $\delta^{1/2}$ | $(-\log \delta)^{-1}$ | 1 | 1 | 1 |
| *PML* | $\delta$ | $(-\log \delta)^{-2}$ | 1 | 1 | 1 |
| *ML* | $\delta^{3/2}$ | $\delta$ | $\delta^{1/2}$ | $(-\log \delta)^{-1}$ | 1 |

*Order of* $\lim_n (n\ \mathrm{Var}\widehat{\beta}_n)$ *for* $\delta = 1 - 4\beta$, $\delta \to 0_+$ *as a function of the dimension* $d$ *of the lattice*

### *5.3.4.2. Ising Model*

We shall examine the isotropic, nearest neighbor Ising model over $\mathbb{Z}^2$ with parameters $(h, \beta)$,

$$\pi(x_s \mid \cdot) = (2ch(h + \beta v_t))^{-1} \exp(x_s(h + \beta v_s)),$$

where $v_s$ is the sum of the four neighboring configurations of $x_s$. When $h = \beta = 0$, the $ML$, $PML$, minimum $\chi^2$ with optimal weights and $CLS$ estimators are asymptotically equivalent; if $\theta = {}^t(h, \beta)$,

$$\sqrt{n}(\widehat{\theta}_n - \theta) \longrightarrow \mathcal{N}_2\left(\begin{pmatrix} 0 \\ 0 \end{pmatrix}, \begin{pmatrix} 1 & 0 \\ 0 & 0.05 \end{pmatrix}\right).$$

For the Coding estimator, there is an efficiency loss when estimating the external field parameter $h$, with $\lim_n (n\ Var\widehat{h}_n) = 2$ instead of 1. If the minimum $\chi^2$ is implemented with the identity weight matrix ($A = Id$, Ordinary Least square ($OLS$)), the asymptotic variance is given by $\begin{pmatrix} 1.71 & 0 \\ 0 & 0.59 \end{pmatrix}$.

If $h = 0$, for dimension 2, the rate of convergence toward zero of $(\widehat{\beta}_{MV} - \beta)$ is greater than $n^{-1/2}$ in $\beta = \beta_c = 0.441$ (cf [15]). Indeed, $\lim\ (n\ Var\widehat{\beta}_n) = \lim_{\beta \to \beta_c} (p(\beta)'')^{-1} = 0$, where $p$ is the pressure given in example 2, § 2.2.2. Several authors ([15], [72], [128], [129]) have studied the nonstandard behavior of the $ML$ estimator at the critical temperature.

## 5.4. Estimation in the Case of Partial Observations

Assume that the field $X = \{X_s, s \in S\}$ with distribution $\mu \in \mathcal{G}(\pi_\alpha)$ is observed through $Y$ with distribution

$$P_\varphi(y_\Lambda \mid x_\Lambda) = \prod_{s \in \Lambda} p_\varphi(y_s \mid x_s), \quad \Lambda \in \mathcal{S}.$$

The joint distribution of $(X, Y)$ is still Markovian, parametrized in $\theta = (\alpha, \varphi)$. Such examples of partial observations are common in image analysis (cf. § 2.4 and § 6.7). More generally, let $\pi_\theta(x, y)$ be a joint Markovian distribution, for example,

$$\pi_\theta(x, y) = Z^{-1}(\theta) \exp - < \theta, H(x, y) > .$$

If only $y = y_0$ is observed, we will say we have a *partial or incomplete observation*; the problem of estimating $\theta$ based only on $y_0$ is important and presents the following new difficulties:

(1) The marginal likelihood of $y$: $\psi_\theta(y) = \sum_x \pi_\theta(x, y)$ in general cannot be calculated numerically or analytically. This leads to the utilization of the Expectation–Maximization ($EM$) algorithm.

(2) In general, it is difficult to prove the identifiability of the model in $\theta$ based only on observation $y$.

(3) If $(X, Y)$ is Markovian, $Y$, in general, is not, and there is no justification of Markovian methods like $PML$ or Coding for $Y$.

(4) In general, we cannot say $\theta \longmapsto \psi_\theta(y_0)$ is concave; there can exist several local maxima.

We shall describe two theoretical approaches for which there exist good asymptotic properties: the $ML$ method and the the marginal contrast or moment method. The algorithmic implementations ($EM$, double stochastic gradient, Gibbsian $EM$) shall be detailed in § 6.7.

*Estimation by Maximum Likelihood.*

Comets and Gidas [37] establish the strong consistency of the $ML$ estimator in the case of incomplete data when $X$ is a translation invariant specification. The techniques they use are based on $LD$ inequalities, analogous to those given in 5.2.2, for the degraded empirical field $Y$. They also establish a new variational principle for the conditional pressure. Bounds for the probability of $\{|\widehat{\theta}_n - \theta_0| > \varepsilon\}$ are also given.

*Estimation by Marginal Contrast: The Case $X$ Ergodic.*

We will consider the $V$-marginal, invariant under translations, additive contrast

$$U_n(\theta) = d_n^{-1} \sum_{D_n} g(Y(V + i), \theta),$$

where $V$ is a finite subset of $S = \mathbb{Z}^d$. If for $\theta \neq \theta_0$,

$$K(\theta, \theta_0) = E_0[g(Y(V), \theta) - g(Y(V), \theta_0)] > 0,$$

and if $(X, Y)$ is ergodic, then with the standard conditions over $g$, the minimum contrast estimator $\widehat{\theta}_n$ is consistent (cf. [88]). In particular, the moments method yields consistent estimators whenever the given moments allow the identifiability of $\theta$. We remark that in general the distribution of $Y$ is not analytically obtainable. The theoretical moments of $\theta$ must be calculated empirically by simulation, for a grid of values of $\theta$, which requires important numerical calculus. The estimation of $\theta$ is then done "graphically" in terms of the observed moment; this method was implemented by Geman and Mac Clure (cf. [66]).

In very particular cases, the functional considered over $Y$ (moments, conditional distributions, marginal distributions) can be calculated analytically; for example, if $y_i = x_i + \varepsilon_i$ is a Gaussian perturbation of Gaussian $X$, $p_i(y_i \mid y_{\partial i})$ can be found explicitly. Another case is that developed by Frigessi and Piccioni [58]: $X$ is an isotropic, four nearest neighbor Ising model, without external field (it has a unique parameter $\beta$), and $Y$ is the degraded field with a channel transmission noise (with error rate $\varepsilon$). In this case and *without the uniqueness condition over* $\mathcal{G}(\beta)$, the authors show the consistency of the estimators of the moments of $(\beta, \varepsilon)$ based on first and second-order moments (nearest neighbor) of $Y$.

If $X$ is weakly dependent, we have (classically) the asymptotic normality of $\widehat{\theta}_n$, and we can easily construct subhypothesis tests as mixtures of $\chi_1^2$, as was developed in 3.2.

*Estimation by Restoration Estimation Method (RE)*, cf. § 6.7).

In § 6.7 we describe a Monte Carlo type Markov Chain method (*MCMC*) for the estimation of incomplete data. This very flexible algorithm is a *Restoration Estimation* method: if $\theta_k$ is the value estimated at step $k$, we restore $X$ following distribution $\mathcal{L}(X \mid Y, \theta_k)$, which gives $X_k$. Having chosen an estimation procedure for complete data, we then estimate $\theta$ as $\theta_{k+1} = h(X_k, Y)$. If the estimation procedure is reasonable and if chain $(\theta_k)$ is ergodic, we estimate $\theta$ by $\hat{\theta} = \sum_{K_0+1}^{K_0+K} \theta_k / K$ for big enough $K_0$ and $K$.

## 5.5. Bibliographical Comments

The results on the estimation of Gibbs fields are relatively new and have developed due to the interest of these models in pattern recognition and image analysis.

Estimation by *PML* or by Coding was introduced in a pioneer article by Besag (1974). The relative efficiency of these methods was studied, in the Gaussian setting, in Besag and Moran (1975), and Besag (1977). A first study on estimation for isotropic Ising models without exterior field appears in a Technical Report of Besag [15]; the asymptotic efficiency of the $ML$ estimator is considered there, as well as its behaviour near the critical temperature $\beta_c$, based on the calculus of the pressure given by Onsager (1944). Several articles of Pickard ([128], [129]) refer particularly to the nonstandard behavior of estimation methods in a phase transition point (cf. also Comets and Gidas [36] and Gidas [72]). We remark also on the work of Strauss (1975) where he approximates the likelihood of a nearest neighbor binary model using truncations of the moment generating function.

The logistic, or minimum $\chi^2$ estimator, was reintroduced in the binary spatial context by Possolo (1976): it is an adaptation of the Berkson method, classical in binary econometrics. The Conditional Least Squares method was introduced by Lele and Ord for analyzing geographical data.

In the stationary case, the idea of decomposing the measure as a linear combination of extremal ergodic elements was considered in several articles; among them, that of Possolo. Here, we followed Guyon and Künsch; the idea of using the variational principle appeared in Künsch (1981). The Large Deviations approach is due to Comets (1991). It also allows us to consider potentials with arbitrary range and yields as a byproduct exponential consistency rates. This approach was generalized by Comets and Gidas to the case of imperfect observations [37].

In the case of a field $\pi_\theta$ which satisfies the unicity condition of Dobrushin and Simon, the techniques are based on mixing (cf. § 2.1.3) and Property 2.2. The study of $PML$ estimators in this context is considered in [80].

When $S$ is arbitrary and if we do not ask anything of $\mathcal{G}(\theta)$, the approach is based on Lemma 5.2.3; the proof of this property of subergodicity is due to Geman and Graffigne [70], in the case of an invariant specification (cf. also Guyon [79], Jensen and Möller [99], and generalized in Guyon and Hardouin [82]). In this last work, we find the construction of a general $\chi^2$ difference of contrast test, which was already mentioned by Besag [14] and used by Cross and Jain [40].

## 5.6. Exercises

(1) *Estimation of the dynamics of a Markovian field.*
$X = \{X_t, t \in \mathbb{N}\}$ is a homogeneous and ergodic Markov chain, $X_t = \{X_i(t), i = 1, n\}$, $X_i(t) \in \{0, 1\}$, with distribution:

$$P(x(t) \mid x(t-1)) = Z^{-1}(\alpha, \beta, \gamma; x(t-1)) \exp\{U(x(t) \mid x(t-1))\}, \text{ where}$$
$$U(x(t) \mid x(t-1)) =$$
$$\sum_{i=1,n} x_i(t)[\alpha x_i(t-1) + \beta(x_{i-1}(t-1) + x_{i+1}(t+1)) + \gamma(x_{i-1}(t) + x_{i+1}(t))]$$

(define $x_{-1} = x_n$ and $x_{n+1} = x_1$). Here, $n$ is fixed, $t = 1, T$ and $T \to \infty$.

(a) What difficulties are related to maximum likelihood estimation? Find conditional distribution $P_i(x_i(t) \mid x_j(t),\, j \neq i,\, x_\ell(t-1),\, \ell = 1, n)$ and the associated conditional pseudolikelihood.

(b) Study the consistency and asymptotic normality of the $PML$ estimator of $(\alpha, \beta, \gamma)$. Test temporal independence: $\alpha = \beta = 0$; test spatial independence: $\beta = \gamma = 0$.

(2) $X$ is a homogeneous texture over $\mathbb{Z}^2$ with $K$ gray levels, Markovian, with eight nearest neighbors and with only singleton and pairwise potentials. Write the Coding tests for the following subhypothesis: $X$ is autobinomial; $X$ is is a reversible texture; $X$ is isotropic.

(3) Simulate the isotropic Ising model with four nearest neighbors ($\alpha = \beta$), over $D = \{1, 2, \ldots, 32\}^2$, without exterior field and with free boundary conditions (set $x_i = 0$ if $i \notin D$).

(a) Estimate $\alpha = \beta$ by Coding and $PML$. In the nonisotropic model, estimate $(\alpha, \beta)$ by Coding and $PML$. Test $\alpha = \beta$.

(b) For a fixed Coding $C$, check experimentally by Monte Carlo methods that $(Var(\widehat{\alpha}^C \mid x(\overline{C})))^{-1/2} \cdot \widehat{\alpha}_C$ is $\mathcal{N}(0,1)$.

(c) Analogously, check that the difference of coding test is a $\chi_1^2$ test. What is the behaviour of the $PML$ test (if the process is mixing)?

(4) *Independence of two spatial characters.*
$U$ and $V$ are two characters observed over $\mathbb{Z}^2$, $u_i = 1$ (resp., $v_i = 1$) if $U$ is present at site $i$ ($V$ is present), 0 if not. Assume the model for $(U, V)$ is invariant under translation, Markovian with four nearest neighbors and isotropic.

(a) Show that the general model $(\Omega)$ depends on 12 parameters, and that its potentials can be expressed as $\phi_1(u_i, v_i) = \alpha u_i + \beta v_i + \gamma u_i v_i$, $\phi_2((u_i, v_i), (u_j, v_j)) = \delta_1 u_i u_j + \delta_2 v_i v_j + \delta_3 u_i v_j + \delta_4 v_i u_j + \delta_5 u_i u_j v_j + \delta_6 v_i u_j v_j + \delta_7 u_i v_i u_j + \delta_8 u_i v_i v_j + \delta_9 u_i v_i u_j v_j$.

(b) Describe the submodel that assumes independence of $U$ and $V$ and test this submodel.

(c) Consider model $(\Omega)$ where only $\delta_1$ and $\delta_2$ are nonzero in the definition of the pair potential $\phi_2$. Defining $U_i$ (resp., $V_i$) as the sum of the four nearest neighbors at site $i$, of $u_j$ (resp., in $i$ of $v_j$), determine the conditional distributions $\pi(u_i, v_i|\cdot)$, $\pi(u_i|v_i, \cdot)$. Based on distributions $\pi(u_i \mid v_i, \cdot)$, construct a difference of coding test for the independence among both characters.

CHAPTER 6

# Stochastic Algorithms

We shall begin by recalling some definitions and results concerning finite state space Markov chains.

## 6.1. Some Results on Nonhomogeneous Markov Chains

Let $\Omega$ be a finite state space, and $P = (P_{ij}, i, j \in \Omega)$ a transition probability matrix. We define the *contraction coefficient* $c(P)$ of $P$ (Dobrushin [49])

$$(6.1) \qquad c(P) = \frac{1}{2} \sup_{i,k \in \Omega} \|P(i,\cdot) - P(k,\cdot)\| = \frac{1}{2} \sup_{i,k} \sum_{j \in \Omega} |P(i,j) - P(k,j)|.$$

Here, $\|\cdot\|$ is the $\ell^1$ norm or variation norm. This coefficient, always bounded by 1, satisfies

$$(6.2) \quad (1)\ c(P) = 1 - \inf_{i,k} \Big( \sum_j \inf\{P_{ij}, P_{kj}\} \Big).$$

$$(6.3) \quad (2)\ c(P \cdot Q) \le c(P)c(Q).$$

(3) If $\nu_1$, and $\nu_2$ are two probabilities over $\Omega$,

$$(6.4) \qquad \|\nu_1 P - \nu_2 P\| \le \|\nu_1 - \nu_2\| c(P).$$

$$(6.5) \quad (4)\ \text{if } \delta = \inf_{i,j} P_{ij} \text{ and } K = |\Omega|,\ c(P) \le 1 - \delta K.$$

Let $X_1, X_2, \ldots$ be a nonhomogeneous Markov chain ($MC$) over $\Omega$ with transition matrix $P_n$ at time $n$; if $k > m$, we call $P^{(m,k)} = P_{m+1} \cdot \cdots . P_k$ the transition from $m$ to $k$.

**(6.1.1) Definition.** Strong and Weak Ergodicity.

(*a*) *The Markov chain is said to be weakly ergodic if, for all* $m \geq 1$,

$$\lim_{k \to \infty} \sup_{\nu_1, \nu_2} \|\nu_1 P^{(m,k)} - \nu_2 P^{(m,k)}\| = 0.$$

(*b*) *The Markov chain is said to be strongly ergodic if there exists a distribution* $\pi_\infty$ *over* $\Omega$ *such that*

$$\lim_{k \to \infty} \sup_{\nu} \|\nu P^{(m,k)} - \pi_\infty\| = 0.$$

Weak ergodicity entails, for large time spans, independence of the distribution with respect to the initial state and strong ergodicity, and the stabilization of this distribution independently of the initial one.

By (4.3), weak ergodicity is equivalent to

$$\text{for all } m, \quad c(P^{(m,k)}) \longrightarrow 0, \quad k \longrightarrow \infty. \tag{6.6}$$

A sufficient condition, by (6.3), is

$$\text{for all } m, \quad \prod_{n \geq m} c(P_n) = 0. \tag{6.7}$$

On the other hand, we have the following sufficient condition for strong ergodicity ([95], [97]):

**(6.1.1) Theorem.** *Assume that for each* $n$, *there exists a probability* $\pi_n$ *invariant by* $P_n$ *that satisfies* $\sum_n \|\pi_n - \pi_{n+1}\| < \infty$. *Then if the Markov chain* $X$ *is weakly ergodic, it will also be strongly ergodic.*

Probability $\pi_\infty$ is just the limit of the Cauchy sequence $(\pi_n)$. In particular, if there exists fixed $\pi$ such that $\pi P_n = \pi$ for each $n$, weak ergodicity will entail strong ergodicity with $\pi_\infty = \pi$. In particular, if $P_n = P$ (the chain is homogeneous) and if there exists $\pi$ s.t. $\pi P = \pi$, weak ergodicity entails strong ergodicity.

*Strong Law of Large Numbers.*

Consider $f : \Omega \longrightarrow \mathbb{R}$; we would like to have

$$\lim_{n \to \infty} \frac{1}{n} \sum_{1}^{n} f(X_i) = \int_\Omega f d\pi_\infty. \tag{6.8}$$

Call $P_\mu$ the distribution of the M.C. starting at $\mu$ and $(c_n)$ the increasing sequence, $c_n = \sup_{1 \le i \le n} c(P_i)$.

**(6.1.2 ) Theorem.** ([62], [95]) *Assume that* $(X_n)$ *is strongly ergodic:*

(*a*) *If* $\lim_n n(1 - c_n) = \infty$, *then limit* (6.8) *will ocurr in* $L^2(P_\mu)$ *and in probability.*

(*b*) *If* $\sum_{n \ge 1} \frac{1}{n^2(1-c_n)^2} < \infty$, *the limit will occur* $P_\mu - a.e.$

REMARK: If the Markov chain is homogeneous, with irreducible transition $P$, and invariant distribution $\pi$, then we have the following variance control:

$$\lim_n n\ Var_\mu\Big(\frac{1}{n}\sum_1^n f(X_i)\Big) = < (I - P)^{-1}(I + P)\widetilde{f}, \widetilde{f} >_\pi,$$

where $< f, g >_\pi = \sum_\Omega f(\omega)g(\omega)\pi_\omega$, $\widetilde{f} = f - \sum_\Omega f(\omega)\pi_\omega$. This result can be used to accelerate simulation for a fixed $f$ [142].

*Central Limit Theorem.*

Let $(f_i)$ be a sequence of real functions over $\Omega$, $S_n = \sum_1^n f_i(X_i)$.

**(6.1.3) Theorem.** ([49]) *Under the following conditions,*

(*a*) $Var(f_i(X_i)) \ge c > 0$ *for all* $i$, (*b*) $\sup_i \|f_i\|_\infty \le C < \infty$, *and*

(*c*) $\lim n^{1/3}(1 - c_n) = \infty$.

*Then*

$$\frac{S_n - ES_n}{(Var\ S_n)^{1/2}} \xrightarrow{\mathcal{D}} \mathcal{N}(0, 1)$$

In $(c)$, power $\frac{1}{3}$ is optimal.

## 6.2. Distribution Simulation Algorithms

Let $\Omega$ be a finite space of configurations and $\pi$ a distribution over $\Omega$ such that for all realization $x$, $\pi(x) > 0$. If we know how to construct a weakly ergodic Markov chain over $\Omega$, with transition matrices $(P_n)$ satisfying $\pi P_n = \pi$, then the chain will be strongly ergodic and for all $x_0$

$$\lim_k P(X_k = x \mid X_0 = x_0) = \pi(x). \tag{6.9}$$

### *6.2.1. The Metropolis Dynamic ([66], [87])*

Construct a stationary chain ($P_n = P$) as follows: choose $Q$, the acceptance transition, to be a symmetric transition matrix over $\Omega$, and define $P$ by

$$(6.10) \qquad P_{xy} = \begin{cases} Q_{xy}\dfrac{\pi(y)}{\pi(x)} & \text{if } \pi(y) < \pi(x), \\ Q_{xy} & \text{if } \pi(y) \geq \pi(x) \text{ and } x \neq y, \\ 1 - \sum\limits_{y:y\neq x} P_{xy} & \text{if } y = x. \end{cases}$$

$P$ is $\pi$-reversible, that is, satisfies the detailed balance equation $\pi_x\ P_{xy} = \pi_y\ P_{yx}$, for all $x, y$. This yields $\pi P = \pi$. In the definition of $P$, only ratios $\frac{\pi(y)}{\pi(x)}$ appear. These ratios are equal to $\exp[U(x) - U(y)]$ if $\pi$ is defined in terms of an energy $U$. If $P$ is irreducible and aperiodic, $P$ is strongly ergodic [97] and thus (6.9) is satisfied. $P$ shall be irreducible whenever $Q$ is.

*Metropolis Algorithm: $X_0 = x(0)$ is chosen arbitrarily.*

*Transition $k \longrightarrow k+1$: If $X_k = x$, choose $y$ according to $Q(x,\cdot)$.*

- *If $\pi(y) \geq \pi(x)$ and $y \neq x$, define $X_{k+1} = y$.*

•• *If not, define* $X_{k+1} = \begin{cases} y \text{ with probability} & \dfrac{\pi(y)}{\pi(x)} \\ x \text{ with probability} & 1 - \dfrac{\pi(y)}{\pi(x)}. \end{cases}$

*If $\pi$ is defined by an energy $U$, then*

$$P_{xy} = Q_{xy} \exp(-[U(y) - U(x)]^+), \quad a^+ = \sup(a, 0).$$

**(6.2.1) Example.** Spin Exchange Dynamic [40].
Assume $\Omega = E^S$, $S = \{1, 2, \ldots, N\}$. Let $x(0)$ be an initial configuration and $\Omega_0 = \Omega_{x(0)}$ the set of configurations with the same "empirical distribution" as $x(0)$; that is, if $n_x(e) = \#\{i : x_i = e, i = 1, N\}$, $x \in \Omega_0 \Longleftrightarrow n_x(e) = n_{x(0)}(e)$, $e \in E$.

We now define the spin exchange dynamic over $\Omega_0$. Choose two different sites, $s$, $t$ of $S$, at random and change the values of the realizations in $s$ and $t$. This transforms configuration $x$ into $y$. This defines a symmetric and irreducible transition $Q$ and we now proceed following (6.10). This algorithm is used by Cross and Jain for the synthesis of Markovian textures with fixed marginals (see 2.2.4).

### *6.2.2. The Gibbs Sampler (Geman[2] [66], [67], [173])*

It is also called Glauber's dynamic. Assume $\Omega = E^S$, $S = \{1, 2, \ldots, N\}$, and $E$ is a finite state space. Let $\{s_1, s_2, s_k, \ldots, k \in \mathbb{N}\}$ be a visiting scheme for $S$ such that each site is visited infinitely often.

*The transition $P_k$ from $X_k$ to $X_{k+1}$ is then defined by*

$$X_s(k+1) = \begin{cases} X_s(k) & \text{if } s \neq s_k, \\ u & \text{if not}, \end{cases}$$

*where $u$ follows the conditional distribution $\pi_{s_k}(\cdot \mid X_{\overline{s}_k}(k))$ with $\overline{s} = S \backslash \{s\}$,*

$$P_k(x,y) = \mathbb{1}(x_{\overline{s}_k} = y_{\overline{s}_k})\pi(y_{s_k} \mid x_{\overline{s}_k}). \tag{6.11}$$

**(6.2.1) Theorem.** *If $\pi(x) > 0$ for each configuration $x \in \Omega$ and if each site of $S$ is visited infinitely often, then*

$$\lim_{n\to\infty} P_n(X(k) = x \mid X(0) = x_0) = \pi(x).$$

*Proof:* We shall limit ourselves to the proof when the sequence of visits is periodic: $s_k = [k] + 1$ (modulo $N$). The condition $\pi$ positive yields $\delta = \inf_{s \in \Omega} \inf_{x \in \Omega} \pi_s(x_s \mid x_{\overline{s}}) > 0$. Let $m \geq 0$ and $Q_m$ be defined by

$$Q_m = P_{mN+1} \cdot P_{nN+2} \cdots P_{(m+1)N}.$$

Then $Q_m$, the transition associated to the $m$-*th* sweep of $S$, is homogeneous, $Q_m = Q$ with

$$Q(x,y) = \prod_{s=1}^{N} \pi_s(y_s \mid y_1, \ldots, y_{s-1}, x_{s+1}, \ldots, x_N),$$

and thus, $\inf_{x,y \in \Omega} Q(x,y) \geq \delta^N > 0$. Then, following (6.5),

$$c(Q) \leq 1 - (L\delta)^N \tag{6.12}$$

if $L = |E|$. $(X_k)$ is weakly ergodic and as $\pi Q = \pi$, $(X_k)$ is strongly ergodic. □

REMARKS:

(1) A general proof that includes the possibility of relaxing the values over subsets $A_k$ (here $A_k = \{s_k\}$), also taking into account constraints over $\pi$, is given Geman [66]. The visiting scheme is not required to be periodic, and can also be stochastic (cf. Exercise 16).

(2) The result is still true if $\Omega = \prod_S E_i$, $E_i$ finite.

(3) In practice, it has been observed that $m = 50$ to 100 sweepings gives good results. If $\lambda_0$ is the subdominating eigenvalue of $Q$, the theoretical convergence rate is $|\lambda_0|^m$, but calculating $\lambda_0$ becomes rapidly impossible due to the cardinality of $\Omega$.

(4) If $P \in \mathcal{G}(\pi)$ is a distribution over $\mathbb{Z}^2$ that we want to simulate over a finite domain $S$, there are two possibilities: either $P$ is the unique element of $\mathcal{G}(\pi)$ and simulation is possible by enlarging $S$ into a big enough $S^K$ and fixing arbitrary conditions over the boundary of $S^K$, or if not, simulation will depend on boundary conditions.

(5) Since $c(Q) < 1$, because of 6.2 (b), we have

$$\lim_{n\to\infty} \frac{1}{n} \sum_{1}^{n} f(X_i) = \int_{\Omega} f d\pi \quad a.e.$$

(6) If each state space is compact, the Gibbs Sampler can still be considered [173]. For $E = \mathbb{R}$ and $X$ Gaussian, the Sampler is valid (cf. Younes, 6.6.1). For a more general result, see 6.6.2.

## 6.3. Optimization by Simulated Annealing

Let $\Omega$ be a finite set, $U : \Omega \longrightarrow \mathbb{R}$. We are interested in the points of $\Omega$ where $U$ reaches its minimum,

$$\Omega_{\text{Min}} = \Big\{x : U(x) = \text{Min}_{y\in\Omega} U(y)\Big\}.$$

Consider $\beta > 0$ and $\pi_\beta$ the distribution over $\Omega$ with energy $\beta U$:

$$\pi_\beta(x) = Z_\beta^{-1} \exp\{-\beta U(x)\}. \tag{6.13}$$

Define $\pi_\infty$ as the uniform distribution over $\Omega_{\text{Min}}$,

$$\pi_\infty(x) = 1(x \in \Omega_{\text{Min}}) \cdot |\Omega_{\text{Min}}|^{-1}. \tag{6.14}$$

Then:

$$\lim_{\beta\to\infty} \|\pi_\beta - \pi_\infty\| = 0. \tag{6.15}$$

The goal of *Simulated Annealing* ($SA$) is to construct a (non homogeneous) Markov chain that converges towards $\pi_\infty$. That is, such that

$$\lim_{k\to\infty} P(X_k \in \Omega_{\text{Min}} \mid X_0 = x) = 1, \quad \text{for all } x.$$

Simulated Annealing with the Gibbs dynamic is well adapted to the case when $\Omega = E^S$ is a product space and $U$ is Markovian: for example, for image reconstruction by $MAP$ (cf. 6.8.2.1). $SA$ with the Metropolis dynamic does not assume any particular structure for the finite space $\Omega$ or for $U$; it can be applied to any combinatorial optimization problem.

## *6.3.1. Simulated Annealing for the Gibbs Sampler Dynamic*

We shall assume $\Omega = E^S$, $E$ finite, $S = \{1, 2, \ldots, N\}$, and shall call $\pi_{\beta,s}(x_s \mid x_{\overline{s}})$ the conditional distribution at $s$ of $\pi_\beta$. We shall continue assuming that the Gibbs Sampler's visiting scheme for $S$ is periodic. If $(\beta(k))$ is a sequence that increases to $+\infty$ (we will say that $T_k = \beta(k)^{-1}$ is a cooling schedule), the $SA$ algorithm, for this scheme, is as follows:

(a) *Site visiting scheme:* $s_k = [k] + 1 \ (N)$.
(b) *Cooling scheme:*

$$\beta(k) = \beta(q) \quad if \quad k \in \{qN, qN+1, \ldots, (q+1)N - 1\}.$$

(c) *The initial value* $X(0) = x$ *is chosen arbitrarily and the transition from* $X(k)$ *to* $X(k+1)$ *is as follows:*

$$X_s(k+1) = \begin{cases} X_s(k) & if\ s \neq s_k, \\ u & if\ s = s_k, \end{cases}$$

*where* $u$ *is chosen at random according to* $\pi_{\beta(k),s_k}(\cdot \mid x_{\overline{s}_k}(k))$, *so that*

$$P_{k,\beta(k)}(x,y) = \mathbb{1}(x_{\overline{s}_k} = y_{\overline{s}_k})\pi_{\beta(k),s_k}(y_{s_k} \mid x_{\overline{s}_k}(k)).$$

Define $\Delta = \sup\{U(x) - U(y),\ x = y \text{ except in one site}\}$.

**(6.3.1) Theorem.** *Choose* $\beta(q) = \gamma \log q$:

(a) *If* $\gamma \le \gamma_0 = (N\Delta)^{-1}$, *then for all* $x$

$$P(X(k) \in \Omega_{\text{Min}} \mid X(0) = x) = 1.$$

(b) *If* $0 < \gamma < \dfrac{\gamma_0}{2}$, *then for all* $x$, $X(k) \longrightarrow 1_{\Omega_{\text{Min}}}$ *a.e.*

*Proof:* Define $Q_\beta = \prod_{k=1,N} P_{k,\beta}$. It can be easily checked that

$$\pi_\beta Q_\beta = Q_\beta. \tag{6.16}$$

Upper bound (6.12) can be written as $c(Q_\beta) \le 1 - (L\delta_\beta)^N$ with $\delta_\beta = \inf_{s\in S}\inf_{x\in\Omega} \pi_{\beta,s}(x_s \mid x_{\overline{s}})$. As $\pi_\beta$ is associated to energy $\beta U$, we have

$$\pi_{\beta,s}(x_s \mid x_{\overline{s}}) = \left[\sum_{u\in E} \exp -\beta[U(u, x_{\overline{s}}) - U(x_s, x_{\overline{s}})]\right]^{-1} \ge \frac{1}{L}e^{-\Delta\beta},$$

which thus yields:

$$c(Q_\beta) \le 1 - e^{-\beta N\Delta}. \tag{6.17}$$

Following Theorem 6.1.1, the strong ergodicity of $(Q_{\beta(k)}, k)$ is assured if $\sum \|\pi_{n+1} - \pi_n\| \leq \infty$, and if the chain is weakly ergodic. Examine the first condition.

Note that we can normalize the energy $U$ in such a way that $U \geq 0$ and the minimum of $U$ is 0. Fix a confuguration $x$ and set $Z_n = Z(\beta(n)), h_n = \exp\{\beta(n)U(x)\}$.

$$\begin{aligned}
|\pi_{n+1} - \pi_n| &= |\frac{h_{n+1}}{Z_{n+1}} - \frac{h_n}{Z_n}| \\
&\leq (Z_{n+1}\, Z_n)^{-1}\, \{h_{n+1}|Z_{n+1} - Z_n| + Z_{n+1}|h_{n+1} - h_n|\} \\
&\leq (\inf_n Z_n)^{-2}\{(\sup_n h_n)|Z_{n+1} - Z_n| + (\sup_n Z_n)|h_{n+1} - h_n|\}
\end{aligned}$$

Since $(\inf Z_n) \geq 1$, $(Z_n)$ is decreasing and positive as $(h_n)$; so, we have

$$\sum_0^\infty |\pi_{n+1}(x) - \pi_n(x)| < \infty.$$

Weak ergodicity follows from $\prod_{n\geq m} c(Q_{\beta(n)}) = 0$, or equivalently,

$$\sum_{n\geq m} (1 - c(Q_{\beta(n)})) = +\infty. \tag{6.18}$$

Now, because of (6.17), $1 - c(Q_{\beta(n)}) \geq \left(\frac{1}{n}\right)^{N\Delta\gamma}$ so that the divergence of the series is guaranteed if $\gamma \leq \gamma_0$; hence, $(a)$.

On the other hand, $n^{-2}(1 - c_n)^{-2} \leq n^{-2+2N\Delta\gamma}$ and the second series is convergent if $\gamma < \frac{\gamma_0}{2}$. Result $(b)$ follows from the strong law of large numbers (Theorem 6.1.2, (b)). □

## *6.3.2. Simulated Annealing for the Metropolis Dynamic*

The results given here are due to Hajek [85]. We assume $\Omega$ is finite, but not necessarily with a product structure. Let $Q$ be a symmetric, irreducible transition matrix over $\Omega$. Define $\beta(k) = \gamma \log k$. Transition at time $k$ is given by

$$P_{\beta(k)}(x, y) = Q_{xy} \exp(-\beta(k)[U(x) - U(y)]^+).$$

Define $M = \inf\{n \geq 1 : Q^n > 0\}$, $\Delta = \mathrm{Max}\{U(x) - U(y), x, y \in \Omega, Q_{xy} > 0\}$, $\gamma_0 = (M\Delta)^{-1}$.

**(6.3.2) Theorem.**

*(a) If* $\gamma \leq \gamma_0$, $\lim_{k\to\infty} Pr(X_k \in \Omega_{\mathrm{Min}} \mid X_0 = x) = 1.$

(b) *If* $\gamma < \frac{\gamma_0}{2}$, $X_k \longrightarrow 1_{\Omega_{\mathrm{Min}}}$ *a.e.*

*Optimal Cooling Scheme.* Define $W_x = \{y \in \Omega : Q_{xy} > 0\}$, $x \in \Omega$. Since $Q$ is symmetric, $W = \{W_x \backslash \{x\}\}$ is a neighborhood system over $\Omega$. $x \in \Omega$ is a local minimum if $U(x) \leq U(y)$ for $y \in W_x$. Define $\Omega_{L\min}$ as the set of local minima ($\Omega_{L\min} \supseteq \Omega_{\min}$). We will say $x, y$ communicate at level $h$ either if $y = x$ and $U(x) \leq h$, or if there exists a sequence $x = x(1), x(2), \ldots, x(k) = y$ of $\Omega$, $x(j+1) \in W_{x(j)}$, with $U(x(j)) \leq h$, for every $j$. The depth $d_x$ of a local minimum $x$ will be the smallest $D > 0$ such that there exists $y \in \Omega$ that communicates with $x$ at level $U(x) + D$ and $U(y) < U(x)$.

**(6.6.3) Theorem.** *Let* $D = \sup\{d_x, x \in \Omega_{L\min} \backslash \Omega_{\mathrm{Min}}\}$. *Then*

$$\lim P(X_k \in \Omega_{\mathrm{Min}} \mid X_0 = x) = 1, \text{ for all } x \in \Omega,$$

*if and only if* $\sum_{k=1}^{\infty} \exp(-\beta(k) \cdot D) = +\infty$. *In particular, if* $\beta(k) = \gamma \cdot \log k$, *this condition is equivalent to* $\gamma \leq \gamma_0 = \frac{1}{D}$.

### *6.3.3. Choosing a Cooling Schedule*

In terms of the temperature, the distribution of the *k-th* sweeping of $S$ is

$$\pi_k(x) = Z^{-1}(T_k) \exp\Big(-\frac{U(x)}{T_k}\Big), \quad T_k = \frac{T_0}{\log(k+1)}, \quad T_0 \text{ big.}$$

The initial temperature must be big enough, then decrease slowly. This gives the analogy with an annealing process: a high initial temperature yields $\pi_0$ that makes possible any state $x$ (particles are randomly distributed in the liquid phase). Then, a slow temperature decrease organizes the solid crystalographic system conveniently.

Unfortunately, this double constraint, $T_0$ big and a slow decrease, cannot be assumed in practice: it is necessary to choose other, more realistic schemes that "mimic" the theoretic scheme (cf. [1]), for example,

$$T_k = T_0 c^k \quad \text{where} \quad 0.8 \leq c \leq 0.99.$$

## 6.4. Sampling and Optimization Under Constraints

$\Omega = E^S$ where $E$ and $S$ are finite. Let $C : \Omega \longrightarrow \mathbb{R}^+$, $\Omega_c = \{x \in \Omega : C(x) = 0\}$ be a set of constraints. Let $U : \Omega \longrightarrow \mathbb{R}$ be an energy function and $\Pi(x) = Z^{-1} \exp -U(x)$ its associated probability.

*First problem:* The restriction of $\Pi$ over $\Omega_c$ is

$$\Pi_c(x) = 1_{\Omega_c}(x)\frac{\exp(-U(x))}{\sum_{y\in\Omega_c}\exp(-U(y))}.$$

How can we obtain a sample of $\Pi_c$?

*Second problem:* Define $\Omega^c_{\mathrm{Min}} = \{x \in \Omega_c : U(x) = \mathrm{Min}_{y\in\Omega_c} U(y)\}$ and $\Pi^c_\infty$ the uniform measure over $\Omega^c_{\mathrm{Min}}$. How can we construct a chain that converges towards $\Pi^c_\infty$?

Consider $\beta > 0$, $\lambda > 0$ and the probability

$$\pi(x,\beta,\lambda) = Z^{-1}(\beta,\lambda)\exp -\beta(U(x) + \lambda C(x)). \tag{6.19}$$

It can be easily checked that for each $x$:

$$(a) \quad \lim_{\lambda\to\infty} \Pi(x,1,\lambda) = \Pi_c(x), \qquad (b) \quad \lim_{\lambda,\beta\to\infty} \Pi(x,\beta,\lambda) = \Pi^c_\infty.$$

Choose a sequence $\{s_k, k \geq 1\}$ of site visits such that for a certain integer $R$, and for all $k$, $\{s_k, s_{k+1}, \ldots, s_{k+R}\} = S$, (for example, the sequence of periodic visits, $R = N$). Define $\Pi_k(x) = \pi(x,\beta_k,\lambda_k)$. Let $X_1, X_2, \ldots$ be the chain over $\Omega$ with transition $P_k(x,y) = 1(x_{\overline{s}_k} = y_{\overline{s}_k}) \cdot \pi_k(y_{s_k} \mid x_{\overline{s}_k})$.

**(6.4.1) Theorem.** ([66], [69])

*(a) If $\beta_k \equiv 1$ and $\lambda_k = \lambda_0 \log k$, with $\lambda_0$ small enough, then for all $x$,*

$$\lim_{k\to\infty} P(X_k = y \mid X_0 = x) = \Pi^c(y).$$

*(b) If $\beta_k \to \infty$, $\lambda_k \to \infty$, and $\lambda_k\beta_k = c_0 \log k$ with $c_0$ small enough, then for all $x$,*

$$\lim_{k\to\infty} P(X_k \in \Omega^c_{\mathrm{Min}} \mid X_0 = x) = 1.$$

This last result is still true if at each time $k$, relaxation is considered over a subset of sites $A_k$, under the following conditions over the sequence of multiple visits:

$$\exists R \in \mathbb{N} \text{ such that } \forall k : A_k \cup A_{k+1} \cup \cdots \cup A_{k+R} = S$$

with transition: $P_k(x,y) = 1(x_{\overline{A}_k} = y_{\overline{A}_k})\pi_k(y_{A_k} \mid x_{\overline{A}_k})$.

# 6.5. Case $X_i$ with Values in $\{0,1\}$: Dynamics of a Birth and Death Process (BDP)

The results described below are due to Preston [132]; they are related to the strong ergodicity of a BDP and to the description of the limiting Markovian distribution when, spatially, the BDP is also Markovian. At time $t$, $t \in \mathbb{R}^+$, the configuration over the finite set $S$ of sites is defined by a subset $A$, $A \subseteq S$: $A = \{s \in S : x_s(t) = 1\}$, $x_t = 0$ over $\overline{A} = S \backslash A$. The temporal dynamic is Markovian:

• *Birth* in $s \notin A$: $P(X_{t+dt} = A \cup \{s\} \mid X_t = A) = \beta(s, A)dt + o(dt)$.
• *Death* in $s \notin A$: $P(X_{t+dt} = A \mid X_t = A \cup \{s\}) = \delta(s, A)dt + o(dt)$.
• *Simultaneous births or deaths*: The probability that there is more than one death or one birth over $[t, t + dt]$ is a $o(dt)$.

As $\beta$ and $\delta > 0$, $\{X_t, t \in \mathbb{R}^+\}$ is an irreducible Markov process with equilibrium state $\pi$: if the process is reversible and if $\beta$ and $\delta$ are Markovian for a neighborhood graph over $S$,

$$\beta(s, A) = \beta(s, A \cap \partial s) \quad \text{if } s \notin A, \quad \delta(s, A) = \delta(s, A \cap \partial s) \quad \text{if } s \in A,$$

the limiting distribution is Markovian and is characterized by the identity

$$\frac{\pi(A \cup \{s\})}{\pi(A)} = \frac{\beta(s, A)}{\delta(s, A)}, \quad s \notin A, \quad A \in \mathcal{P}(S).$$

Conversely, equations

$$\beta(s, A) = \frac{\pi(A \cup \{s\})}{\pi(A)} \quad \text{and} \quad \delta(s, A) = 1 \quad \text{if } s \notin A \tag{6.21}$$

define a BDP with equilibrium state $\pi$. We have the following simulation algorithm for $\pi$:

*Choose $\beta$ and $\delta$ following* (6.21):

(1) $t_0 = 0$, *choose arbitrary* $x_0 = A_0$.

(2) *Let $t_k$ be the $k$-th birth or death instant and $A$ the new configuration.*

• *At each site $s$ of $S \backslash A$, simulate independent exponentials with parameters $\beta(s, A)$.*

•• *At each site $s$ of $A$, simulate independent exponentials with parameters $\delta(s, A \backslash \{s\})$.*

*Let $\Delta_k$ be the* inf *of these exponentials, $s_{k+1}$ the associated site, $t_{k+1} = t_k + \Delta_k$, then*

$$x(t_{k+1}) = \begin{cases} A \cup \{s_{k+1}\} & \textit{if } s_{k+1} \notin A \quad \textit{(Birth)}, \\ A \backslash \{s_{k+1}\} & \textit{if not} \quad \textit{(Death)}. \end{cases}$$

Then the process $x(t) = \{x(t_k), t_k \leq t < t_{k+1}, k \geq 0\}$ has $\pi$ as its equilibrium distribution.

## 6.6. Simulation Algorithms when the State Space Is Nonfinite

Condition (6.6), using the contraction coefficient (6.1), is sufficient but not necessary for ergodicity. On the other hand, (6.5), which gives upper bounds for the contraction coefficients, is well adapted to chains in a finite state space. In this section we shall give two ergodicity results when the state space is nonfinite as well as a asymptotic stability result for Markov chains.

### *6.6.1. Gibbs Sampler in the Gaussian Case*

Consider $X = (X_1, X_2, \ldots, X_N) \sim \mathcal{N}_N(0, \Sigma)$ such that $\Sigma^{-1} = Q$ exists. Observe that in the Markovian case, $q_{i,j} = 0$ if $i$ and $j$ are not neighbors. The distribution of $X_i$, conditional to $X_j$, $j \neq i$ is

$$\mathcal{N}(-q_{ii}^{-1} \sum_{j \neq i} q_{ij} X_j,\ q_{ii}^{-1}). \tag{6.22}$$

Assume $\{s_k, k \geq 1\}$ is a visiting scheme for $S$ such that for a given $R$, $\{s_k, s_{k+1}, \ldots, s_{k+R}\} = S$ for all $k$ (e.g. the periodic sequence), $X(0) = x_0$ is arbitrary, and the transition at time $k$:

$$X_s(k+1) = \begin{cases} X_s(k) & \text{if } s \neq s_k, \\ \xi & \text{if } s = s_k \end{cases}$$

where $\xi$ is chosen at random following (6.22) with $i = s_k$.

**(6.6.1) Theorem.** (Younes [173]) $X(k) \xrightarrow{\mathcal{D}} \mathcal{N}_N(0,\Sigma)$.

*Proof:* For all $k$, $X(k)$ is Gaussian. Define $M_k = EX(k)$, $\Sigma_k = Var\ X(k)$; we will check that $M_k \longrightarrow 0$, $\Sigma_k \longrightarrow \Sigma$. The distribution of $X(k+1)$ conditional to $X(k)$ is

$$\mathcal{N}(A_{s_k} X(k), S_{s_k}), \quad A_i = I - B_i, \tag{6.23}$$

where only row $i$ $(q_{ii}^{-1} q_{i,j}, j = 1, N)$ of $B_i$ is nonzero and the only nonzero of $S_i$ is $q_{ii}^{-1}$ at position $(i, i)$. We have the following recurrence relationship:

$$M_{k+1} = A_{s_k} M_k, \quad \Sigma_{k+1} = S_{s_k} + A_{s_k} \Sigma_k A'_{s_k}. \tag{6.24}$$

As distribution $\mathcal{N}(0, \Sigma)$ is invariant under the sampler, it satisfies (6.24) so that

$$\Sigma_{k+1} - \Sigma = A_{s_k}(\Sigma_k - \Sigma) A'_{s_k}. \tag{6.25}$$

Since $Q$ is positive definite, we can consider the associated norm $\| \cdot \|_Q$ over $\mathbb{R}^N$. Let $e(i)$ be the *i-th* vector of the canonical basis of $\mathbb{R}^N$; $B_i$ turns out to be the $Q$ orthogonal projection over $e(i)$. $A_i$ is the complementary orthogonal projection.

Let $n_0 = 0$, and $n_1, n_2, \ldots$ be the successive sweeping times,

$$n_{p+1} = \inf\{k > n_p,\ \{s_{n_p+1}, \ldots, s_k\} \supseteq S\}.$$

Define $K_p = A_{s_{n_{p+1}}} \cdot \cdots \cdot A_{s_{n_p+1}}$. $\|K_p\|_Q \le 1$ as $K_p$ is the product of orthogonal projections. Assume that for a certain $x$, $\|K_p x\|_Q = \|x\|_Q$. Then $\|A_k x\|_Q = \|x\|_Q$ for all $k \in S$; that is, $x = 0$ and thus as $S$ is finite, $\|K_p\|_Q < 1$. Because $(n_{p+1} - n_p) \le R$, there is only a finite number of $K_p$ so that there exists $\rho < 1$ such that for all $p$, $\|K_p\|_Q \le \rho < 1$.

It follows from (6.24) and (6.25) that $M_k \longrightarrow 0$ and $\Sigma_k \longrightarrow \Sigma$. □

REMARK: For certain classes of variance $\Sigma$, Barone and Frigessi [9] propose an acceleration of this Gaussian Gibbs sampler.

### *6.6.2. Tierney's Ergodicity Result*

The result obtained for the Gibbs sampler in the Gaussian case is a particular case of a more general result due to Tierney ([161]). Let $P$ be a transition over $(E, \mathcal{E})$ for which we have an invariant probability $\pi : \pi P = \pi$.

$P$ is *irreducible* if for all $A \in \mathcal{E}$ such that $\pi(A) > 0$, and for all $x \in E$, there exists an integer $n$ such that $P^n(x, A) > 0$.

$P$ is *periodic* if there exists $d \ge 2$ and a partition of $E$ in $d$ events $E_1, E_2, \ldots, E_d$ such that if $x \in E_j$, $x$ can only communicate with an element $y \in E_{j+1}$ (defining $E_{d+1} = E_1$). If not, $P$ is *aperiodic*. We shall then write

(H1) $P$ is irreducible;
(H2) $P$ is aperiodic; and
(H3) for all $x \in E$, $P(x, \cdot)$ is absolutely continuous with respect to $\pi(\cdot)$.

**(6.6.2) Theorem.** *Let $P$ be a transition and $\pi$ an invariant under $P$ probability. Under (H1), (H2), and (H3) we have*

$$\lim_{n \to \infty} \|P^n(x, \cdot) - \pi(\cdot)\| = 0.$$

For the Gaussian Gibbs sampler, the three conditions are satisfied: as the transition density is strictly positive, the chain is irreducible and aperiodic and transitions are absolutely continuous with respect to $\pi$.

Another example where this Theorem can be applied is that of the bivariate process $Y = (X, \Lambda)$, $X_i \in \mathbb{R}$ (gray level) and $\Lambda_i \in S$ (label), such that, for example, $\Lambda$ is a Gibbs field and $(X|\Lambda)$ is a Markovian, Gaussian field.

### *6.6.3. Lyapunov's Stability Criteria and Existence of an Invariant Distribution (Duflo [54])*

Let $(X_n)$ be a Markov chain over $(E, \mathcal{E})$ where $E = \mathbb{R}^d$ and $\mathcal{E}$ is the Borel $\sigma$-algebra. The concept of stability is weaker than that of ergodicity, such as it is defined in (6.1). It is useful and easily manageable when studying the asymptotic stability of Monte Carlo algorithms via Markov chains (*MCMC*).

We will say the chain is *stable* if there exists an invariant probability $\pi$ such that, a.e., for any initial distribution $\nu$ for $X_0$, we have for all $\Gamma \in \mathcal{E}$:

$$\lim_{n\to\infty} \frac{1}{n+1} \sum_{k=0}^{n} 1_{(X_k \in \Gamma)} = \pi(\Gamma).$$

We will say transition $P$ is Fellerian if for any continuous and bounded variable $g : E \to \mathbb{R}$, variable $\pi g$ defined by

$$\pi g(x) = \int_E g(y)\pi(x, dy)$$

is also bounded and continuous.

**(6.6.3) Theorem.** Lyapunov's Stability Criteria. *Consider a Markov chain over $\mathbb{R}^d$ with transition $P$ that satisfies the following conditions*

*(D1) $P$ is Fellerian.*
*(D2) $P$ has, at most, one invariant probability.*
*(D3) There exists a continuous "Lyapunov function" $V : E \to \mathbb{R}^+$ that satisfies*

$$\lim_{\|x\|\to\infty} V(x) = +\infty \quad \text{and } \limsup_{\|x\|\to\infty} \frac{\pi V(x)}{V(x)} < 1.$$

*Then the chain is stable.*

Condition (D2) is satisfied if, for example, the transition function is absolutely continuous with respect to the Lebesgue measure with a strictly positive density. In order to check (D3), it is necessary to try different functions $V$, for example, $\|x\|$ or $\|x\|^\alpha$. If $P$ admits a strictly positive density, then (D2) is satisfied.

This Theorem works in two directions: in the first place it assures the existence of an invariant probability for $P$, and in the second, it gives the asymptotic behaviour of the chain.

## 6.7. Stochastic Algorithms for Estimation of Gibbs Models

If $X$ is a Gibbs field over $\Omega = E^S$, $E$, $S$ finite, the likelihood of a Gibbs parametric model includes the normalization constant $Z(\theta)$,

$$Z(\theta) = \sum_{\omega \in \Omega} \exp\{-U_\theta(\omega)\},$$

which is uncalculable given the large size of $\Omega$. The following section describes a stochastic gradient algorithm which allows us to avoid this difficulty when maximizing the likelihood in terms of $\theta$.

## *6.7.1. Likelihood Maximization Algorithm (Younes [172])*

$X = \{X_s, s \in S\}$, $S = \{1, 2, \ldots, N\}$, $X_s \in E$ finite, $|E| = m$ is a Gibbs field which belongs to an exponential model with distribution

$$\pi_\theta(x) = \exp[- < \theta, U(x) > - \log Z(\theta)], \quad \theta \in \mathbb{R}^p.$$

If we observe $x_0$, the ML equation is given by $U(x_0) = E_{\widehat{\theta}}[U(X)]$. We shall assume that $\log \pi_\theta(x)$ is strictly concave: $-Var_\theta[U(X)] = [\log \pi_\theta(x)]^{(2)}$ is definite negative.

*Description of the Stochastic Gradient Algorithm.*

- We observe $x_0$ and initialize the parameter at arbitrary $\theta_0$.

•• Step $n \longrightarrow n+1$: Let $\theta_n$ be the value calculated at step $n$.

$$\theta_{n+1} = \theta_n + [U(X_{n+1}) - U(x_0)]/(n+1)\Delta$$

Here, $X_{n+1}$ is obtained by the Gibbs Sampler at site $s_n$ visited at time $n$ under $\theta_n$ and $X_n$, $P_n(X_{n+1} = y \mid X_n = x) = P_{\theta_n}(x, y)$ (cf. 6.11).

$(X_n)$ thus obtained is an inhomogeneous Markov chain and choosing $\Delta$ big enough assures the a.s. convergence of the algorithm.

**(6.7.1) Theorem.** *Let $\theta_*$ be the unique solution of the ML equation. If $\Delta > \Delta_0$ is big enough, then $\theta_n \longrightarrow \theta_*$ a.s. and there exists $\varepsilon > 0$ such that a.s. and for all $\varepsilon < \varepsilon_0$,*

$$\|\theta_n - \theta_*\| = o(n^{-\varepsilon}).$$

*$\Delta_0$ can be chosen to be $2\mu N\varphi_0$, where*

$$\mu = \operatorname*{Max}_{s \in D, x_s, x'_s, x_{\overline{s}}} \|U(x_s, x_{\overline{s}}) - U(x'_s, x_{\overline{s}})\|_\infty, \quad \varphi_0 = \operatorname{Max} \|U(x) - U(x_0)\|.$$

REMARKS:

(1) This result is related, on one hand, to the control of process $(X_n)$ in terms of kernels $(P_{\theta_n})$, and on the other, to the dynamics of the mean differential equation [11].

(2) This algorithm, equipped with a stopping rule, can be used for practical purposes. A comparative study with other estimation methods (for example, $PML$) is considered in [173] where the superiority of the $ML$ is shown.

(3) At each step, this algorithm can be combined with a simulated annealing restoration algorithm: if $(Y_n)$ is the Markov chain,

$$P(Y_{n+1} = y \mid Y_n = x) = P_{\frac{\theta_n}{T_n}}(x, y),$$

where $\theta_n \longrightarrow \theta_*$ fast enough and $T_n = \dfrac{c}{\log n}$ with $c$ big enough, then $(Y_n)$ is strongly ergodic with limiting distribution given by the uniform distribution over

$$\Omega_{\max,\theta^*} = \{\omega : \pi_{\theta^*}(\omega) = \operatorname*{Max}_x \pi_{\theta^*}(x)\}.$$

(4) If $\pi_\theta(x, y) = Z^{-1}(\theta) \exp[- < \theta, U(x, y) >]$ is the joint distribution at $(x, y)$ and if the partial observation is $y_0$, the ML estimator is given by

$$\frac{d}{d\theta} \log \psi_\theta(y_0) = 0, \quad \text{where} \quad \psi_\theta(y) = \sum_x \pi_\theta(x, y).$$

Assume also $E_\theta\left[\frac{d}{d\theta} \log \pi_\theta(\cdot, y_0) \mid y_0\right] = 0$; that is, $E_\theta[U(\cdot, y_0) \mid y_0] = E_\theta(U)$.

In order to solve this equation, Younes [175] proposes a stochastic algorithm which considers two parallel simulations. One, to simulate $\pi_\theta$, and the other to simulate $\pi_\theta(\cdot, y_0)$ by Gibbs' sampling:

$$\text{(S1)} \qquad (X_1^{n+1}, Y_1^{n+1}) \sim \pi_{\theta_n}, \qquad\qquad \text{(S2)} \qquad X_2^{n+1} \sim \pi_{\theta_n}(\cdot, y_0)$$

and

$$\theta_{n+1} = \theta_n + \frac{c}{n+1}[U(X_1^{n+1}, Y_1^{n+1}) - U(X_2^{n+1}, y_0)].$$

If $c$ is small enough, and if $\theta$ belongs to a compact, the algorithm converges to a solution of the $ML$ equation. This algorithm, as the $ML$ one for complete data, can be combined with a $MPM$ reconstruction algorithm (cf. 7.8.2).

In this same partial data context, we shall describe the Gibbsian $EM$ algorithm due to Chalmond.

## *6.7.2. The Gibbsian EM Algorithm (Chalmond [30])*

Let $X = \{X_s, s \in S\}$, $S = \{1, 2, \ldots, n\}$ be a Gibbs field with state space $E = \{0, 1, \ldots, m-1\}$, and distribution $\pi_\alpha(x) = \exp\{-U_\alpha(x) - \log Z(\alpha)\}$. We shall assume that we only observe a degraded image $Y = \{Y_s, s \in S\}$, with distribution

$$\pi_\varphi(y \mid x) = \prod_{t=1}^{N} P_\varphi(y_t \mid x_t).$$

$X$, $(X,Y)$, and $(X \mid Y)$ are Markov fields, whereas $Y$ is not. If $P_\varphi(y \mid x)$ is proportional to $\exp\{-g_\varphi(x,y)\}$, $(X \mid Y)$ is a Gibbs field with energy

$$U_\theta(x \mid y) = U_\alpha(x) + \sum_S g_\varphi(x_t, y_t), \quad \theta = (\alpha, \varphi).$$

*Standard EM Algorithm.* Define $z = (x,y)$ the complete observation with distribution $P_\theta(z)$, and $y$ the incomplete observation. Starting with an initial value $\theta_0$, stage $n \to n+1$ of the $EM$ algorithm with current parameter value $\theta_n$, is given by

$$\theta_{n+1} = \operatorname*{Arg\ Max}_\theta E[\log P_\theta(X,y) \mid y, \theta_n].$$

*Step E* (Expectation): Expectation calculation under $y$ and $\theta_n$.
*Step M* (Maximization): Choosing $\theta_{n+1}$ that maximizes the expectation in terms of $\theta$.

The popularity of the algorithm is due to the following property:

THEOREM. — *Let $\ell_\theta(y)$ be the likelihood of the incomplete observation $y$ for $\theta$. Then the sequence $(\ell_{\theta_n}(y))$ is increasing over $(\theta_n)$.*

We point out that we don't know beforehand if the $ML$ limit for $y$ is finite, or if the sequence depends on the choice of $\theta_0$, the initial estimator.

*Chalmond's Gibbsian EM Algorithm.* $\pi_\alpha(x)$ is replaced by the pseudo-likelihood $P^*_\alpha(x) = \prod_{i \in S} \pi_i(x_i \mid x_{\partial i}, \alpha)$, and $P_\theta(z)$ by the pseudo-likelihood $P^*_\theta(x,y) = \prod_{i\in S} P_\varphi(y_i \mid x_i)\pi_i(x_i \mid x_{\partial i}, \alpha)$. If $\theta_n$ is the current value, $\theta_{n+1} = \operatorname*{Arg\ Max}_\theta E[\log P^*_\theta(X \mid y) \mid \theta_n, y]$; the Gibbs sampler can be used to calculate the expectation by Monte Carlo methods.

**(6.7.1) Example.** *Images with m-Levels and Degradation by Channel Noise.*

- $X$ is modeled by $p_{ij} = P(X_t = i \mid X_{\partial i} = j)$, $i = 0, m-1$ and $j$ in $E^{\partial i}$.
- Channel noise: $Y$ has $K$ levels with $p_i(k) = P(Y_t = k \mid X_t = i)$. We have

$$P^*_\theta(x,y) = \prod_{i,k} p_i(k)^{m_{ik}(x,y)} \cdot \prod_{i,j} p_{ij}^{n_{ij}(x,y)} \quad \text{with}$$

$$m_{ik} = \sum 1((x_s, y_s) = (i,k)), \quad n_{ij} = \sum 1((x_s, x_{\partial s}) = (i,j))$$

We update $\theta_n$ at step $n+1$ by

$$\begin{aligned} p_{i,j}^{(n+1)} &= E[n_{ij}|y,\theta_n]/\textstyle\sum_i E[n_{ij}|y,\theta_n], \ p_i^{(n+1)}(k) \\ &= E[m_{ik}|y,\theta_n]/\textstyle\sum_k E[m_{ik}|y,\theta_n] \end{aligned}$$

Since $E[n_{ij} \mid y, \theta] = \sum_x n_{ij}(x) P_\theta(x \mid y)$, $E[m_{ik} \mid y, \theta] = \sum_x m_{ik}(x,y) P_\theta(x \mid y)$, the Gibbs sampler is used to estimate both expectations. For example, for the first: $\widehat{E}[n_{ij} \mid y, \theta] = \frac{1}{s_n - s_0} \sum_{s_0}^{s_n} n_{ij}(x(s))$; $x(s)$ is the configuration at step $s$, $s_0$ is the initial time required in order for the sampler to be stationary, and $s_n$ is the duration of the simulation. Analogous equations that describe $\theta_{n+1}$ in terms of $\theta_n$ are also given in [30] for a gray level degradation, $y_t \in \mathbb{R}$: $p(y_t \mid x_t = i) \sim \mathcal{N}(\mu_i, \sigma^2)$.

### *6.7.3. Markov Chain Monte Carlo Method (MCMC) for the Estimation of Partially Observed Markov Models*

Let $(X, Y)$ be a Markovian model whose distribution depends on a parameter $\theta$. We shall assume that only $Y$ is observed and that the marginal distribution of $Y$ allows the identification of $\theta$. The idea for the estimation of $\theta$ is as follows: assume that we reconstruct the missing observation, $\widehat{X}$, and let $\theta = h(X, Y)$ be a good estimator of $\theta$ in a complete observation context; we will choose $\widehat{\theta} = h(\widehat{X}, Y)$. Such algorithms are studied in the image context by Quian and Titterington [134] and Lavielle [112]: the first studies estimation-restoration coupling, wile the second, based on the ideas of Celeux and Diebolt [28], constructs a homogeneous Markov chain $(\widehat{\theta}(k),\ k \geq 1)$ as follows: $\widehat{\theta}(0)$ is arbitrary; if $\widehat{\theta}(k-1)$ is the estimator at time $k$, we simulate $\widehat{X}(k-1)$ according to $P(X \mid Y = y,\ \widehat{\theta}(k-1))$ and $\widehat{\theta}(k) = h(\widehat{X}(k-1), y)$. Once the chain is ergodic, we choose $\widehat{\theta}$ as the mean of the successive values of $\widehat{\theta}(k)$. We call such algorithms Restoration-Estimation algorithms ($RE$).

An example of simulated data in [112] shows the good qualities of this algorithm: $Y = f * X + \varepsilon$ is a $40 \times 40$ image, resulting in a convolution plus an additive noise; $(X \mid \Lambda)$ is an independent field, a mixture of Gaussian distributions conditional to a label process $\Lambda$; then, $\theta = (\theta_\Lambda, \theta_{X|\Lambda}, \theta_{Y|X})$, the parameter we want to estimate, defines the label process, the convolution filter, and the mixture of Gaussians. This estimation, based only on $Y$, is correct for all parameters.

### *6.7.4. Gibbs Sampler and Bayesian Statistics*

Assume we observe a realization of a r.v. $Y$ with density $f(y, \theta)$, where $\theta$ is a parameter of $\mathbb{R}^m$. The Bayesian statistics principle is the following: we fix *a prior distribution* $g(\theta, \mu)$ for $\theta$ with $\mu$ a known parameter. The Bayesian estimator for $\theta$, observed $y$, is based on the conditional distribution

$$g(\theta|y) = \frac{f(y, \theta)\ g(\theta, \mu)}{h(y)}.$$

The numerator contains the joint distribution of $(y, \theta)$, and the denominator the marginal distribution of $y$. Except for very particular cases this marginal cannot be given analytically, as it is based on an $m$ dimensional integral of $\theta$. This difficulty has lead Bayesians to develop numerical techniques in order to give approximate values for $h(y)$, and thus of $g(\theta|y)$.

The Gibbs sampler has opened a new and very flexible methodology in Bayesian statistics, even for relatively complex models (cf., e.g., Gelfand et al. [65], Gelfand and Smith [64], Smith and Roberts [150]). In many situations (cf. Exercises 13 and 15), we have an analytic expression for the conditional densities $g_i(\theta_i \mid \theta_j,\ j \neq i,\ y)$ $i = 1, \ldots, m$. If for all $i, \theta, y$ we have $g_i > 0$, then we can obtain the distribution of $g(\theta \mid y)$ *empirically* by using the Gibbs sampler for the conditional distributions $g_i$ and periodic scanning $[i] = i \bmod m$. If $\theta(k)$ is the realization after the $k$-*th* scanning, $(\theta(k),\ k \geq 1)$ is an ergodic chain and, in particular,

$$P(\theta \in A|y) = \lim_{k \to \infty} \tfrac{1}{K} \textstyle\sum_{k=1,K} 1(\theta(k) \in A \mid y).$$

## 6.8. Stochastic Algorithms for Image Reconstruction

The basic problem is reconstructing an image $x = \{x_s, s \in S\}$ from an observation $y$,

$$y = \phi(H(x), b). \tag{6.26}$$

Some examples are given in § 2.4: $x_s \in E$ qualitative, quantitative, or mixed. $H$ is a local degradation, $b$ is a noise factor, and $\phi$ is associated to the transmission process.

### 6.8.1. *Bayesian Methods*

Reconstruction $\widehat{x} = \widehat{x}(y)$ must satisfy the following two constraints:

(1) $\widehat{x}$ must satisfy the prior assumptions on $x$, which are expressed in terms of prior energy.

(2) $\widehat{x}(y)$ must "agree" with observation $y$.

Both constraints are exclusive in a certain sense, and $\widehat{x}$ shall minimize a cost function,

$$U(x \mid y) = U_1(x) + U_2(x \mid y).$$

In the Markovian setting, $U_1(x)$ corresponds to the prior energy for $x$, and $U_2(x \mid y)$ to the posterior energy conditional to $y$. The posterior energy $U(x \mid y)$ is the key criteria that leads to the different Bayesian methods of global reconstruction.

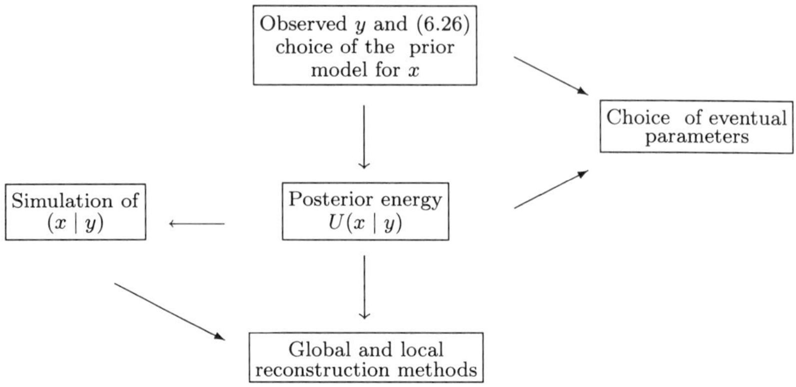


## 6.8.2. *Global Reconstruction Methods: MAP, ICM, and MPM*

We shall consider three important reconstruction methods: the $MAP$, the $ICM$, and the $MPM$ The local version of $MPM$, the contextual classification, is given in 6.8.3. The interest of these methods is illustrated by several examples. The richness (but also the complexity) of these methods only becomes clear when dealing with them in practice or by reading the image literature, some of which is given in the references.

Generally, the energy that is used in the reconstruction criteria $U_\theta(x \mid y)$ depends on a parameter that must be "chosen." We shall not consider this problem in great depth here, and limit ourselves to certain comments:

(1) Generally, reconstruction iterative algorithms are coupled with iterative "choice" algorithms, or parameter "estimation." The initialization of these requires an initial point $\widehat{x}(0)$.

(2) The choice of a "good" parameter is a difficult problem that depends on a qualitative criteria (visual or any other) over $\widehat{x}_\theta(y)$. This "good" choice also depends on the method and on the noise levels [48]. If we dispose of an "expert" who can judge the quality of the reconstruction, it is reasonable to let him decide, among several reconstructions, that which he considers the "best."

(3) There are three families of methods for the choice of $\theta$:

(*i*) the statistical methods which interpret $\theta$ as the parameter of a statistical model and then use good estimation functionals;

(*ii*) the "nonparametric" methods as "cross-validation" which automatically incorporate the reconstruction method;

(*iii*) the fitting of parameters by means of conditional probabilities which control the modelization accurately ([8], [136]).

*6.8.2.1. Maximum a Posteriori (MAP; Geman, [67])*

It is the solution associated to the global cost function $c(x,\widehat{x}) = \mathbb{1}(x \neq \widehat{x})$, that is,

$$\widehat{x}(y) = \operatorname{Arg}\operatorname*{Max}_{x} P(x \mid y). \tag{6.27}$$

This global cost function is very penalizing: it is enough that at one site $\widehat{x}_i \neq x_i$ for the reconstruction to be penalized by 1; this explains, in part, why the $MAP$ algorithm yields such "smooth" results.

As $x$ belongs to a product space $\Omega = E^S$, we could consider a Simulated Annealing ($SA$) algorithm with a Gibbs dynamic if $P_s(x_s \mid x_{\overline{s}}, y)$ depends locally on $x_{\overline{s}}$. If not, we will choose the Metropolis dynamic for an adequate choice of the acceptance transition $Q$. The cooling schedule shall be realistic, for example, with a slow exponential decay (cf. 6.3.3).

Visually, the $MAP$ is very sensible to variations of the regularization parameters that appear in the prior energy: a value that is too big can lead, in the case of a binary image, to a one-color reconstruction with least conditional energy ([48], [118], [130]).

In order to get a good reconstruction, it is necessary to scan the image many times. However, the complexity of this stochastic algorithm grows linearly in $|\Omega|$, the size of the space of configurations, whereas the complexity of combinatorial algorithms is exponential. In spite of this, the $SA$ algorithm is still expensive: a present-day research area is the parallelization of this algorithm (at a given time, the $SA$ can be done at several sites) and the implementation of special architectures. Theoretical results show, however, that in general these synchronous algorithms do not converge towards the uniform distribution over $\Omega_{\text{Min}}$ ([4], [163], [164]).

In the case of binary images, Porteons, Greig and Seheult [130] implemented an exact $MAP$ algorithm which allows us to compare exact $MAP$ and $SA$ for a particular cooling scheme. The conclusion (see below Figure 6.4.) is that $SA$ could be preferred to exact $MAP$.

To illustrate $MAP$ reconstruction, we present two examples considered in the Geman brothers' article [67].

**(6.8.1) Example.** (Figure 6.1) Consider $x$, a $64 \times 64$ hand-drawn image with three gray levels ($a$), degraded by an additive Gaussian noise $\mathcal{N}(0, \sigma = 0.7)$($b$). We assume an isotropic, eight nearest neighbors, Markovian prior model for $x$ with 2 point cliques and, for $s$ and $t$ neighbors, we define $U_{\{s,t\}}(x_s, x_t) = \frac{1}{3}$ if $x_s = x_t$, $-\frac{1}{3}$ if not.

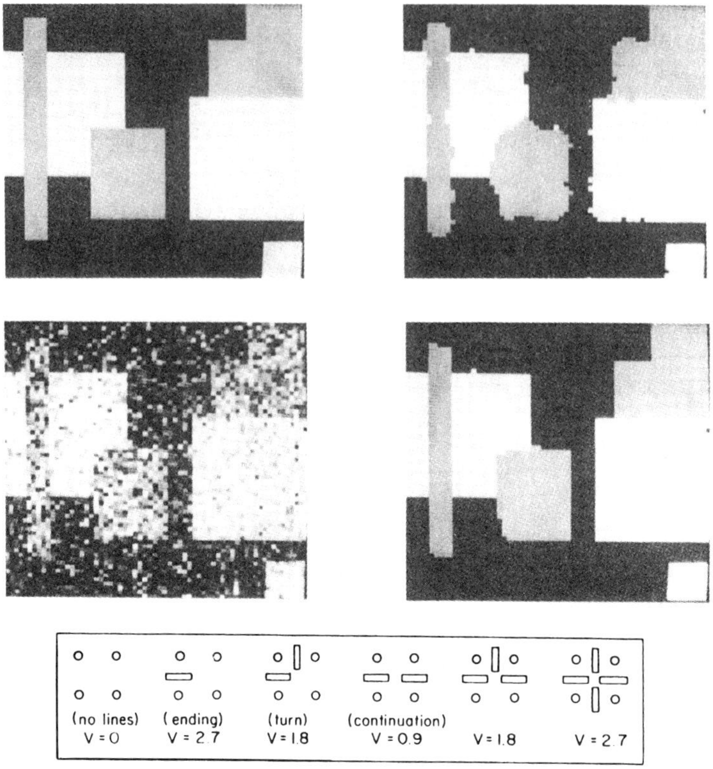


*Figure* 6.1: *Reconstruction by MAP.*

*(a)* 64 × 64 *Original image with* 3 *levels; (b) Noisy image;*

*(c) Reconstruction after* 1000 *scans over the gray levels only;*

*(d) Reconstruction after* 1000 *scans including an edge process;*

*(e) Edge potentials.*

*Source: Geman [67], IEEE PAMI, 1984.*

A first reconstruction is obtained after scanning the image 1000 times $(c)$. A second reconstruction, this time including a line (edge) process, improves these results $(d)$. The energy for the Gray × Lines model is $U(x \mid \ell) + U(\ell)$: if there is a line between neighbors $s$ and $t$, we define the corresponding potential to be zero; if not, we maintain the old structure. $U(\ell)$ organizes contours, according to the chosen potential. In this same setting of a gray level image with edges, another model, as well as the problem of choosing the corresponding parameters, was studied by Chalmond [29] (cf. 2.4, Example 3).

**(6.8.2) Example.** (Figure 6.2) The $x$, a $64 \times 64$ image to be restored, is that of a road signal, with two levels $0, 1$. Thus, we seek to reconstruct a binary image on the basis of an image $y = Hx + b$ degraded ($(a)$ by a convolution $Hx$, and $(b)$ by an additive noise $b \sim \mathcal{N}(0, \sigma = 0.5)$). Two reconstructions are given: one, after 100 scans of an $SA$ algorithm ($c$); the other after 1000 scans ($d$). In both cases, the energy function (Gray $\times$ Edge) is used, with two possible edge directions.

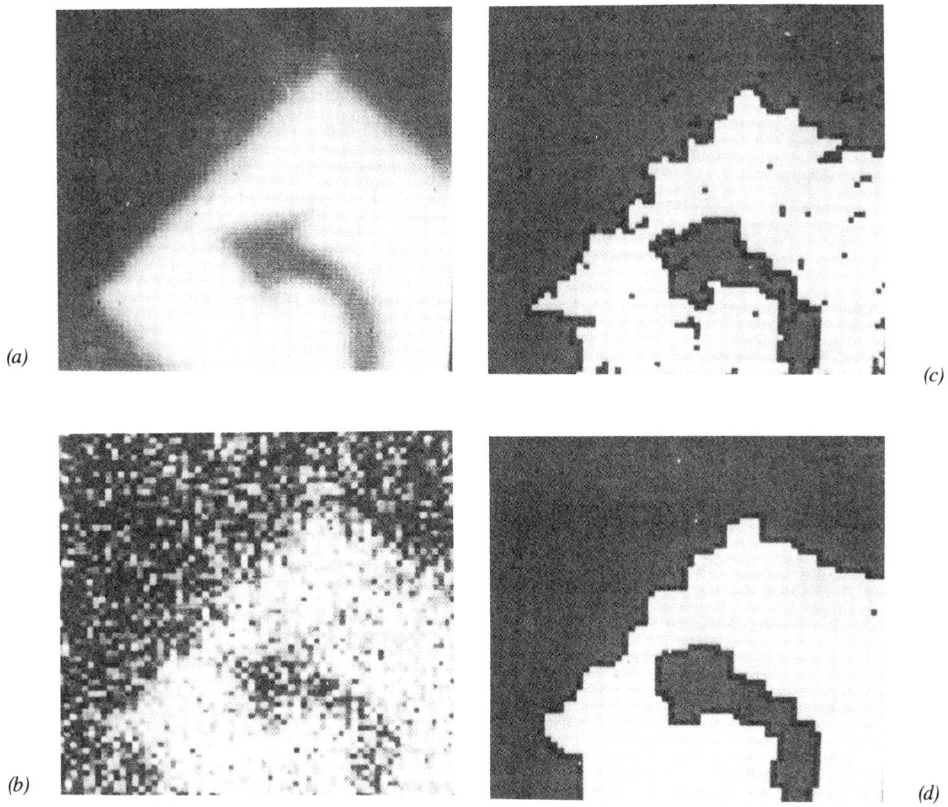

*Figure 6.2: Restoration by MAP including an edge process.*
*(a) Noisy image of a road sign; (b) with additive noise;*
*(c) Restoration after 100 scans; (d) after 1000 scans;*
*Source: Geman [67], IEEE PAMI, 1984.*

*6.8.2.2. ICM (Iterative Conditional Mode, Besag [18])*

$ICM$ is a *deterministic* fast algorithm which mimics $MAP$ but that requires a *good initialization.* The implementation of the algorithm can be understood from the following equality:

$$P(x \mid y) = P_s(x_s \mid x_t,\ t \neq s, y)\ P(x_t,\ t \neq s \mid y). \tag{6.28}$$

By choosing the value at site $s$ that maximizes the conditional probability (conditional mode), $P(x \mid y)$ increases. The $ICM$ algorithm is then as follows: assume we have a visiting sequence that passes infinitely often by each site;

*(i) Choose the initial value $\widehat{x}(0)$ conveniently.*

*Step $k \longrightarrow k+1$ is given by*

*(ii) If $s = s(k+1)$ is the visited site at time $k+1$, choose $\widehat{x}(k+1)$*

$$\widehat{x}(k+1) = \begin{cases} \widehat{x}_t(k) & \text{if } t \neq s, \\ \underset{x_s \in E}{\text{Arg Max}}\, P_s(x_s \mid \widehat{x}_t(k),\ t \neq s, y) & \text{if not.} \end{cases}$$

Throughout $ICM$, $P(\widehat{x}(k) \mid y)$ is increasing. This deterministic algorithm converges towards a local extremum that depends on the initial value $\widehat{x}(0)$. This algorithm performs well and is fast, which explains its success. Experience shows that it is enough to scan an image between 5 and 10 times in order to obtain a good restoration. More scans may perhaps degrade the reconstruction.

By analogy with $SA$, we can, during the implementation, vary the regularization parameter; for example, in a segmentation (cf. (2.47)), varying parameter $\beta$ from a small value (configurations are almost free) toward a reasonable value ($\beta = 1.5$) organizes $x$ well.

**(6.8.3) Example.** (Figure 6.3) *A* $90 \times 98$ *scene with 3 colors* [18].
The original scene $x(a)$ has three colors. The degraded image, which is not represented here, is originated by independent, circular, Gaussian distributions which treat the three values of $x_i$ in an exchangeable fashion: $y_i \sim \mathcal{N}C(\mu(x_i), \sigma^2)$ ($\mu(1) = 0°$, $\mu(2) = 120°$, $\mu(3) = 240°$). The choice of $\sigma^2$ is such that pixel by pixel maximum likelihood reconstruction (that is, contextual blind discriminant analysis) yields a reconstruction with 40% error rate $(b)$.

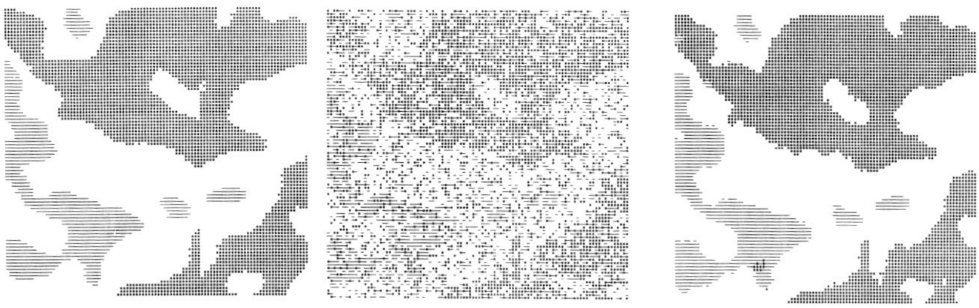

*(a)* *(b)* *(c)*

*Figure 6.3: Reconstruction by ICM of a three-level image.*
*(a) Original image;*
*(b) Restoration by ML (e = 40%);*
*(c) Restoration by ICM after 6 scans (e = 5%).*
*Source: Besag, JRSS B., 1986 [18].*

The prior model for $x$ is the exchangeable model (2.47):

$$P(x) = Z^{-1}(\beta)\ \exp(\beta n(x)), \tag{6.29}$$

for the eight nearest neighbors relationship, $n(x) = \sum_{<i,j>} \mathbb{1}(x_i = x_j)$ and $\beta = 1.5$. The $ICM$ reconstruction, with $(b)$ as initial value, after six scannings, is given in $(c)$: the error classification rate for this reconstruction is less than 5%.

**(6.8.4) Example.** (Figure 6.4) *Exact $MAP, SA, ICM$ and influence of $\beta$* [130]. The original $64 \times 64$ binary image $x$ $(a)$ is degraded by a channel transmission noise with 25% error rate (noisy image $y$, $(b)$). We consider the prior model defined by (6.29) for $x$. Then $x$ is reconstructed by three methods: exact $MAP$ (first column), $SA$ (second column) and $ICM$ (third column) for three values of $\beta : 0.3, 0.7, 1.1$. For the $ICM$, $\widehat{x}(0) = y$. Exact $MAP$ was obtained using a Ford–Fulkerson algorithm.

*Figure* 6.4: *Exact MAP, SA and ICM for several values of the regularization parameter* $\beta$.
(a) *original image;* (b) *noisy image* (25% *error rate*):
*column* 1: *MAP* (c,f,i); *column* 2: *SA* (d,g,j); *column* 3: *ICM* (e,h,k);
*line* 1: $\beta = 0.3$; *line* 2: $\beta = 0.7$; *line* 3: $\beta = 1.1$.
*The cooling scheme is* $\mathrm{T_k} = \mathrm{A} \cdot \rho^{\mathrm{k}}$, $\mathrm{k} = 1, 564$, $\rho = 0.99$, *and* $\mathrm{A} = 2.9$.
*Source: Porteons et al., JRSS B.* [130].

These simulations call for the following comments:

(1) For $\beta = 0.7$ and 1.1, $SA$ is better "to the eye" than exact $MAP$;

(2) For "big" values of $\beta$, images restored by $MAP$ are too smooth: $MAP$ is not robust in terms of $\beta$;

(3) $ICM$ is correct; and

(4) In spite of this, it is $MAP$ with $\beta = 0.3$ which has the smallest error rate.

*6.8.2.3.Marginal Posterior Mode (MPM) (Marroquin, Mitter, Poggio [118])*
An example of cost function that penalizes less an error in only one site is, for example, the additive cost error function, $c(\widehat{x}, x) = \sum_S 1(x_i \neq \widehat{x}_i)$. The Bayesian solution associated to the distribution $P(x \mid y)$ is the $MPM$, the *marginal posterior mode* at each site

$$(6.30) \qquad \widehat{x}(y) = \{\widehat{x}_s(y) = \operatorname*{Arg\ Max}_{x_s \in E} P_s(x_s \mid y),\ s \in S\}.$$

$P_s(x_s \mid y)$ is the marginal distribution of $x$ at $s$, conditional to the complete observation $y$. For a Gibbs model, this law cannot be given analytically. Thus, it shall be evaluated by a Monte Carlo procedure, simulating (for example, by means of the Gibbs sampler) the conditional joint distribution $P(x \mid y)$, whose conditional distributions $P_s(x_s \mid x_t, t \neq s, y)$ we know explicitly. If $x_s \in E$ has $G$ states, $S = \{1, \ldots, n\}$, we shall require $G \times S$ memory places. $G$ memory places are associated to each site $s$ and permit the evaluation of the empirical frequency of $(X_s \mid y)$.

Calculation time shall be relatively significant, as a significant number of simulations of $(x \mid y)$ shall be required (for the Monte Carlo), each simulation requiring several complete scannings (for the Gibbs sampler). Restoration time is thus comparable to that of $MAP$.

In terms of the regularization parameter, $MPM$ is more robust than $MAP$, but less than $ICM$. As opposed to $MAP$, the $MPM$ is not constrained to a particular algorithmic schedule. These remarks explain why $MPM$ is often prefered to $MAP$.

When the state space $E$ is $\mathbb{R}$ or a finite subset of $\mathbb{R}$, computation is easier when we use the mean of the marginal in each site: this is the Threshold Posterior Mean ($TPM$); threshold, because for $E$ a discrete space, we take in $E$ the nearest value from the sample mean.

**(6.8.5) Example.** (Figure 6.5) *Segmentation of an Aerial Image* [22].
The observed $64 \times 64$ image $y$ $(a)$, taken from an aerial image, is a texturized image that shows different types of land plots, separated by roads. The label image $x$ that is to be reconstructed has three levels. A Markovian nearest neighbors model is assumed for $x$ and for $y$ a texture model composed of independent variables $\mathcal{N}(\mu(x_i), \sigma^2)$. Parameter estimation and the reconstruction of $x$ are done simultaneously and iteratively by a Gibbsian $EM$ algorithm and $MPM$.

An initial reasonable partition $\widehat{x}(0)$ is obtained from the gray level histogram of $y$ $(b)$ based on quantiles at 33% and 67%, which separate the population in three equal subsets. That is not the initial value that is given here: $(c)$ corresponds to an initial choice of $\widehat{x}(0)$ by $ML$ for a bad initial value of parameters, $\mu_1 = 0$, $\mu_2 = \mu_3 = 0.2$, $\sigma^2 = 0.4$, whereas the same values calculated over the quantiles are respectively $-0.8$, $-0.35$, 1 and 0.4. In spite of this bad initial choice $(c)$, the partition organizes itself well during the iterative process. Figure 6.5.$(e)$ represents $\widehat{x}$ in the 30-*th* iteration.

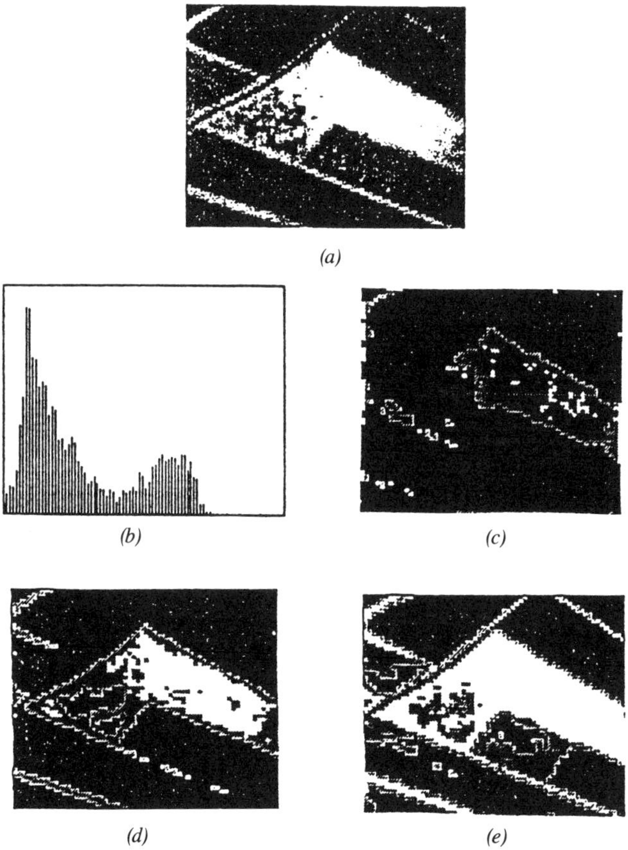


*Figure* 6.5: *Segmentation of an aerial image; the MPM is coupled with a Gibbsian EM for the estimation of the model's parameters; (a) observed image, and (b) its gray level histogram; (c) ML reconstruction; (d) $\longrightarrow$ (e): iterative reconstruction-estimation; (e) corresponds to iteration n*$^{\text{o}}$ 30.

*Source: Chalmond, Pattern Recognition [30].*

### *6.8.3. Local Reconstruction Methods*

Let us consider, for example, the image formation model $y = (y_s)$ where $y_s \sim \mathcal{L}(x_s)$ are independent, pixel by pixel. If $x_s$ is an $m$–level process, this example can correspond to a quantitative Gaussian texture model (teledetection problem) or to a channel transmission deformation.

The *contextual blind* reconstruction is ordinary discriminant analysis, or classification by *maximum likelihood*:

$$(6.31)\ (ML) \qquad \widehat{x}_s = \operatorname*{Arg\ Max}_{x_s} P(y_s \mid x_s).$$

In the absence of information regarding the frequencies of the different levels of $x$, we choose *blind classification* $(BC)$

$$(6.32)\ (BC) \qquad \widehat{x}_s = \operatorname*{Arg\ Max}_{x_s} P(x_s \mid y_s), \quad s \in S.$$

$MPM$ reconstruction is a site by site global reconstruction

$$(6.30)\ (MPM) \qquad \widehat{x}_s = \operatorname*{Arg\ Max}_{x_s} P(x_s \mid y), \quad s \in S.$$

We also have the related *contextual discriminant analysis* (or *contextual classification* $(CC)$) where we choose $\widehat{x}_s$ in terms of a local context $y(V_s)$,

$$(6.33)\ (CC) \qquad \widehat{x}_s = \operatorname*{Arg\ Max}_{x_s} P(x_s \mid y(V_s))), \quad s \in S.$$

For example, consider $V_s = V + s$ where $V$ is a five site window formed by the origin and its four nearest neighbors. Bayes formula yields

$$P(x_s = j \mid y(V_s)) = \frac{1}{P(y(V_s))} \sum_{x_{\partial s}} P(y(V_s) | x(V_s) = (j, x_{\partial s})) \times P(x(V_s) = (j, x_{\partial s}))$$

where $\partial s = V_s \backslash \{s\}$. Our *prior information* over the partition $x$ we want to reconstruct is the geometrical regularity of the constant label zones. A simple way to translate this prior information is to limit the number of possible configurations $(j, x_{\partial s})$ to the most probable $x(V_s)$: for example, for the *simple contextual classification (SCC)*, only the $m$ constant configurations are kept:

$$(6.34)\ (SCC) \qquad \widehat{x}_s = \operatorname*{Arg\ Max}_{j} P(y(V_s) \mid x(V_s) \equiv j).$$

In this example, all the prior information over $x$ is translated by this simplification: *no modeling is required for $x$.*

The *geometrical contextual classification (GCC*, cf. [92]) authorizes a few other specifications for $x(V_s)$, for example, the $m$ constant configurations of type 1, and the $8m(m-1)$ two level configurations of type 2 and 3:

*Three possible configurations for* x($\mathrm{V_s}$) *in GCC.*

**(6.8.6) Example.** (Figure 6.6) *Comparison of global and local methods* [170]: The $32 \times 32$ original image is a drawn $\beta$ with two levels $\{0, 1\}(a)$; the texture differentiation over the observation corresponds to the variance model: $y_s \sim \mathcal{N}(0, \sigma^2(x_s))$ with $\sigma^2(0) = 1$ and $\sigma^2(1) = 6$. After such a degradation, it is very hard to distinguish the $\beta$.

*(a) Original 64 × 64 image.*
*(b) Simulated textures based on*
*(a) $\sigma_D^2 = 1$, $\sigma_1^2 = 6$; and*
*on segmentations of (b):*
*(c) CA, (d) CCS,*
*(e) CCG, (f) MPM,*
*(g) MAP, and (h) ICM.*

*Figure 6.6: Comparison of global and local methods.*
*Source: Yao, Cahier du CERO [170].*

Visual segmentation over observation (*b*) is very difficult. This is confirmed by the very bad quality of the blind classification ($BC$). If the global methods $MAP$ and $MPM$ are the best, among the fast methods ($ICM$, $SCC$, and $GCC$), both contextual methods are better than $ICM$ in this example.

## 6.9. Bibliographical Comments

The contraction coefficients for a Markov chain and its properties were introduced by Dobrushin in the context of $CLT$ for a nonhomogeneous chain. For general results concerning Markov chains, the reader can consult Isaacson and Madsen, Iosifescu and Theodorescu.

Monte Carlo methods and in particular the Metropolis dynamic are presented by Hammersley and Handscomb, see also Hastings. The Gibbs sampler is due to Geman [67], as well as simulated annealing for this dynamic. Younes extended this result to the case where the state space is compact, and to the Gaussian case. Tierney generalizes this result to more general infinite state spaces. Lyapunov's criteria for Markov chains is given in Duflo. The simulated annealing algorithm was proposed by Kirkpatrick, Gellatt, and Vecchi. For a state space $\Omega$ which is not necessarily a product space and the Metropolis dynamic, Hajek gives an optimal cooling scheme.

The results concerning relaxation and cooling under constraints are given in [69]. D. Geman [66] gives a very elegant and general version of all these algorithms including relaxation and cooling over subsets. Another version of these results for more complicated state spaces is given in Ligget (1985). Parallelization techniques and synchronization of S.A. are studied by Trouvé (cf. also Azencott et al. [4]).

Convergence towards a Gibbs state of a Birth and Death process with a local evolution law was established by Preston.

The stochastic gradient algorithm for the calculation of the maximum likelihood estimator is due to Younes, and the Gibbsian $EM$ was studied by Chalmond in the context of estimation-reconstruction of an image model. Qian and Titterington [134] and Lavielle [112] develop Monte Carlo methods for the estimation of hidden Markovian models.

The utilization of stochastic algorithms for images developed significantly after the pioneer article of the Geman brothers, where they considered $SA$ for the $MAP$ reconstruction. Besag proposed $ICM$ as a fast version of $MAP$, and the $MPM$ was introduced by Marroquin, Mitter, and Poggio.

Contextual methods are studied by many authors; we cite among others, Switzer, Mardia, Yjort, Haslett, and Yao.

## 6.10. Exercises

(1) *Simulation of an SAR model.* $S$ is a finite set, $\{Y_i, i \in S\}$ and $\{a_{ij}, i, j \in S\}$ are given real numbers, $a_{i,i} = 0$ if $i \in S$ and $\alpha = \sup_i \sum_j |a_{ij}| < 1$.

(a) Show that the system $X_i = \sum_j a_{ij} X_j + Y_i$, $i \in S$ has as its unique solution, the limit of the sequence $X^{(n)}$ defined by

$$X^{(0)} = 0, X_i^{(n+1)} = \textstyle\sum_j a_{ij} X_j^{(n)} + Y_i, i \in S.$$

(b) Under the above conditions, deduce a simulation method for a $SAR$ model where $Y = \varepsilon$ is a Gaussian white noise.

(c) What difficulties arise if we want to apply this same method to a $CAR$ model with $Y = e$ a Markovian residual?

(2) In order to simulate an isotropic, with four nearest neighbors, binary texture (with values $\{0, 1\}$) with *equidistributed marginals,* we can proceed in the two following ways:

(a) Use the Gibbs sampler for the conditional distribution $\pi(x_i|\cdot) = Z^{-1}(b, v_i) \exp bx_i(v_i - 2)$, where $v_i$ is the sum of the four nearest neighbors (see 2.2.4).

(b) Use the Metropolis dynamic for the spin exchange dynamic (cf. § 6.2.1) and the energy $U(x) = \alpha \sum_i x_i + \beta \sum_{<i,j>} x_i x_j$ provided that the initial configuration has as many zeros as ones. Check visually that simulations are "analogous" if $\beta = b$.

(3) Over the torus $T^2 = \{1, 2, \ldots, 32\}^2$ and with the four nearest neighbors relationship, we consider the Markovian dynamic for a binary field with states $\{0, 1\}$ and energy:

$$\begin{aligned} &U(x(t) \mid x(t-1)) \\ &= \sum_i x_i(t)[\alpha + \beta x_i(t-1) + \gamma(x_{i+(1,0)}(t-1) + x_{i+(0,1)}(t-1)) + \delta v_i(t)]. \end{aligned}$$

$v_i(t)$ is the sum of the four nearest neighbors with configuration $i$ at time $t$. Simulate this dynamic for $\alpha = -2\delta = -5$, $\beta = \gamma = 1$.

(4) Over the torus $T^2 = \{1, 2, \ldots, 32\}^2$, let $\pi$ be the Markovian binary distribution, with states $\{0, 1\}$ and energy $(\alpha \sum_i x_i + \beta \sum_{<i,j>} x_i x_j)$ with $\alpha = -2\beta = -5$ and the four nearest neighbors relationship. What is a birth/death process associated by (6.21)? Implement the algorithm that simulates $\pi$ by a birth/death dynamic and give the simulation's evolution every 10 events.

(5) Let $X = \{X_s, s \in S\}$ be a multidimensional Markovian field, $X_s \in E^K$. We shall consider simulating $X$, limiting ourselves to coordinate-by-coordinate simulation.

*Examples*:

(a) Let $S = \{1, 2, \ldots, n\}$ be the unidimensional torus equipped with the two nearest neighbors relationship. Simulate the Gaussian Markovian field with energy (cf. Exercise 12, Chapter 2, $0 \le \alpha < \frac{1}{2}$):

$$-U(X, Y) = \sum_i (X_i^2 + Y_i^2) - 2\alpha \sum_{<i,j>} X_i Y_i.$$

(b) Let $S$ be a finite subset of $\mathbb{Z}^2$ equipped with the four nearest neighbor relationship. Simulate the bivariate process $Z_i = (U_i, V_i) \in \{0, 1\}^2$ with potentials (cf. Exercise 4, Chapter 5):

$$\phi_1(u_i, v_i) = \alpha u_i + \beta v_i + \gamma u_i v_i,$$

$$\phi_2((u_i, v_i), (u_j, v_j)) = \delta_1 u_i u_j + \delta_2 v_i v_j \quad \text{if } i, j \quad \text{are neighbors}$$

(i) either directly based on the conditional distribution

$$\mathcal{L}((u_i, v_i) \mid (u_j, v_j), j \neq i)$$

(ii) or by duplicating $S$ as $S^* = S_U \cup S_V$ and then considering the conditional distributions $\mathcal{L}(u_i \mid u_j, j \neq i, V)$ and $\mathcal{L}(v_i \mid v_j, j \neq i, U)$.

(6) Given a real image of "homogeneous" textures with $G$ gray levels:

(a) Identify a model of Markovian textures by penalized pseudo-likelihood, for example, the autobinomial model (cf. § 2.2.4) (cf. § 3.4.3 and Theorem 5.3.5).

(b) Considering the model with estimated parameters, simulate the texture and compare with the original texture (cf. [40] and § 2.2.4, Figure 4).

(7) *Simulation of an edge process.* Consider the model of Example 3, § 2.4. We shall assume we have a binary edge process with potentials defined by the five parameters: $\alpha_1$ indicates a horizontal edge, $\alpha_2$ a vertical edge; $\beta_1$, $\beta_2$ indicate prolongation of horizontal and vertical edges, respectively; $\gamma$ for corners.

(a) Give the joint distribution and discuss the effect of each parameter.

(b) Simulate the process for different parameter values: isotropic or not, with weak or strong "turning" rates, and weak or strong border rates.

(8) *Segmentation of a texturized image.* $S = \{1, 2, \ldots, 32\}^2$ is divided in three regular areas with labels $\lambda = 1, 2, 3$.

(a) If $\lambda_i = k$, we observe $Y_i \sim \mathcal{N}_3(\mu_k, I)$, site-by-site independent variables, $\mu_1 = {}^t(0.5, 0, 0)$, $\mu_2 = {}^t(0, 0.5, 0)$ and $\mu_3 = {}^t(0, 0, 0.5)$. Reconstruct partition $\lambda$: (1) by $ML$, or Blind Classification, (2) by Contextual Simple Classification using a local window with nine points; (3) by $ICM$, with the solution of (1) as its initial value; (4) by $MPM$ (for (3) and (4), use the prior model (2.47) for $\lambda$).

(b) $\lambda$ is degraded into $\delta$ by a channel noise: $P(\delta_i = \lambda_i) = 0.4$, $P(\delta_i \neq \lambda_i) = 0.6$ with the same probability for the two different possibilities for $\delta_i$. Restore $\lambda$ using $MAP$.

(9) A $32 \times 32$ image is formed by two regular areas with the same surface. This image is degraded by a channel noise with 0.4 error rate. Restore the image using $MAP$ under the constraint that both surfaces are equal.

(10) *Matrices with minimum band size.* $A_0 = (a_{ij})_{1 \leq i,j \leq n}$ is a matrix of zeros and ones. By successive of row and column permutations, we create a class $\mathcal{A}$ based on $A_0$ and we wish to minimize $U(A) = \sum_{i,j} d(i,j) \mathbb{1}(a_{ij} = 1)$ where $d(i,j)$ is a distance. Let $Q$ be the transition over $\mathcal{A}$ which consists in interchanging at random (uniform probability) two columns of $A$ and then two rows of $A$. Show that $Q$ is irreducible and symmetric. Describe the Metropolis algorithm and apply it to minimizing $U$ for $A_0 = 50 \times 50$, $a_{ij}$ chosen to be one with probability 0.1, $d(i,j) = |i - j|^4$.

(11) *Maximal segmentation of a graph.* Over $S = \{1, 2, \ldots, n\}$, we define a graph where connection $\{i,j\}$ has an associated symmetrical weight $w_{ij}$. For a given subset $A$ of $S$ ($A \in \Omega = \mathcal{P}(S)$), we measure the degree of attachment between $A$ and $S \backslash A$ by $U(A) = \sum_{i \in A, j \notin A} w_{ij}$. Our goal is to maximize this expression. Let $Q$ be a dynamic over $\Omega$ defined as follows: we choose at random a point $s \in S$; if $s \in A$, we change $A$ for $A \backslash \{s\}$; if not, $A$ for $A \cup \{s\}$. Show that $Q$ is irreducible and symmetric. Describe the Metropolis algorithm for the maximization of $U$.

(12) *The Traveling Salesman's problem.* A salesman wants to complete his tour passing once and only once by each of the $n$ cities $S = \{1, 2, \ldots, n\}$. If $\Omega$ is the set of all such tours (starting and ending at 1), $\pi \in \Omega$ is represented by a cyclic permutation $\pi(1), \ldots, \pi(n)$, $\pi(1) = 1$, $\pi(j) \in S \backslash \bigcup_{\ell=1,j-1} \{\pi(\ell)\}$, $j = 2, n$: we visit $1 = \pi(1)$, then $\pi(2), \ldots, \pi(n)$ and return to 1 (called $\pi(n+1)$) and the covered distance is $U(\pi) = \sum_{i=1,n} d(\pi(i), \pi(i+1))$, where $(d(i,j))$ is a matrix of distances among the $n$ cities.

Numerate $\Omega$. Let $Q$ be the dynamic over $\Omega$ defined as follows: two values $p$ and $q$, $2 \leq p < q \leq n$, are chosen at random, and we exchange, in $\pi$, the values of $\pi(p)$ and $\pi(q)$. Show that $Q$ is irreducible and symmetric. Give the maximization algorithm for $U$.

Is the dynamic where only $p$, $2 \leq p \leq n-1$, is chosen at random and where $\pi(p), \pi(p+1)$ are exchanged, acceptable?

(13) *Bayesian estimation and Gibbs sampling* (Gelfand and Smith [65]). We observe $Y = (Y_1, \ldots, Y_p)$, $Y_i \sim \mathcal{N}(\theta_i, \sigma_i^2)$ and independent with known $\sigma_i^2$ and wish to estimate $\theta = (\theta_1, \ldots, \theta_p)$ under the nonlinear constraint

$$\theta_1 \leq \theta_2 \leq \ldots \leq \theta_p. \qquad (OC)$$

(a) Assume, to begin with, that the $\theta_i$ are i.i.d. $\mathcal{N}(\mu, \tau^2)$, without constraints. Write the joint distribution of $(Y, \theta)$ and check that $(\theta \mid Y)$ is the product of independent Gaussian distributions

$$\mathcal{L}(\theta_i \mid y_i) \sim \mathcal{N}\Big(\frac{\mu\, \sigma_i^2 \, + y_i\, \tau^2}{\sigma_i^2 \, + \, \tau^2}, \frac{\sigma_i^2\, \tau^2}{\sigma_i^2 \, + \, \tau^2}\Big). \qquad (C)$$

(b) Now assume that $\theta_i$ satisfy the order constraint $(OC)$. Check that the analytic form of the density of $(\theta \mid Y)$ is not calculable, whereas the conditional distribution $(\theta_i \mid Y, \theta_j,\ j \neq i)$ is just distribution $(C)$ restricted to the interval $[\theta_{i-1}, \theta_{i+1}]$, defining $\theta_0 = -\infty$ and $\theta_{p+1} = \infty$. How can this distribution be simulated? (See Exercise 15(1c)).

(c) Give a procedure for calculating empirically the distribution of the Bayesian estimator of $\theta$ given $Y$ and under the order constraint $(OC)$.

(14) *Restoration-estimation algorithm for incompletely observed Markov chains.* Let $X = \{X_k, k \in \mathbb{Z}\}$ be a Markov chain with states $\{-1, +1\}$ and transition matrix $P = \begin{pmatrix} p & 1-p \\ 1-p & p \end{pmatrix}$. $X$ is also a Markovian field with two nearest neighbors where the only nonzero potential is the pair potential $\phi(x_i, x_j) = \beta x_i x_j$ if $|i - j| = 1$ with $\beta = \frac{1}{2} \log \frac{p}{1-p}$, (cf. §2.2.2). We shall assume that $X$ is observed only at times $0, s, 2s, \ldots, Ns$ $(s > 1)$ and define $X' = (X_0, X_s, \ldots, X_{Ns})$, $X^* = (X_k, 0 \leq k \leq Ns$ and $k$ not a multiple of $s)$.

(a) Check that $(X^* \mid X')$ is a Markov field whose potentials can be identified in terms of $p$ and $X'$. Give an algorithm for simulating $X^*$ for known $p$ and observed $X'$.

(b) Show that on the basis of $X'$, $p$ is identifiable if $s$ is odd, but it is not anymore if $s$ is even (examine cases $s = 2$ and $s = 3$).

(c) Implement the algorithmic procedure for estimating $p$ described in Section 6.7.3. which can be summarized as follows: if $\hat{p}_k$ is the estimated value at step $k$, we simulate $X^*(k)$ according to distribution $(X^* \mid X', \hat{p}_k)$; we then estimate $\hat{p}_{k+1}$ by $ML$ over the complete observation $(X', X^*(k))$.

(d) In this example, since $X$ is a Markov chain, $p$ can be directly estimated by $ML$. Compare numerically this estimator to that of (c). What are its advantages and its limitations?

(e) Recalling Exercise (11-(b)), Chapter 2, show that the estimation method described in (c) can be extended to a Markov field with four nearest neighbors over $\mathbb{Z}^2$ using the pseudo-likelihood as the estimation functional.

(15) *Parameter estimation for incompletely observed Gaussian data by Monte Carlo Markov chain (MCMC).* Let $-\infty = a_0 < a_1 \ldots < a_p = +\infty$ be

$(p+1)$ given real numbers and the $p$ associated classes $C_j = (a_{j-1}, a_j], j = 1,p$. An $n$-sample of a Gaussian distribution $\mathcal{N}(m, \sigma^2)$ is observed only in terms of classes $C_j$, $j = 1,p$: we observe $n_j$ = number of observations in $C_j$, $j = 1,p$ $(n = n_1 + \cdots + n_p)$.

(1a) Give a reasonable initial estimator $\hat{\theta}_0$ for $(\theta = m, \sigma^2)$.

(1b) Implement the $MCMC$ estimation procedure described in Section 6.7.3: if $\hat{\theta}_k$ is the current value of the estimator, describe the two steps of iteration $k \to k+1$.

(1c) Check that if $U$ has a uniform distribution over $[0,1]$, variable $Y$ defined by

$$Y = m + \sigma \Phi^{-1}(p(U; a, b, m, \sigma)), \text{ with}$$

$$p(U; a, b, m, \sigma) = \Phi(\frac{a-m}{\sigma}) + U.\left[\Phi(\frac{b-m}{\sigma}) - \Phi(\frac{a-m}{\sigma})\right]$$

with $\Phi$ a standard normal, is a Gaussian $\mathcal{N}(m, \sigma^2)$ variable restricted to $[a, b], a \le b$ (cf. Devroye [46]). Implement the algorithm discused in (1b).

(2) Find the likelihood of the partial observation $(n_1, n_2, \ldots, n_p)$ of a Gaussian distribution. Compare (2) and (1b).

(16) *Gibbs sampler with random scanning.* Consider $S = \{1, 2, \ldots, n\}$ and $\pi$ a positive distribution over $\Omega = E^S$, $E$ finite. Let $\{p_s > 0\}$, $\sum_S p_s = 1$ be a distribution over $S$. Consider the Gibbs sampler (cf. 6.2.2) with random scanning: at step $k$, $s_k$ is chosen according to distribution $\{p_s, s \in S\}$.

(a) Show that the transition of this dynamic is homogeneous given by

$$P = \textstyle\sum_S p_s P_s \text{ where } P_s(x,y) = \pi_s(y_s \mid x_s, j \neq s) 1(y_j = x_j, j \neq s).$$

(b) Check that $P$ is irreducible and give a lower bound in terms of $(x, y)$ of $P^n(x, y)$.

(c) Show that the sampler is $\pi$-ergodic.

(17) *Synchronous and asynchronous sampling (cf. 2.2.6, [176], [177]).* Let $\Omega = E^S$ be a finite configuration state space over $S = \{1, 2, \ldots, n\}$ and $p_s(x, y_s) > 0, s \in S, x \in \Omega, y_s \in E$ s.t. each $p_s(x, \cdot)$ is a probability over $E$.

(a) Show that the *synchronous kernel* $R(x,y) = \prod_S p_s(x, y_s)$ is ergodic and there exists a unique invariant distribution $\mu$ (defined as the *synchronous distribution*) for $R$.

(b) Let $\pi$ be a positive measure over $\Omega$, $\pi(x) > 0, x \in \Omega$, and consider the associated family $p_s(x, y_s) = \pi_s(y_s \mid x_t, t \neq s)$. Show that the argument which assures that $\pi$ is invariant for the sequential asynchronous kernel $Q$ (cf. Theorem 6.2.1) is no longer valid for the synchronous distribution $\mu$.

(c) *Example*: Consider $E = \{0, 1\}$, and $\pi$ defined by the quadratic (asynchronous) energy, $U(x) = \sum_{<s,t>} w_{st} x_{st}$; that is,

$$\pi_s(y_s \mid x_{\partial_s}) = e^{y_s v_s(x)}(1 + e^{v_s(x)})^{-1}, \quad \text{where } v_s(x) = \textstyle\sum_t w_{st} x_t.$$

Check that the distribution defined by

$$\mu(x) = \Gamma^{-1} \prod_S (1 + e^{v_s(x)})$$

satisfies $\mu R = \mu$. Finding the (synchronous) energy for $\mu$, show that the neighborhood system for $\mu$ is given by

$$V^{(2)}(s) = \{t \in S, \exists u \in S, \ < u, s > \ \text{and} \ < u, t >\},$$

and check that the synchronous dynamic is reversible, that is, $\mu(x)Q(x,y) = \mu(y)Q(y,x)$ (cf. Remark 3 of Theorem 2.2.3).

# Bibliography


[1] E. Aarts and T.J. Kors. — 1989, Simulated Annealing and Boltzman Machines: Stochastic approach to combinatorial optimization and neural computing, Wiley

[2] R. Azencott. — 1988, Simulated Annealing, Séminaire Bourbaki 1987–88, n° 697

[3] R. Azencott. — 1988, Image analysis and Markov fields, Proceed. Int. Conf. Ind. Appl. Math., SIAM Philadelphia

[4] R. Azencott, Ed. — 1992, Simulated annealing: Parallelization techniques, Wiley

[5] R. Azencott and D. Dacunha-Castelle. — 1984, Séries d'observations irrégulières, Masson

[6] R. Azencott, Th. Carlier, J.L. Guiziou and J.F. Yao. — 1992, Automated detection of geological horizons using M.R.F. techniques, Proceed. of the 6-*th* Conf. in Image Analysis, Come, Italy, 705–713

[7] R. Azencott, B. Chalmond and P. Julien. — 1992, Bayesian 3-*D* path search and its applications to focusing seismic data, L.N.S. n° 74, Springer, eds. Barone, Frigessi, Piccioni, 46–74

[8] R. Azencott, C. Graffigne and C. Labourdette. — 1992, Edge detection and segmentation of textxtured plane image, L.N.S. n° 74, Springer, eds. Barone, Frigessi, Piccioni, 75-88

[9] P. Barone and A. Frigessi. — 1990, Improving stochastic relaxation for gaussian random fields, Proba. in the Engineering and Informational Sciences, n° 4, 369–389

[10] B. Beghin, Ch. Gourieroux and A. Montfort. — 1979, Identification of ARMA model: The corner method, in Time Series Analysis, Ed. O. Anderson, North–Holland

[11] A. BENVENISTE, M. METIVIER AND P. PRIOURET. — 1987, Algorithmes adaptatifs et approximations stochastiques; Théorie and applications, Masson

[12] I. BERKES AND G.J. MORROW. — 1981, Strong invariance principles for mixing random fields, Z.f.W., 57, 15–37

[13] J. BESAG. — 1972, On the correlation structure of some two dimensional stationary processes, Biometrika, 59, 43–48

[14] J. BESAG. — 1974, Spatial interaction and the statistical analysis of lattice systems, J.R.S.S. B 36, 192–236

[15] J. BESAG. — 1976, Parameter estimation for Markov Fields, Tech. report, series 2, Dept. Stat., Univ. Princeton

[16] J. BESAG. — 1977, Efficiency of pseudo likelihood estimation for simple Gaussian fields, Biometrika, 64, 616–618

[17] J. BESAG. — 1981, On a system of two dimensional recurrence equations, JRSS B., 43, n° 3, 302–318

[18] J. BESAG. — 1986, On the statistical analysis of dirty pictures, JRSS B., 48, 259–302

[19] J. BESAG AND P.J.GREEN. — 1993, Spatial statistics and Bayesian computation, JRSS(B), 55, n° 1, 25-37

[20] J. BESAG AND P.A.P. MORAN. — 1975, On the estimation and testing of spatial interaction for Gaussian lattice processes, Biometrika, 62, 3, 555–562

[21] E. BOLTHAUSEN. — 1982, On the C.L.T. for stationary mixing random fields, Ann. Proba., vol. 10, n° 4, 1047–1050

[22] L. BREIMAN. — 1968, Probability, Addison-Wesley

[23] D.R. BRILLINGER. — 1975, Times Series: Data Analysis and Theory, Holt, Rinehart and Winston

[24] P. BRODATZ. — 1966, Textures, Dover N.Y

[25] D. BROOK. — 1964, On the distinction between the conditional and the joint probability approaches in the specification of nearest–neighbor systems, Biometrika 51, 481–483

[26] O. CATONI. — 1990, Etude asymptotique des algorithmes de recuit simulé, Thèse Université Orsay

[27] O. CATONI. — 1992, Image restoration by stochastic dichotomic reconstruction of contour line, L.N.S. n° 74 Springer, Eds. Barone, Frigessi, Piccioni, 101–116

[28] G. CELEUX, J. DIEBOLT. — 1988, A random imputation principle: The Stochastic E.M. algorithm, I.N.R.I.A., preprint n° 901

[29] B. CHALMOND. — 1988, Image restoration using an estimated Markov model, Signal processing, 15, 115–129

[30] B. CHALMOND. — 1989, An iterative Gibbsian technique for reconstruction of M-ary image, Pattern Recognition, 22, (6), 747–761

[31] B. CHALMOND. — 1991, Traitement de différents cas de restauration d'imagerie: Construction de modèle, détection, estimation et reconstruction Bayesienne, Habilitation à la direction de recherche, Univ. d'Orsay

[32] S.C. CHAY. — 1972, On quasi-Markov random fields, J. Multi An., 2, 14–76

[33] A.D. Cliff and J.K. Ord. — 1981, Spatial Autocorrelation, Pion, London, Second Ed

[34] F. Comets. — 1986, Grandes déviations pour les champs de Gibbs sur $\mathbb{Z}^d$, CRAS Paris, t. 303, Série I n° 11, 511–513

[35] F. Comets. — 1992, On consistency of a class of estimators for exponential family of Markov random fields on the lattice, Ann. Stat., Vol 20, n° 1, 455–468

[36] F. Comets and B. Gidas. — 1991, Asymptotics of M.L.E. for the Curie–Weiss model, Ann. of Stat. Vol. 19, n° 2, 557–578

[37] F. Comets and B. Gidas. — 1992, Parameter estimation for Gibbs distributions from partially observed data, Ann. Appl. Proba, Vol 2, 142–170

[38] J. Coursol and D. Dacunha-Castelle. — 1982, Remarques sur l'approximation de la vraisemblance des processus Gaussiens stationnaires, Teor. Veroy. Prim., 27, 155–159

[39] N. Cressie. — 1991, Statistics for Spatial Data, Wiley

[40] G.R. Cross and A.K. Jain. — 1983, Markov random field texture models, IEEE Trans. PAMI, 5, n° 1, 25–39

[41] D. Dacunha-Castelle and M. Duflo. — 1983, Probabilités et statistiques, tome 2: Problèmes à temps mobile, Masson

[42] R. Dahlhaus. — 1983, Spectral analysis with tapered data, J.T.S.A., 4 (3), 163–175

[43] R. Dahlhaus. — 1984, Parameter estimation of stationary processes with spectra containing strong peaks, dans "Robust and Nonlinear Times Series Analysis," Eds. Franke, Hardle and Martin, L.N.S. n° 26, 50–67

[44] R. Dahlhaus and H.R. Künsch. — 1987, Edge effect and efficient parameter estimation for stationary random fields, Biometrika, 74, (4), 877–882

[45] A. Devijver. — 1988, Image segmentation using Causal Markov Random Field models, Lect. Notes in Comp. Sci., Pattern Recognition, Proced. 4th Int. Conf., Ed. J. Kittler, Springer, 131–143

[46] L. Devroye. — 1986, Non-uniform random variate generation, Springer

[47] J.M. Dinten. — 1990, Tomographie à partir d'un nombre limité de projections: régularisation par des champs markoviens, Thèse Univ. d'Orsay

[48] J.M. Dinten, X. Guyon and J.F. Yao. — 1991, On the choice of the regularization parameter: The case of binary images in the Bayesian restauration framework, 55–77, Spatial Statistics and Imaging, Possolo Ed., Proceed. AMS–IMS–SIAM joint summer conf., 1988, Lect. Notes, Inst. Math. Stat., Hayward

[49] R.L. Dobrushin. — 1956, Central limit theorems for nonstationary Markov chains, Th. Proba. Appl., vol. 1, n° 4, 329–383

[50] R.L. Dobrushin. — 1968, The description of a random fields by mean of conditional probabilities and condition of its regularity, Th. Proba. Appl., 13, 197–224

[51] R.L. Dobrushin. — 1970, Prescribing a system of random variables by conditional distribution, Th. Proba. Appl., 15, 458–486

[52] P. Doukhan. — Mixing: Properties and examples, 142 p., L.N.S. n° 85, Springer

[53] P. Doukhan and X. Guyon. — 1991, Mélange pour les champs linéaires, C.R.A.S. t. 313, Série I, 465–470

[54] M. DUFLO. — 1990, Méthodes récursives alétoires, Masson

[55] W. FELLER. — 1943, The general form for the so called Law of Iterated Logarithm, Trans. Am. Soc., 54, 373–402

[56] H. FÖLLMER AND S. OREY. — 1988, Large Deviations for the empirical field of Gibbs measure, Ann. Proba., Vol. 16, 961–977

[57] E. FRANCOIS AND P. BOUTHEMY. — 1991, Multiframe based identification of mobile components of a scene with a moving camera, Rapport Rech. n° 1368, INRIA Rennes

[58] A. FRIGESSI AND M. PICCIONI. — 1991, Parameter Estimation for two–dimensional Ising fields corrupted by noise, 78–89, Spatial Statistics and Imaging, Possolo Ed., Proceed. AMS–IMS–SIAM joint summer conf., 1988, Lect. Notes, Inst. Math. Stat., Hayward

[59] A. FRIGESSI, C.R. HWANG AND L. YOUNES. — 1990, Optimal spectral structure of reversible stochastic matrices, Monte Carlo methods and the simulation of Markov Random Fields, L.N.S. n° 74, Springer, Eds. Barone, Frigessi, Piccioni

[60] J.P. GAMBOTTO. — 1981, Two dimensional time series for textures, in Digital image processing, Eds. Simon and Haralick, Reidel Prob. Company 211–229

[61] T.W. GAMELIN. — 1969, Uniform Algebras, Prentice-Hall

[62] N. GANTERT. — 1990, Laws of large Numbers for the Annealing Algorithm, Stoch. Proc. and Appl., 35, 309–313

[63] D.D. GARBER. — 1981, Computational models for texture analysis and texture synthesis, Thèse, Univ. South. California

[64] A.E. GELFAND AND A.F.M. SMITH. — 1990, Sampling Based approaches to calculating marginal densities, JASA, Vol 85, n° 4, 398–409

[65] A.E. GELFAND, S.E. HILLS, A. RACINE-POON AND A.F.M. SMITH. — 1990, Illustration of Bayesian inference in normal data models using Gibbs sampling, JASA, Vol. 85, n° 412, 972–985

[66] D. GEMAN. — 1990, Random Fields and Inverse problem in Imaging, L.N.M. n° 1427, Springer

[67] D. GEMAN AND S. GEMAN. — 1984, Stochastic relaxation, Gibbs distributions and the Bayesian restoration of Images, IEEE–PAMI–6, 721–741

[68] D. GEMAN, S. GEMAN AND CH. GRAFFIGNE. — 1987, Locating Textures and Objects Boundaries, Patt. Rec. Th. and Appl., Devijver and Kitter Eds., Heidelberg, Springer

[69] D. GEMAN S. GEMAN, CH. GRAFFIGNE AND P. DONG. — 1990, Boundary detection by constrained Optimization, IEEE–PAMI, vol. 12, 7

[70] S. GEMAN AND C. GRAFFIGNE. — 1987, M.R.F. image models and their applications to computer vision, Proceed. Int. Congress Math., 1986, A.M.S., Providence, Ed. M. Gleason

[71] H.O. GEORGII. — 1988, Gibbs measure and phase transitions, De Gruyter

[72] B. GIDAS. — 1991, Parametric estimation for Gibbs distributions, I: Fully observed data, in "Markov random fields: Theory and Applications", Eds. Jain and Chellapa, Academic Press

[73] V.V. GORODETSKII. — 1977, On the strong mixing property for linear sequences, Th. Proba. Appl., 22, 411–413

[74] C. GRAFFIGNE. — 1987, Experiments in texture analysis and segmentation, Ph. D., Division App. Maths, Brown Univ

[75] U. GRENANDER. — 1989, Advances in Pattern Theory, Ann. of Stat., vol. 17, n° 1, 1–30

[76] U. GRENANDER AND B.E. OSBORN. — Parameter estimation in Pattern theory, to appear JASA

[77] I.M. GUELFAND AND N.YA. VILENKIN. — 1964, Generalized functions, 4, Applications of Harmonic Analysis, Academic Press

[78] X. GUYON. — 1982, Parameter estimation for a stationary process on a $d$-dimensional lattice, Biometrika 69, 95–105

[79] X. GUYON. — 1986, Estimation d'un champ de Gibbs, Preprint Univ. Paris 1

[80] X. GUYON. — 1987, Estimation d'un champ par pseudo–vraisemblance–conditionnelle: Etude asymptotique et application au cas markovien, in Spatial Proc. and Spatial Times Series Analysis, Ed. F. Droesbeke, Pub. Fac. Univ. St. Louis, Bruxelles, 16–62

[81] X. GUYON. — 1990, Champs stationnaire sur $\mathbb{Z}^2$: Modèles, statistiques et simulations, Contribuciones en Proba. y Estadistica Matematica, n° 1, Ed. Acta Cien. Venez. y Soc. Bernoulli Latinoamericana, Caracas

[82] X. GUYON AND C. HARDOUIN. — 1992, The chi2 difference of coding test for testing Markov Random Field hypothesis, L.N.S. n° 74, Springer, Eds. Barone, Frigessi, Piccioni, 165–176

[83] X. GUYON AND H.R. KÜNSCH. — 1992, Asymptotic comparison of estimators in the Ising model, L.N.S. n° 74, Springer, Eds. Barone, Frigessi, Piccioni, 177–198

[84] X. GUYON AND S. RICHARDSON. — 1984, Vitesse de convergence du T.C.L. pour des champs faiblement dépendants, Z.f.W., 66, 297–314

[85] B. HAJEK. — 1988, Cooling schedules for optimal annealing, Math. of Operations Research, vol. 13, n° 2, 311–329

[86] P. HALL AND C.C. HEYDE. — 1980, Martingale limit theory and its applications, Acad. Press

[87] J.M. HAMMERSLEY AND D.C. HANDSCOMB. — 1964, Monte Carlo Methods, London, Methuen Company

[88] C. HARDOUIN. — 1992, Quelques résultats nouveaux en statistique des processus: Contraste fort, régressions à résidus à longue portée et estimation par log–périodogramme, Thèse Univ. Paris VII

[89] J. HASLETT. — 1985, Maximum likelihood discriminant analysis on the plane using a Markovian model of spatial context, Pattern Recognition, 18, 287–296

[90] W.K. HASTINGS. — Monte Carlo sampling methods using Markov Chains and their applications, Biometrika, 57, 97-109

[91] H. HELSON AND D. LOWDENSLAGER. — 1958, Prediction theory and Fourier series in several variables, Acta Mathematica, n° 99, 165–202

[92] N.L. HJORT. — 1985, Neighbourhood based classification of remotely sensed data based on geometric probability models, Tech. report 10, Stanford

[93] I.A. IBRAGIMOV AND YU.V. LINNIK. — 1971, Independant and stationary sequences of random variables, Wolters–Nordhoff

[94] I.A. IBRAGIMOV AND YU.A. ROZANOV. — 1974, Processus Aléatoires Gaussiens, Ed. MIR, Moscou

[95] M. IOSIFESCU AND R. THEODORESCU. — 1969, Random Processes and Learning, Grundlehren der Math., Wissenschaften, Bd. 150, Springer

[96] S. IOVLEFF. — 1994, Reconstruction Bayesienne en Tomographie Geophysique et approximation d'equation d'onde, Thèse Université Paris 1

[97] D.L. ISAACSON AND R.Q. MADSEN. — 1976, Markov chains: Theory and Application, Wiley

[98] M. JANZURA. — 1988, Statistical analysis of Gibbs Random fields, 10-*th* Prague Conf. 1986, "Inf. Th. Stat. decis. funct., Random Proc.," Reidel Publishing Comp., 429–438

[99] J.L. JENSEN AND J. MÖLLER. — 1991, Pseudolikelihood for exponential family models of spatial point processes, The Ann. Applied Proba., vol. 1, n° 3, 445–461

[100] S. KIRKPATRICK, C.D. GELLATT JR. AND M.P. VECCHI. — 1983, Optmisation by Simulated annealing, Science, 220, 671–680

[101] H. KOREZLIOGLU AND P. LOUBATON. — 1986, Spectral factorisation of wide sense stationary processes on $\mathbb{Z}^2$, J. Multi. Anal., vol. 19, n° 1, 26–47

[102] O. KOSLOV AND N. VASILYEV. — 1980, Reversible Markov chains with local interaction, in Multicomponent random systems, Eds. Dobrushin and Sinai, Dekker

[103] M.G. KREIN. — 1940, Sur le problème du prolongement des fonctions hermitiennes positives et continues, C.R. Acad. Sci. URSS, vol. 26, 17–22

[104] H.R. KÜNSCH. — 1980, Reelwertige Zufallsfelder anf einem Gitter: Interpolation, Variations Sprinzip und Stat. Analyse, Thèse E.T.H. Zurich

[105] H.R. KÜNSCH. — 1981, Thermodynamics and Statistical analysis of Gaussian Random Fields, Z.f.W., 58, 407–421

[106] H.R. KÜNSCH. — 1981, Almost sure entropy and variational principle for random fields with unbounded state space, Z.f.W., 58, 69–85

[107] H.R. KÜNSCH. — 1982, Decay of correlations under Dobrushin's uniqueness condition and its applications, Comm. Math. Phys., 84, 207–222

[108] H.R. KÜNSCH. — 1984, Time reversal and Stationary Gibbs measures, Stoch. Proc. and Applications, 17, 159–166

[109] H.R. KÜNSCH. — 1987, Intrinsic Auto Regressions and related models on the two dimensional lattice, Biometrika, 74, 3, 517–524

[110] M. LAVIELLE. — 1990, Deconvolution $2D$ and détection de ruptures. Application en géophysique, Thèse Univ. d'Orsay

[111] M. LAVIELLE. — 1991, $2D$ Bayesian deconvolution, Geophysics, Vol. 56, n° 12, 2008

[112] M. LAVIELLE. — 1992, A stochastic algorithm for parametric and non–parametric estimation in the case of incomplete data, to appear in Signal Processing

[113] S.R. LELE AND J.K. ORD. — 1986, Conditional least square estimation for spatial processes: Some asymptotic results, Tech. report n° 65, Univ. of Pennsylvania

[114] T. LIGGET. — 1985, Interacting Particle Systems, G.M.W. 276, Springer

[115] P. LOUBATON. — 1988, Prédiction et représentation markovienne des processus stationnaires vectoriels sur $\mathbb{Z}^2$. Utilisation des techniques d'estimation spectrale $2-D$ en traitement d'antenne, Thèse ENST, Paris

[116] K.V. MARDIA. — 1984, Spatial discrimination and classification maps, Comm. Stat. Theor. Math., 13 (18), 2184–2197

[117] K.V. MARDIA. — 1988, Multidimensional Multivariate Gaussian M.R.F. with application to Image Processing, J. Mult. An., 24, 265–284

[118] J. MARROQUIN, S. MITTER AND T. POGGIO. — 1987, Probabilistic solution of ill posed problems in computational vision, JASA 82, 76–89

[118′] R.J. MARTIN. — 1990, The use of Time–series models and methods in the analysis of agricultural fields trials, Comm. Stat. Theor. Meth., 19, 55–81

[119] G. MATHERON. — 1973, The Intrinsic random functions and their applications, Adv. Appl. Prob. 5, 439–468

[120] P.A.P. MORAN. — 1950, Notes on continuous stochastic phenomena, Biometrika, 37, 17–23

[121] C.M. NEWMAN. — 1980, Normal fluctuations and the FKG inequalities, Comm. Math. Phys. 74, 119–128

[122] C.M. NEWMAN. — 1983, A general central limit theorem for FKG systems, Comm. Math. Phys. 91, 75–80

[123] C.M. NEWMAN. — 1984, Asymptotic independence and limit theorems for positively and negatively dependent variables; Inequalities in Stat. and Proba., IMS Lecture Notes-Monograph Series, vol. 5, 127–140

[124] X.X. NGUYEN AND H. ZESSIN. — 1979, Ergodic theorem for general dynamic systems, Z.f.W, 48, 159–176

[125] S. OLLA. — 1988, Large deviations for Gibbs random fields, Proba. Th. Rel. Fields, 77, 343–357

[126] L. ONSAGER. — 1944, Crystal statistics I: A two dimensional model with order–disorder transition, Phys. Rev., 65, 117–149

[127] D.K. PICKARD. — 1980, Unilateral Markov fields, Adv. Appl. Proba., 12, 655–671

[128] D.K. PICKARD. — 1982, Inference for general Ising Model, J.A.P., 19 A, 345–357

[129] D.K. PICKARD. — 1987, Inference for Markov Fields: The simplest nontrivial case, JASA, vol. 82, n° 397, 90–96

[130] B.T. PORTEONS, D.M.GREIG AND A.H. SEHEULT. — 1989, Exact M.A.P. estimation for binary Images, JRSS B., 51, n° 2, 271–279

[131] A. POSSOLO. — 1986, Estimation of binary Markov Random Fields, Tech. Report n° 77, Dept. of Stat., Univ. of Washington

[132] C.J. PRESTON. — 1974, Gibbs States on countable sets, Cambridge Univ. Press

[133] B. PRUM. — 1986, Processus sur un réseau et mesures de Gibbs. Applications, Masson

[134] W. QUIAN AND D.M. TITTERINGTON. — 1991, Estimation of parameters for hidden Markov models, Phil. Trans. R. Soc. Lond. A, 337, 407–428

[135] B. RIPLEY. — 1981, Spatial Statistics, Wiley

[136] B. RIPLEY. — 1986, Statistics, images and pattern recognition, The Canadian J. of Stat., vol. 14, n° 2, 83–111

[137] B. RIPLEY. — 1988, Statistical inference for spatial processes, Cambridge Univ. Press

[138] R.T. ROCKFELLAR. — 1970, Convex Analysis, Princeton Univ. Press

[139] M. ROSENBLATT. — 1985, Stationary sequences and Random Fields, Birkhaüser

[140] YU. A. ROZANOV. — 1967, Stationary Random Processes, Holden Day

[141] YU. A. ROZANOV. — 1967, On Gaussian field with given conditional distribution, Th. of Proba. and Its Applications, vol. XII, n° 3, 381–391

[142] R. RUBINSTEIN. — 1983, Simulation and the Monte Carlo methods, Wiley

[143] W. RUDIN. — 1962, Fourier Analysis on Group, Interscience

[144] W. RUDIN. — 1963, The extension problem of positive definite function, Ill. J. Math., 7, 532–539

[145] B. SEBASTIEN. — 1993, Modèles Lineaires avec résidus spatialement autocorrélés: Application à l'expérimentation agricole, Thèse INAPG, Paris

[146] A. SEGHIER. — 1991, Extension des fonctions de type positif and entropie associée – Cas multidimensionnel, Ann. IHP, vol. 8, n° 6, 651–675

[147] A. SEN. — 1976, Large Sample Size distribution of statistics used in testing spatial correlation, Geographical Analysis, vol. 8, 175–184

[148] R. SENOUSSI. — 1990, Statistique asymtotique presque sûre de modèles statistiques convexes, Ann. I.H.P., vol. 26, n° 1, 19–44

[149] B. SIMON. — 1979, A remark on Dobrushin's uniqueness Theorem, Comm. Math. Phys., 68, 183–185

[150] A.F.M. SMITH AND G.O. ROBERTS. — 1993, Bayesian Computation via the Gibbs Sampler and related Markov Chain Monte Carlo Methods, JRSS-B, 55, n° 1, 3–23

[151] F. SPITZER. — 1971, Random fields and interacting particule systems, L.N. on M.A.A. summer seminar

[152] C. STEIN. — 1972, A Bound for the error in normal approximation of a sum of dependant random variables, Proc. Berkeley Sympos., M.S.P.2, 583–603

[153] D.J. STRAUSS. — 1975, Analysing Binary lattice data with nearest–neighbor property, J. Appl. Proba., 12, 702–715

[154] D.J. STRAUSS. — 1977, Clustering on coloured lattice, J. Appl. Proba., 14, 135–143

[155] D.J. STRAUSS. — 1986, On a general class of models for interaction, SIAM Rev., vol. 28, 4, 513–527

[156] W.G. SULLIVAN. — 1973, Potentials for almost Markovian random fields, Comm. Math. Phys., 33, 61–74

[157] H.J. SUSSMAN. — 1988, On the convergence of learning algorithms for Boltzman Machines, preprint, Rutgers Univ

[158] P. SWITZER. — 1980, Extension of linear discriminant analysis for statistical classification of remotely sensed sattelite imagery, J. Intern. Math. Geo., 12, 367–376

[159] H. TAKAHATA. — 1983, On the rates in the C.L.T. for weakly dependent random fields, Z.f.W. 64, 445–456

[160] A.A. TEMPEL'MAN. — 1972, Ergodic theorems for general dynamical systems, Trans. Moscow Math. Soc., 26, 94–132

[161] J.L. TIERNEY. — 1992, Exploring posterior distributions using Markov chains, in Computer Science and Statistics: $23^{rd}$ Symposium on the interface, Ed. Keramidas, p 563-570, and Ann. of Stat. (to appear)

[162] D. TJÖSTHEIM. — 1978, Statistical Spatial Modeling, Adv. Appl. Proba., 10, 130–154

[163] A. TROUVÉ. — 1988, Problèmes de convergence et d'ergodicité pour les algorithmes de récuit parallélisés, CRAS, t. 307, Série 1, 161–164

[164] A. TROUVÉ. — 1991, Noyaux partiellement synchrones et parallélisation du recuit simulé, CRAS, t. 312, 155–158

[165] A. TROUVÉ,. — 1993, Parallélisation massive du Recuit Simulé, Thése, Université d'Orsay

[166] J.W. TUKEY. — 1967, An introduction to the calculation of numerical spectrum analysis, dans "Spectral Analysis Times Series", Ed. Harris, Wiley, 22–46

[167] P.J.M. VAN LAARHOVEN AND E.H.L. AARTS. — 1987, Simulated Annealing: Theory and Applications, Dordrecht: D. Reidel

[168] P. WHITTLE. — 1954, On stationary processes in the plane, Biometrika, 41, 434–449

[169] C.S. WITHERS. — 1981, Conditions for linear processes to be strong–mixing, Z.f.W., 57, 477–480

[170] J.F. YAO. — 1989, Segmentation bayésienne d'Images: comparaison des méthodes contextuelles and globales, Cahier du Centre d'Etudes and de Rech. Op. 30 (4), 269–290

[171] J.F. YAO. — 1990, Méthodes Bayésiennes en segmentation d'Image and Estimation par rabotage des modèles spatiaux, Thèse Univ. d'Orsay

[172] L. YOUNES. — 1988, Estimation and Annealing for Gibbsian Fields, Ann. I.H.P. vol. 2, 269–294

[173] L. YOUNES. — 1988, Problème d'estimation paramétrique pour les champs de Gibbs Markoviens – Application au traitement d'Images, Thèse Univ. d'Orsay

[174] L. YOUNES. — 1989, Parametric Inference for Imperfectly observed Gibbsian fields, Proba. Th. Rel. Fields, 82, 625–645

[175] L. YOUNES. — 1991, Parameter estimation for imperfectly observed Gibbs fields and some comments on Chalmond's EM Gibbsian algorithm, LNS, n° 47, p 240–258, Springer, Eds. Barone, Frigessi, Piccioni

[176] L. YOUNES. — 1993, Synchronous random field and Image restoration, preprint, DIAM, ENS Cachan

[177] L. YOUNES. — 1993, Synchronous Boltzman Machines can be universal approximators, preprint, DIAM, ENS Cachan

[178] J. YUAN AND T. SUBBA RAO. — 1990, Non parametric Estimation of the second order spectrum of a stationary process on $\mathbb{Z}^d$, Tech. Report n° 196, Univ. Manchester

# Index